*Trends in Mathematics* is a series devoted to the publication of volumes arising from conferences and lecture series focusing on a particular topic from any area of mathematics. Its aim is to make current developments available to the community as rapidly as possible without compromise to quality and to archive these for reference.

Proposals for volumes can be sent to the Mathematics Editor at either

Birkhäuser Verlag
P.O. Box 133
CH-4010 Basel
Switzerland

or

Birkhäuser Boston Inc.
675 Massachusetts Avenue
Cambridge, MA 02139
USA

Material submitted for publication must be screened and prepared as follows:

All contributions should undergo a reviewing process similar to that carried out by journals and be checked for correct use of language which, as a rule, is English. Articles without proofs, or which do not contain any significantly new results, should be rejected. High quality survey papers, however, are welcome.

We expect the organizers to deliver manuscripts in a form that is essentially ready for direct reproduction. Any version of $T_EX$ is acceptable, but the entire collection of files must be in one particular dialect of $T_EX$ and unified according to simple instructions available from Birkhäuser.

# Algebra
# Some Recent
# Advances

I.B.S. Passi
Editor

Birkhäuser Verlag
Basel · Boston · Berlin

Editor's address:

Centre for Advanced Study in Mathematics
Punjab University
Chandigarh
India


1991 Mathematical Subject Classification 13-02




A CIP catalogue record for this book is available from
the Library of Congress, Washington D.C., USA


Deutsche Bibliothek Cataloging-in-Publication Data
Algebra : some recent advances / I.B.S. Passi ed. – Boston ; Basel ;
Berlin : Birkhäuser, 1999
  (Trends in mathematics)

  ISBN-13: 978-3-0348-9998-7          e-ISBN-13: 978-3-0348-9996-3
  DOI: 10.1007/978-3-0348-9996-3

©1999 Hindustan Book Agency (India) and Indian National Science Academy
Authorized edition by Birkhäuser Verlag, P.O. Box 133, CH-4010 Basel, Switzerland,
for exclusive distribution worldwide except India

Softcover reprint of the hardcover 1st edition 1999


9 8 7 6 5 4 3 2 1

# Contents

# Preface

The Indian National Science Academy has planned to bring out monographs on special topics with the aim of providing accessible surveys/reviews of topics of current research in various fields. Prof. S. K. Malik, FNA, Editor of Publications INSA asked me in October 1997 to edit a volume on algebra in this series. I invited a number of algebraists, several of them working in group rings, and it is with great satisfaction and sincere thanks to the authors that I present here in *Algebra: Some Recent Advances* the sixteen contributions received in response to my invitations.

I. B. S. Passi

# On Abelian Difference Sets

*K.T. Arasu\* and Surinder K. Sehgal*

## 1. Introduction

We review some existence and nonexistence results – new and old – on abelian difference sets. Recent surveys on difference sets can be found in Arasu (1990), Jungnickel (1992*a, b*), Pott (1995), Jungnickel and Schmidt (1997), and Davis and Jedwab (1996). Standard references for difference sets are Baumert (1971), Beth *et al.* (1998), and Lander (1983). This article presents a flavour of the subject, by discussing some selected topics.

Difference sets are very important in combinatorial design theory and in communication engineering while designing sequences with good correlation properties. Our extended bibliography covers a wide variety of papers written in the area of difference sets and related topics.

In section 2, we define difference sets and provide some classical examples of them. Section 3 deals with the important and useful concept of multipliers introduced by Hall (1947). They can be used to investigate the existence question. In section 4, we introduce group rings and characters, the basic algebraic tools in the study of difference sets. Section 5 is devoted to the tools from algebraic number theory which can be used to study difference sets. Extensions of the so-called Mann test are given in section 6. These results impose severe restrictions on the parameters of a difference set. In section 7, we summarize all the known examples of difference sets satisfying certain parameter conditions. The remaining sections deal with difference sets with 'extremal' parameter values – these classes are known as planar difference sets, Paley-Hadamard difference sets and Menon-Hadamard difference sets.

We close by listing all the open cases of $(v, k, \lambda)$ difference sets (up to our knowledge) for $k \leq 150$.

---

\*Work partially supported by NSA grant # MDA 904-97-1-0012 and by AFOSR grant F49620-96-1-0328.

## 2. Difference Sets

Let $G$ be a multiplicatively written group of order $v$. A subset $D$ of $G$ of cardinality $k$ is said to be a $(v, k, \lambda)$ difference set if the multiset $(d_1 d_2^{-1} : d_1, d_2 \in D, d_1 \neq d_2)$ contains each nonidentity element of $G$ exactly $\lambda$ times. A difference set is called cyclic, abelian, or nonabelian if the underlying group has the corresponding property. The parameter $n = k - \lambda$ is called the order of $D$. In this article, we confine ourselves to abelian difference sets. Unless otherwise specified, we write the underlying abelian group $G$ additively. An easy counting argument shows:

**Proposition 2.1.** *Let $D$ be a $(v, k, \lambda)$ difference set in $G$. Then $\lambda(v - 1) = k(k - 1)$.*

The following is easy to see:

**Proposition 2.2.** *If $D$ is a $(v, k, \lambda)$ difference set in $G$, then $D' = G \backslash D =$ the complement of $D$ in $G$ is a $(v, v - k, v - 2k + \lambda)$ difference set in $G$.*

**Remark**   $D$ and $D'$ have the same order $n = k - \lambda = (v - k) - (v - 2k + \lambda)$.

**Proposition 2.3.** *Let $D$ be a $(v, k, \lambda)$ difference set in a group $G$. Then for any automorphism $\sigma$ of $G$ and any $g \in G$,*

$$D' = \sigma(D) + g = \{\sigma(d) + g : d \in D\}$$

*is a $(v, k, \lambda)$ difference set in $G$.*

**Remark**   If $\sigma$ is the identity map in Proposition 2.3, then $D' = D + g$ is called a translate of $D$.

**Example 2.4. (Paley 1993)** Let $q$ be a prime power, $q \equiv 3 \pmod 4$. Let $D$ consist of all the nonzero squares in $F_q$, the finite field with $q$ elements. Then $D$ is a $(q, (q - 1)/2, (q - 3)/4)$ difference set in the additive group of $F_q$.

**Remark**   Paley obtained these using a different formulation in terms of Hadamard matrices.

**Example 2.5. (Singer difference sets)** Let $\alpha$ be a generator of the multiplicative group of $F_{q^{d+1}}$. Then the set of integers $\{i : 0 \leq i < \frac{q^{d+1}-1}{q-1}, \text{trace}_{(d+1)/1}(\alpha^i) = 0\} \bmod (q^{d+1} - 1)/(q - 1)$ form a (cyclic) difference set with parameters

$$\left( \frac{q^{d+1} - 1}{q - 1}, \frac{q^d - 1}{q - 1}, \frac{q^{d-1} - 1}{q - 1} \right).$$

Here the trace denotes the usual trace function $trace_{(d+1/1)}(\beta) = \sum_{i=0}^{d} \beta^{q^i}$ from $F_{q^{d+1}}$ to $F_q$.

**Example 2.6. (Cyclotomic Difference Sets)** The following subsets of $F_q$ are difference sets in the additive subgroup of $F_q$:

$$
\begin{aligned}
F_q^{(2)} &= \{x^2 : x \in F_q \setminus \{0\}\}, q \equiv 3 \pmod{4} \ (\textit{quadratic residues, Paley} \\
&\quad \textit{difference sets}); \\
F_q^{(4)} &= \{x^4 : x \in F_q \setminus \{0\}\}, q = 4t^2 + 1, t \ \textit{odd}; \\
F_q^{(4)} &\cup \{0\}, q = 4t^2 + 9, t \ \textit{odd}; \\
F_q^{(8)} &= \{x^8 : x \in F_q \setminus \{0\}\}, q = 8t^2 + 1 = 64u^2 + 9, t, u \ \textit{odd}; \\
F_q^{(8)} &\cup \{0\}, q = 8t^2 + 49 = 64u^2 + 441, t \ \textit{odd}, u \ \textit{even}; \\
H(q) &= \{x^i : x \in F_q \setminus \{0\}, i \equiv 0, 1 \ \text{or} \ 3 \pmod 6)\}, q = 4t^2 + 27, \\
&\quad q \equiv 1 \pmod 6) \ (\text{Hall difference sets}).
\end{aligned}
$$

The above difference sets are called cyclotomic.

**Example 2.7. (Twin prime power difference sets)** Let $q$ and $q + 2$ be prime powers. Then the set $D = \{(x, y) : x, y$ are both nonzero squares or both non-squares or $y = 0\}$ is a twin prime power difference set with parameters

$$
\left( q^2 + 2q, \frac{q^2 + 2q - 1}{2}, \frac{q^2 + 2q - 3}{4} \right)
$$

in the group $(F_q, +) \oplus (F_{q+2}, +)$.

Thus the history of difference sets dates back to Paley (1933). But the systematic study of these did not begin until Hall (1947) started investigating cyclic difference sets arising from finite projective planes. Hall's paper was motivated by the classical paper of Singer (1938). Difference sets in arbitrary groups were formally introduced by Bruck (1955).

## 3. Multipliers

The fundamental notion of multipliers, due to Hall (1947) is very useful in the study of difference sets. Let $D$ be a $(v, k, \lambda)$ difference set in an abelian group $G$. An automorphism $\sigma$ of $G$ is said to be a multiplier of $D$ if $\sigma(D) = D + g$ for some $g \in G$. An integer $t$ relatively prime to the order of $G$ is said to be a *numerical multiplier* or simply, a multiplier, if the automorphism $\sigma : x \to tx$, is a multiplier of $D$. As we shall see soon, multipliers can be used to investigate the existence or nonexistence of a hypothetical $(v, k, \lambda)$ difference set. The first multiplier theorem was proved by Hall (1947) for the case $\lambda = 1$; the straightforward generalization of Hall's multiplier theorem is due to Chowla and Ryser (1950):

**Theorem 3.1. (First multiplier theorem)** *Let $D$ be an abelian $(v, k, \lambda)$-difference set. Let $p$ be a prime divisor of $n = k - \lambda$. Suppose that g.c.d.$(p, v) = 1$. If $p > \lambda$, then $p$ is a multiplier of $D$.*

In order to use the multipliers to study the existence question, we also require results similar to the one mentioned below.

**Theorem 3.2. (McFarland and Rice, 1978)** *Let D be an abelian $(v, k, \lambda)$-difference set in G. Then there exists a translate of D fixed by every numerical multiplier of D.*

**Example 3.3.** Consider a $(13, 4, 1)$ difference set $D$ in $Z_{13}$. By theorem 3.1, 3 is a multiplier of $D$. We may assume that $D$ is a union of orbits of $Z_{13}$ under $x \rightarrow 3x$ (by theorem 3.2). $D$ must be formed from the orbits $\{0\}, \{1, 3, 9\}, \{2, 6, 5\}, \{4, 12, 10\}, \{8, 11, 7\}$. $D_1 = \{0, 1, 3, 9\}$, $D_2 = \{0, 2, 6, 5\}$, $D_3 = \{0, 4, 12, 10\}$, $D_4 = \{0, 8, 11, 7\}$ all work.

**Example 3.4.** Proceeding as in example 3.3, one can show that $D_1 = \{3, 6, 7, 12, 14\}$ and $D_2 = \{7, 9, 14, 15, 18\}$ both are $(21, 5, 1)$ difference sets in $Z_{21}$.

**Example 3.5.** There do not exist $(1975, 141, 10)$ difference sets in $Z_5 \times Z_5 \times Z_{79}$ or $Z_{25} \times Z_{79} = Z_{1975}$.

**Proof.** Let $D$ be a hypothetical $(1975, 141, 10)$ difference set in $G$, where $G = K \times Z_{79}$, and $K = Z_5 \times Z_5$ or $Z_{25}$. By theorem 3.1, $t = 131$ is a multiplier of $D$. We work with the multiplier $t = 131^5$. Note: $131^5 \equiv 1 \pmod{\text{exponent of } G}$, hence $t$ is identity of $K$. Its orbits on $Z_{79}$ are: a single orbit of size 1 and 6 orbits of size 13. By theorem 3.2, we may, without loss of generality, assume that $D$ is a union of these orbits, say $a$ orbits of size 1 and $b$ of size 13.
    Then

$$a + 13b = 141$$

and $a \in [0, 25]$. Note that $a \equiv 11 \pmod{13}$.
    Hence

$$a = 11 \text{ or } 24.$$

The $b$ orbits of size 13 yield $(13)(12) b$ differences of $D$ all of which lie in $Z_{79}$. Hence $(13)(12)b \le (79 - 1)(10)$, using $\lambda = 10$ so

$$b \le 5,$$

contradicting $a + 13b = 141$. Hence, $D$ cannot exist.                              ∎

**Remark** The above example 3.5 settles an open case in a recent table of Lopez and Sanchez (1997). The proof given here is due to Arasu and Sehgal (1998).

**Example 3.6.** $(31, 10, 3)$ difference sets do not exist in $Z_{31}$. For otherwise, 7 would be a multiplier for such a putative difference set. But the orbits of $Z_{31}$ under $x \rightarrow 7x$ have sizes 1, 15, and 15.

**Conjecture 3.7. (The multiplier conjecture)** Theorem 3.1 holds without the assumption that $p > \lambda$.

All known multiplier theorems may be viewed as an attempt to prove the above conjecture. We now state a multiplier theorem of Menon.

**Theorem 3.8. (Second multiplier theorem)** *(Menon 1960) Let D be an abelian $(v, k, \lambda)$-difference set in G, and let $m > \lambda$ be a divisor of n which is co-prime with v. Moreover, let t be an integer co-prime with v satisfying the following condition: For every prime p dividing m there exist a nonnegative integer f with $t \equiv p^f \pmod{v^*}$, where $v^*$ denotes the exponent of G. Then t is a numerical multiplier of D.*

We state another multiplier theorem due to McFarland (1970). We first define a function $M$ as follows:

$$M(2) = 2 \cdot 7, \quad M(3) = 2 \cdot 3 \cdot 11 \cdot 13, \quad M(4) = 2 \cdot 3 \cdot 7 \cdot 31$$

Recursively, $M(z)$ for $z \geq 5$ is defined to the product of the distinct prime factors of the numbers

$$Z, M(z^2/p^{2e}), p - 1, p^2 - 1, \dots, p^{u(z)} - 1,$$

where $p$ is a prime dividing $m$ with $p^e \| m$ and where $u(z) = (z^2 - z)/2$. (The notation $p^a \| m$ means that $p^a | m$ but $p^{a+1} \nmid m$; we then say that $p^a$ strictly divides $m$.)

**Theorem 3.9.** *Theorem 3.8 remains true if the assumption $m > \lambda$ is replaced by $M(n/m)$ are co-prime.*

Our next result is due to Turyn (1964).

**Theorem 3.10.** *Let D be an abelian $(v, k, \lambda)$-difference set in G, and let $n = 2m$, where m is odd and co-prime with v. Moreover, let t be an integer co-prime with v satisfying the following condition: For every prime p dividing m, there exists a nonnegative integer f with $t \equiv p^f \pmod{v^*}$ where $v^*$ denotes the exponent of G. Finally, assume that t is a quadratic residue modulo 7 if v is divisible by 7. Then t is a numerical multiplier for D.*

As a consequence of theorem 3.9, Turyn (1964) obtained:

**Theorem 3.11.** *Let D be an abelian $(v, k, \lambda)$-difference set and assume that $n = 2p^a$ for some odd prime not dividing v. Then p is a numerical multiplier for*

*D, provided that one of the following conditions holds:*

*a is odd;*

*v is not divisible by 7;*

*7|v and p is a quadratic residue modulo 7.*

A stronger version of theorem 3.10 is given by Muzychuk (1998):

**Theorem 3.12.** *Let D be a $(v, k, \lambda)$ difference set in an abelian group G. Suppose $n = 2p^a$ for some odd prime p with $(p, |G|) = 1$. Then p is a multiplier of D.*

In a series of papers, Qiu has recently made slight improvements of all of the above multiplier theorems – in an attempt to settle the multiplier conjecture. Arasu and Xiang (1995) provide a unified multiplier theorem, in a more general setting.

## 4. Algebraic Tools

In this section we discuss the most important technical tools that are useful in the study of difference sets. We warn the reader that we switch to 'multiplicative' notation when we deal with group rings, i.e., we shall write the underlying abelian group G of a given difference set multiplicatively.

Let G be a multiplicatively written group of order $v$ and R a commutative ring with unity 1. Then the group ring $RG$ is the free R-module with elements of G as basis equipped with the following multiplication:

$$\left(\sum_{g} a_g g\right)\left(\sum_{h} b_h h\right) = \sum_{k}\left(\sum_{g,h:gh=k} a_g b_h\right) k.$$

We shall identify the unities of $R, G$, and $RG$ and denote them by 1. We will also use the obvious imbedding of R into RG. For each subset S of G, by abuse of notation we let S also denote the group ring element $S = \sum_{g \in S} g$.

For $A = \sum_{g \in G} a_g g \in RG$ and any integer $t$, we define $A^{(t)} = \sum_{g \in G} a_g g^t$.

The following result follows readily from the definition of a difference set:

**Proposition 4.1.** *Let D be a subset of cardinality k of a group G of order v. Let R be a commutative ring with 1. Suppose that D is a $(v, k, \lambda)$ difference set in G. Then the identity*

$$D \cdot D^{(-1)} = (k - \lambda) + \lambda G \tag{1}$$

*holds in the group ring RG. If R has characteristic 0, then the converse also holds.*

We shall mostly be working with the group ring $ZG$, where Z is the ring of rational integers and G is an abelian group containing a difference set. For each positive integer $t$, we let $\xi_t$ denote a primitive $t^{\text{th}}$ root of unity. A character $\chi$ of an abelian group G is a homomorphism from G to the multiplicative group $C^*$ of complex numbers. If $G$ has exponent $e$, then $\chi$ maps $G$ to the group of $e^{\text{th}}$ roots of unity. Each character $\chi$ of $G$ can be linearly extended to a ring homomorphism from $ZG$ to $Z[\xi_e]$, the ring of algebraic integers in the $e^{\text{th}}$ cyclotomic field $Q(\xi_e)$. The set of all characters of $G$ is denoted by $G^*$. Obviously $G^*$ is a group under pointwise multiplication. $G^*$ is called the character group of $G$.

The following result is well known and is a consequence of the orthogonality relations for characters. For a proof, we refer the reader to Mann (1965).

**Proposition 4.2.** *Let* $A = \sum_g a_g g \in ZG$. *Then* $a_g = \frac{1}{|G|} \sum_{\chi \in G^*} \chi(A)\chi(g^{-1})$. *Hence, if* $A, B \in ZG$ *satisfy* $\chi(A) = \chi(B)$ *for all characters* $\chi$ *of* $G$, *then* $A = B$.

Applying characters $\chi$ of $G$ to both sides of (1), we get

$$\chi(D)\overline{\chi(D)} = \begin{cases} k - \lambda \text{ if } \chi \text{ is nonprincipal} \\ k^2 \text{ if } \chi \text{ is principal} \end{cases} \qquad (2)$$

A subset $D$ of $G$ satisfying (2) above is a $(v, k, \lambda)$ difference set of $G$, in view of proposition 4.2. Thus, once a candidate subset had been identified as a difference set, formal verification is carried out by simply calculating the character sum $\chi(D)$, proving that its modulus is $\sqrt{k - \lambda}$ for all nonprincipal characters of $G$.

The idea of using characters to investigate difference sets is now very standard— it was formally developed in the papers of Turyn (1965) and Yamamoto (1953).

The following well-known theorem entails restrictions on the parameters of a $(v, k, \lambda)$ difference set. It was proved using algebraic techniques.

**Theorem 4.3.** *(Bruck and Ryser (1949) and Chowla and Ryser (1950)) Let $D$ be a $(v, k, \lambda)$ difference set in a group $G$.*

(i) *If $v$ is even, then $n = k - \lambda$ is a square.*

(ii) *If $v$ is odd, then there exist integers $x$, $y$, and $z$, not all zero, such that*

$$x^2 = (k - \lambda)y^2 + (-1)^{\frac{v-1}{2}}\lambda z^2.$$

**Remarks**

(1) Theorem 4.3 is actually applicable to a class of widely studied combinatorial objects – called symmetric designs. $(v, k, \lambda)$ difference sets give rise to symmetric designs.

(2) Part $(i)$ of theorem 4.3 is due to Schutzenberger (1949).

## 5. Number Theoretic Tools

Let $D$ be a $(v, k, \lambda)$ difference set in an abelian group $G$. The corresponding group ring element $D$ of $ZG$ satisfies

$$DD^{(-1)} = (k - \lambda) + \lambda G \qquad (5.1)$$

Hence, $\chi(D)\chi(D) = k - \lambda = n \qquad (5.2)$

for all nonprincipal characters $\chi$ of $G$. Thus $\chi(D)$ is an algebraic integer in $Q(\zeta_e)$ where $\zeta_e$ is a primitive $e^{\text{th}}$ root of unity, $e = $ exponent of $G$.

(5.2) says that the existence of a difference set $D$ implies that of an algebraic integer of a certain modulus. Most of the non-existence results on abelian difference sets rely on this number theoretic restriction.

**Proposition 5.1.** *Let $p$ be a prime and $\zeta_e$ a primitive complex $e^{\text{th}}$ root of unity; write $e = p^\ell e'$ where $e'$ is an integer relatively prime to $p$. The multiplicative order of $p$ modulo $e'$ is denoted by $f$. Let $\Phi(x)$ be the Euler phi function. Then the following identity for ideals holds in $Z[\zeta_e]$:*

$$(P) = (p_1 \cdots P_g)^{\Phi(p^\ell)},$$

*where the $P_i's$ are distinct prime ideals and $g = \Phi(e')/f$. If $t$ is an integer relatively prime to $p$ such that $t \equiv p^s \bmod e'$, then the Galois automorphism $\zeta_e \to \zeta_e^t$ fixes the ideals $P_i$. If $e' = 1$ then $g = 1$ and $P_1 = (1 - \zeta_e)$.*

We now introduce the notion of self-conjugacy due to Turyn (1965). A prime $p$ is said to be self-conjugate modulo a positive integer $m$, if there exists an integer $j$, such that $p^j \equiv -1 \pmod{m'}$, where $m'$ is the $p$-free part of $m$. In the study of abelian difference sets, we say that self-conjugacy assumption is satisfied, if every prime divisor of $n = k - \lambda$ is self-conjugate modulo $\exp(G)$.

As a consequence of proposition 5.1, we get:

**Proposition 5.2. (Mann test)** *Let $p$ be self-conjugate modulo $w$, and let $D$ be a difference set of order $n$ in a group $G$ whose exponent is divisible by $w$. The $p$ cannot divide the square-free part of $n$, i.e., $p^{2a}$ is the exact $p$-power dividing $n$. In particular, for each character $\chi$ of order $w$, we have $\chi(D) \equiv 0 \bmod p^a$.*

**Remark** The proof of proposition 5.2 uses the prime ideal factorization of $(p)$ in $Z[\zeta_w]$. Better results can be proved by exploiting the condition $\chi(D) \equiv 0 \pmod{p^a}$ more carefully. We quote a lemma in this regard.

**Lemma 5.3. (Arasu, et al. (1996))** *Let $p$ be a prime, and let $G$ be an abelian group with a cyclic Sylow $p$-subgroup of order $p^s$. If $Y \in Z[G]$ is an element such that $\chi(Y) \equiv 0 \bmod p^a$ for all character, then we can write*

$$Y = p^a X_0 + p^{a-1} P_1 X_1 + \ldots + p^{a-r} P_r X_r,$$

*$r = \min(a, s)$, where $P_i$ denotes the unique subgroup of order $p^i$. Moreover, if the coefficients of $Y$ are nonnegative, the coefficients of the $X_i$ can be chosen to be nonnegative, too.*

Using lemma 5.3, one can show, for instance, that there are no abelian difference sets with parameters $(4p^2, 2p^2 - p, p^2 - p)$ in groups of order $4p^2$ if $p \equiv 3 \pmod{4}$ or if the Sylow 2-subgroup is elementary abelian.

Recent research activities of Ma (1996) and Schmidt (1997) attempt to overcome the 'self-conjugacy' assumption. An example of such a result is:

**Proposition 5.4.** *Let* $d = p^a m$, *where* $p$ *is an odd prime and* $d > 0$ *is an odd integer relatively prime to* $p$. *If* $X \in Z[\zeta_d]$ *satisfies* $X\bar{X} = p$, *then with suitable* $j$ *either* $\zeta_d^j X \in Z[\zeta_m]$ *or* $X = \pm \zeta_d^j Y$, *where* $Y$ *is a generalized Gauss sum.*

Using this result, further necessary conditions on the existence of abelian difference sets can be obtained.

We refer the reader to Schmidt (1997, 1998), Arasu & Ma (1998) for related results.

## 6. Extensions of Mann Test

We now give various extensions of proposition 5.2, the Mann test. We start with a definition:

Let $G$ be a group and $N$ a normal subgroup of $G$. Let $H = G/N$. For any subset $D$ of $G$, let $s_g = |D \cap gN|$. The numbers $s_g$ are called the intersection numbers of $D$ relative to $N$.

**Theorem 6.1.** *Let* $D$ *be a* $(v, k, \lambda)$ *difference set in a (not necessarily abelian) group* $G$ *of order* $v > k$. *Furthermore, the* $u \neq 1$ *be a divisor of* $v$, *let* $U$ *be a normal subgroup of order* $s$ *and index* $u$ *of* $G$; *put* $H = G/U$, *and assume that* $H$ *is abelian and has exponent* $u^*$. *Finally, let* $p$ *be a prime not dividing* $u^*$ *and assume that* $tp^f \equiv -1 \pmod{u^*}$ *for some suitable nonnegative integer* $f$ *and some numerical* $G/U$-*multiplier* $t$. *Then the following hold:*

(i) *$p$ does not divide the square-free part of* $n = k - \lambda$, *say,* $p^{2j} \| n$ *(with* $j \geq 0$*);*

(ii) $p^j \leq s$;

(iii) *$u > k$ implies that* $p^j | k$.

(iv) *All intersection numbers of* $D$ *relative to* $U$ *are congruent modulo* $p^j$, *say,* $s_x \equiv y \pmod{p^j}$ *for all* $x \in H$.

(v) *One has* $yu \equiv k \pmod{p^j}$; *if we choose* $y_0$ *as the smallest non-negative solution of this congruence, we also have* $y_0 u \leq k$.

**Remark** (i) is due to Mann (1964); (i)-(iii) are due to Jungnickel & Pott (1988); and (iv) & (v) were proved by Arasu *et al.* (1990).

**Example 6.2.** There cannot exist an abelian $(704, 38, 2)$ difference set in a group $G$ whose Sylow 2-subgroup has exponent $u^* = 2$ or $4$. (*Reason:* Apply theorem 6.1 with $p = 3$ using the fact that $3 \equiv -1 \pmod{u^*}$. Then (*i*) holds with $j = 1$, but then (*iii*) gives $3 | 38$, a contradiction.

**Remark** The above proof is due to Jungnickel & Pott (1988), even though these cases were previously ruled out by Arasu (1988) using tedious arguments. These cases answered some open problems in Lander's book (1983).

The following is also due to Mann (1964).

**Theorem 6.3.** *Let $D$ be an abelian $(v, k, \lambda)$ difference set in $G$, and let $w$ be a divisor of $v$. If there exists a multiplier $m$ of $D$ satisfying $m^h \equiv -1 \pmod{w}$ for some positive integer $h$, then either $n = k - \lambda$ is a square or $n$ is of the form $n = a^2 q^3$, where $q$ is a prime and $w$ a power of $q$. In the latter case, every multiplier of $D$ has odd order mod $r$ for all prime divisors $r \neq q$ of $v$; also $q \equiv 1 \pmod 4$, $v$ is odd and $m$ is a quadratic residue mod $q$.*

## 7. Known Families of Difference Sets with $(v, n) > 1$

All known examples of abelian difference sets with $(v, n) = 1$ have parameters as given in examples 2.4–2.7. In this section we discuss the case when $(v, n) > 1$.

**Example 7.1. (McFarland, 1973)** Let $q$ be a prime power and $d$ a positive integer. Let $G$ be an abelian group of order $v = q^{d+1}(q^d + \ldots + q^2 + q + 2)$ which contains an elementary abelian subgroup of order $q^{d+1}$. Identify $E$ as the additive group of $F_q^{d+1}$.

Let $r = \frac{q^{d+1}-1}{q-1}$ and $H_1, H_2, \ldots, H_r$ be the hyperplanes of order $q^d$ of $E$. If $g_0, g_1, \ldots, g_r$ are distinct coset representatives of $E$ in $G$, then

$$D = (g_1 + H_1) \cup (g_2 + H_2) \cup \ldots \cup (g_r + H_r)$$

is a McFarland difference set with parameters

$$\left( q^{d+1} \left( 1 + \frac{q^{d+1}-1}{q-1} \right), q^d \left( \frac{q^{d+1}-1}{q-1} \right), q^d \left( \frac{q^d-1}{q-1} \right) \right)$$

A variation of McFarland's construction by Spence (1977) is given below:

**Example 7.2. (Spence difference sets)** Let $E$ be the elementary abelian group of order $3^{d+1}$ and $G$ a group of order $v = 3^{d+1} \left( \frac{3^{d+1}-1}{2} \right)$ containing $E$. Let $m = \frac{3^{d+1}-1}{2}$ and $H_1, H_2, \ldots, H_m$ denote the subgroups of $E$ of order $3^d$. If $g_1, \ldots, g_m$ are distinct coset representatives of $E$ in $G$, then

$$D = ((g_1 + (E \backslash H_1)) \cup (g_2 + H_2) \cup (g_3 + H_3) \cup \ldots \cup (g_m + H_m))$$

is a Spence difference set with parameters

$$\left( 3^{d+1} \left( \frac{3^{d+1}-1}{2} \right), 3^d \left( \frac{3^{d+1}+1}{2} \right), 3^d \left( \frac{3^d+1}{2} \right) \right)$$

**Example 7.3. (Menon-Hadamard difference sets)** A difference set whose parameters are $(4u^2, 2u^2 - u, u^2 - u)$ is called a Menon-Hadamard difference set.

All of section 11 will be devoted to studying this important class of so-called Menon-Hadamard difference sets.

**Example 7.4. (Davis-Jedwab difference sets)** A difference set with parameters

$$\left(2^{2d+4}\left(\frac{2^{2d+2}-1}{3}\right), 2^{2d+1}\left(\frac{2^{2d+3}+1}{3}\right), 2^{2d+1}\left(\frac{2^{2d+1}+1}{3}\right)\right)$$

is called a Davis-Jedwab difference set. (Here $d$ is any nonnegative integer.)

These difference sets exist in all abelian groups of order $2^{2d+4}\left(\frac{2^{2d+2}-1}{3}\right)$ which have a Sylow 2-subgroup $S_2$ of exponent at most 4, with the single exception $d = 1$ and $S_2 \cong Z_4^3$.

**Remark** The above family is contained in the brilliant unifying work of Davis and Jedwab (1997). Their work has sparked a keen interest in the area of difference sets, leading to new directions of research. It motivated Chen (1997) to obtain

**Example 7.5. (Chen difference sets)** A difference sets with parameters

$$\left(4q^{2d+2}\left(\frac{q^{2d+2}-1}{q^2-1}\right), q^{2d+1}\left(\frac{2(q^{2d+2}-1)}{q+1}+1\right),\right.$$
$$\left. q^{2d+1}(q-1)\left(\frac{q^{2d+1}+1}{q+1}\right)\right)$$

is called a Chen difference set. (Here $d$ is a nonnegative integer and $q$ a prime power.)

**Remark** As observed by Jungnickel & Schmidt (1997), all the known examples of abelian difference sets with $(v, n) > 1$ share the following interesting property: the character sum $\chi(D)$ of such a difference set for a nontrivial character $\chi$ of the underlying group $G$ is divisible by $\sqrt{n}$.

We close this section by briefly introducing the marvellous notion of the so-called "Covering extended building set" due to Davis & Jedwab (1997). This clever idea was introduced by Davis & Jedwab in an attempt to unify all the constructions of abelian difference sets with $(v, n) > 1$.

An $(a, m, h, \pm)$ covering extended building set in an abelian group $G$ is a family $\{D_1, \ldots, D_h\}$ of subsets of $G$ satisfying $(i)$ $|D_1| = a \pm m$ and $|D_i| = a$ for $i = 2, \ldots, h$ and $(ii)$ for each nonprincipal character $\chi$ of $G$ there is exactly one $i$ with $|\chi(D_i)| = m$ and $\chi(D_j) = 0$ if $j \neq i$.

Every covering extended building set gives rise to a difference set, as shown in:

**Theorem 7.6. (Davis & Jedwab (1997))** *Suppose there exists an $(a, m, h, \pm)$ covering extended building set in an abelian group $G$. Then there exists an*

$$(h|G|, ah \pm m, ah \pm m - m^2)$$

*difference set in any abelian group containing $G$ as a subgroup of index $h$.*

**Remarks**

(1) Examples 7.4 and 7.5 are by-products of theorem 7.6, by constructing suitable covering extended building sets.

(2) The case $d = 0$ of example 7.5 reduces to example 7.3; the case $q = 2$ of example 7.5 boils down to example 7.4; the case $q = 3$ of example 7.5 covers example 7.2. For historical reasons, we have decided to discuss these examples individually.

## 8. Difference Sets with Extremal Parameters

Recall that the parameters of a $(v, k, \lambda)$ difference set $D$ satisfy:

$$k(k - 1) = \lambda(v - 1) \text{ and } n = k - \lambda = k^2 - \lambda v.$$

Since $k \equiv n \pmod{\lambda}$ it follows that $k(k - 1) \equiv n(n - 1) \pmod{\lambda}$

$$\equiv \lambda(v - 1) \equiv 0 \pmod{\lambda}.$$

Write $n(n - 1) = \lambda\mu$. But $v = \frac{k(k-1)}{\lambda} + 1 = \frac{(n+\lambda)(n+\lambda-1)}{\lambda} + 1 = 2n + \lambda + \frac{n(n-1)}{\lambda} = 2n + \lambda + \mu$. Thus the parameters $(v, k, \lambda)$ can be rewritten as $(2n + \lambda + \mu, n + \lambda, \lambda)$ where $n(n - 1) = \lambda\mu$.

**Proposition 8.1.** *Let $(v, k, \lambda)$ be a parameter triple of a difference set. Then $4n - 1 \le v \le n^2 + n + 1$.*

**Proof.** As seen above, we can rewrite the parameters as $(2n + \lambda + \mu, n + \lambda, \lambda)$, where $n(n - 1) = \lambda\mu$. We shall first maximize and minimize $f(\lambda, \mu) = \lambda + \mu$ subject to $n(n - 1) = \lambda\mu$. Eliminating $\mu$,

$$f(\lambda, \mu) = f(\lambda) = \lambda + n(n - 1)/\lambda \text{ for } \lambda \ge 1.$$

Treat this as a function of the real variable $\lambda$ on $(1, \infty)$. Then $f'(x) = 1 - n(n - 1)/\lambda^2$ and so $\lambda = \sqrt{n(n - 1)}$ is a critical point on $(1, \infty)$. Also $f'(x) > 0$ for $\lambda$ in $[\sqrt{n(n - 1)}, \infty)$ and $f'(x) < 0$ for $\lambda$ in $[1, \sqrt{n(n - 1)}]$. Thus $f(x)$ has absolute maximum at $\lambda = 1$ and absolute minimum at $\lambda = \sqrt{n(n - 1)}$ for $\lambda$ in $[1, n(n - 1)]$. But our interest is optimizing $f(x)$ for 'discrete' values of $\lambda$. Thus $f(x)$ for integral values of $\lambda$, $\lambda \ge 1$, has absolute minimum at $\lambda = (n - 1)$. The corresponding absolute minimum for $f(\lambda) = \lambda + \mu = \lambda + n(n - 1)/\lambda$ is

$$f(1) = 1 + n(n - 1) = n^2 - n + 1$$

and the absolute minimum is $f(n - 1) \doteq 2n - 1$. Hence, the largest possible value of $v$ is $n^2 + n + 1$ and the smallest possible value of $v$ is $4n - 1$. The corresponding parameter triples are: $(n^2 + n + 1, n + 1, 1)$ and $(4n - 1, 2n - 1, n - 1)$. This completes the proof of proposition 8.1. ∎

We will refer to the parameters $v = n^2 + n + 1$; $v = 4n - 1$ and $v = 4n$ as extremal parameters. Difference sets satisfying these parameter conditions have drawn considerable attention. We devote the next three sections to discussing these three families of difference sets.

## 9. Difference Sets with Largest Possible $v$, For A Given $n$

Let $D$ be an abelian $(v, k, \lambda)$ difference set with $v = n^2 + n + 1$. As seen in section 8, difference sets with the largest values of $v$, for a given $n$ have parameters

$$(v, k, \lambda) = (n^2 + n + 1, n + 1, 1).$$

If we regard the elements of the underlying group $G$ as 'points' and each translate $D + g$ of $D$ for $g \in G$ as 'lines' and the relation 'belongs to' as incidence, then we obtain a finite projective plane of order $n$. For this reason this family is usually referred to as 'planar difference sets'. Thus planar difference sets are precisely those for which $\lambda = 1$.

From example 2.5, planar difference sets of order $n$ exist whenever $n$ is a prime power.

**Conjecture 9.1. (The prime power conjecture, PPC)** If there exists an abelian planar difference set of order $n$, then $n$ is a prime power.

In 1951, Evans and Mann verified the PPC in the cyclic case for $n \leq 1600$; Keiser (Unpublished) extended this for $n \leq 3600$. Gordon (1994) uses all the known nonexistence results and a computer to verify the PPC for $n \leq 2,000,000$.

The simplest nonexistence test for the planar case that would use the fact that $\lambda = 1$ is given in

**Lemma 9.2.** *Let $D$ be a planar abelian difference set of order $n$ in the group $G$. If $t_1, t_2, t_3,$ and $t_4$ are numerical multipliers such that*

$$t_1 - t_2 \equiv t_3 - t_4 \bmod (\exp G),$$

*then $\exp G$ divides the least common multiple of $t_1 - t_2$ and $t_1 - t_3$.*

Hall (1947) proved lemma 9.2 for the cyclic case; Lander (1983) obtained the abelian case. Using lemma 9.2, the following is easily proved:

**Proposition 9.3.** *Let $D$ be a planar abelian difference set of order $n$. Then $n$ cannot be divisible by any of of the numbers 6, 10, 14, 15, 21, 22, 26, 33, 34, 35, 38, 39, 46, 51, 55, 57, 58, 62, or 65.*

Using the Mann test (see section 5), further nonexistence tests can be obtained:

**Theorem 9.4.** *Let $D$ be a planar abelian difference set of order $n$. Moreover, let $p$ and $q$ be prime divisors of $n$ and of $v = n^2 + n + 1$, respectively. Then each of the following conditions implies that $n$ is a square:*

*D has a multiplier which has even order* (mod $q$).

*p is a quadratic nonresidue* (mod $q$).

$n \equiv 4 \; or \; 6$ (mod 8).

$n \equiv 1 \; or \; 2$ (mod 8) *and* $p \equiv 3$ (mod 4).

$n \equiv m \; or \; m^2$ (mod $m^2 + m + 1$) *and p has even order* mod($m^2 + m + 1$).

*The next two results are due to Ostrom (1953).*

**Theorem 9.5.** *Assume the existence of a planar abelian difference set of order* $n = m^2$ *in G. Then there also exists a planar difference set of order m in some subgroup H of G.*

**Theorem 9.6.** *Assume the existence of a planar cyclic difference set D of order* $n = m^s$ *in G, where* $\gcd(s, 3) = 1$. *Then, there also exists a planar difference set* $D'$ *of order m in some subgroup H of G. Moreover, every multiplier of D is also a multiplier of* $D'$.

**Remark**  Theorem 9.5 was proved by Ostrom only for the cyclic case; the abelian version is due to Jungnickel and Vedder (1984).

The best known nonexistence theorems are summarized in:

**Theorem 9.7.** *Let D be an abelian planar difference set.*

*(i) If n is even, then* $n = 2$ *or* $n = 4$ *or* $n \equiv 0$ (mod 8).

*(ii) If* $n \equiv 0$ (mod 3), *then* $n = 3$ *or* $n \equiv 0$ (mod 9).

**Remark**  Item (i) is due to Jungnickel and Vedder (1984) and (ii) is due to Wilbrink (1985).

Nonexistence theorems that specifically apply to square order planar difference sets are due to Arasu (1989):

**Theorem 9.8.** *Let D be a cyclic planar difference set of order n, where* $n = m^{2s}$ *with* $\gcd(s, 3) = 1$, *and let t be a multiplier of D. Assume the existence of a prime q dividing* $m^2 - m + 1$ *such that t has odd order modulo q. Then t has odd order modulo* $m^4 + m^2 + 1$.

**Theorem 9.9.** *Let D be an abelian planar difference set of order n, where* $n = m^{2^r}$ *for some positive integer r, and let t be a multiplier of D. Assume the existence of a prime q dividing* $m^2 - m + 1$ *such that t has odd order modulo q. Then t has odd order modulo* $m^4 + m^2 + 1$.

In a series of papers, Ho has studied the multiplier groups $M$ of abelian planar difference sets. We summarize some of his results below:

**Theorem 9.10. (Ho, 1993, 1994, 1995)** *Let $\Pi$ be a projective plane of order $n$ with Singer group $G$ (not necessarily abelian) and difference set $D \subset G$, and let $M$ be the multiplier group of $D$. Then the Sylow 2-subgroup $S$ of $M$ is a cyclic direct factor of $M$, and hence $M$ is solvable. Moreover, the following hold:-*

a) *Write $n = m^{2^a}$ where $m$ is not a square. Then $|S| \le 2^a$; if $M$ is abelian, then actually, $|S| = 2^a$ and $|M| \le (m+1)2^a$.*

b) *$M$ fixes a line of $\Pi$.*

c) *If $M$ has even order, then each subgroup of $G$ is invariant under the unique involution is $M$, except possibly if $n = 16$ and $G$ is nonabelian.*

d) *Let $H$ be an abelian subgroup of $M$. If $H$ has odd order, then $|H| \le n+1$. If $|H| = n+1$, then $n^2 + n + 1$ is a prime.*

e) *If $M$ is abelian, then either $|M| \le n+1$ or $n$ is a square.*

f) *If $G$ is abelian, then $|M| \le n+1$ except for $n = 4$, where $|M| = 6$.*

g) *If $M$ is abelian and $n$ is a square, then the Sylow 3-subgroup of $M$ is cyclic.*

More interesting results that relate these planar difference results to desarguesian projective planes are in the literature. We refer the reader to Beth, Jungnickel, and Lenz (1998) for further details on these goemetric ideas, conjectures, and the current status of the conjectures.

We finally remark that the only known infinite series of difference sets for a given $\lambda$ is for the case $\lambda = 1$. In this context we mention:

**Conjecture 9.11. (Hall's conjecture)** For a fixed $\lambda > 1$ there exist only a finite number of $(v, k, \lambda)$ difference sets.

**Remark** The above conjecture is originally made for any $(v, k, \lambda)$ symmetric designs.

## 10. Difference Sets with the Smallest Possible $v$, For A Given $n$

As seen in section 8, difference sets with least $v$ for a given $v$ must satisfy: $v = 4n - 1$, and the corresponding triples are:

$$(v, k, \lambda) = (4n - 1, 2n - 1, n - 1).$$

Some authors refer to these as Hadamard difference sets; to avoid confusion with another family of difference sets (See section 11) that share this name by some authors, we shall refer to this family as the Paley-Hadamard family. Example 2.5 with $q = 2$; examples 2.4 and 2.7 and some families covered in example 2.6 are all Paley-Hadamard difference sets.

Paley-Hadamard difference sets with $v = 4n - 1$ give rise to Hadamard matrices of order $4n$. (NOTE: A Hadamard matrix $\mathbf{H}$ is a $t$ by $t$ matrix with entries from $\{1, -1\}$ satisfying $HH' = I_t$ where $\mathbf{H}'$ is the transpose of $\mathbf{H}$, and $I_t$ is the $t$ by $t$ identity matrix. An easy combinatorial argument shows that if there is a Hadamard matrix of order $t$, then $t = 1, 2$ or $t \equiv 0 \pmod 4$. The converse is conjectured to be true, but is still open. Several infinite families of these matrices exist; the smallest open case is $t = 428$. We refer the reader to Seberry and Yamada (1992) for further details.)

**Conjecture (Song and Golomb, 1994)** If there exists a cyclic Paley-Hadamard difference set in a group of order $v$, then $v = 4n - 1$ must be either a prime, or a product of 'twin primes,' or one less than a power of 2.

Song and Golomb (1994) have verified this conjecture for $v < 10,000$, except for the following 17 cases:

$$1295, 1599, 1935, 3135, 3439, 4355, 4623, 5775, 7395, 7743, 8227,$$
$$8463, 8591, 8835, 9135, 9215, \text{ and } 9423$$

For further results on these, refer Golomb and Song (1997).

## 11. Difference Sets Satisfying *v = 4n*

"$v = 4n$" is one more than the smallest allowed value "$v = 4n - 1$" for the order of a group admitting a $(v, k, \lambda)$ difference set, for a given $n = k - \lambda$. This family is very rich and is of considerable interest in communication engineering. Since Menon (1962) was the first to investigate these systematically, these are called the Menon difference sets by some authors. Some others call these 'Hadamard difference sets,'[1] since these difference sets give rise to Hadamard matrices. We shall refer to these as Menon-Hadamard difference sets. Menon showed that they have parameters of the form $(4N^2, 2N^2 \pm N, N^2 \pm N)$.

In this section, we survey recent progress made in the study of Menon-Hadamard difference sets.

We begin with the following composition theorem of Menon (1962).

**Theorem 11.1.** *The existence of Menon-Hadamard difference sets with $N = u_1$ and $N = u_2$ implies that of a Menon-Hadamard difference set with $N = 2u_1u_2$.*

Note that any singleton is a $(4, 1, 0)$ Menon-Hadamard difference set in $Z_4$ or $Z_2 \times Z_2$. This rather trivial example, in conjunction with theorem 1, provides an infinite family of these difference sets in abelian 2-groups of exponent 4 and order an even power of 2.

Turyn (1984) used elementary methods and proved.

**Theorem 11.2.** *Menon-Hadamard difference sets exist in $Z_2^2 \times Z_3^{2b}$ and $Z_4 \times Z_3^{2b}$.*

Turyn's seminal paper (1965) establishes an exponent bound on certain Sylow $p$-subgroups of an abelian group containing a Menon-Hadamard difference set. Using Turyn's results (see Turyn (1965)) one can show: if $G$ is an abelian group of order $2^{2a+2}$ containing a Menon-Hadamard difference set, then $\exp(G) \leq 2^{a+2}$

Using the results of Davis (1991) and Dillon (1990a, b), Kraemer (1993) proved the following theorem:-

**Theorem 11.3.** *There exists a Menon-Hadamard difference set in an abelian group $G$ of order $2^{2a+2}$ if and only if $\exp(G) \leq 2^{a+2}$.*

Until 1992, all known examples of Menon-Hadamard difference sets satisfied $N = 2^r 3^s$ for some integers $r$ and $s$. Xia's (1992) sensational construction methods yield

**Theorem 11.4.** *Menon-Hadamard difference sets exist in $H \times Z_{p1}^4 \times Z_{p2}^4 \times \ldots \times Z_{pt}^4$ where each $p_i$ is a prime, $p_i \equiv 1 \pmod 4$ and $H$ is any group of order 4.*

Xiang and Chen (1996) provided a more transparent proof of Xia's results.

Using the notion of the so-called "binary supplementary quadruple systems" (BSQ) introduced by Jedwab (1992), theorem 4 can be strengthened by replacing $H$ by any abelian 2-group that meets the exponent bound of Turyn, viz. $\exp(H) \leq 2\sqrt{|H|}$. BSQ of size $3^b \times 3^b$ for each $b \geq 1$ are due to Arasu, Davis, Jedwab, and Sehgal (1993). These, coupled with Xia's BSQ asnd Jedwab's (1992) results yield.

**Theorem 11.5.** *Menon-Hadamard difference sets exist in $H \times Z_{3^{b1}}^2 \times \ldots \times Z_{3^{bs}}^2 \times Z_{p_1}^4 \times \ldots \times Z_{p_t}^4$, where $H$ is an abelian 2-group satisfying $\exp(H) \leq 2\sqrt{|H|}$ and each $p_i$ is a prime, $p_i \equiv 3 \pmod 4$.*

The obvious question was: what can be said about primes $p$, $p \equiv 1 \pmod 4$? i.e., can such a prime divide the order of an abelian group $G$ which contains a Menon-Hadamard difference set?

The smallest case $p = 5$ was solved by Van Eupen and Tonchev (1997):

**Theorem 11.6.** $Z_2 \times Z_2 \times Z_5^4$ *contains a Menon-Hadamard difference set.*

Van Eupen and Tonchev obtained this result using a computer search, which was motivated by a result of Ray-Chaudhuri and Xiang (1997) that provided a connection of these difference sets to certain projective 2-weight codes. Wilson and Xiang (1997) provide a very general construction method for Menon-Hadamard difference sets in groups $H \times Z_p^4$, where $H$ is any group of order 4 and $p$ is prime, $p \equiv 1 \pmod 4$. They succeeded in giving such difference sets for the cases $p = 13$ and $p = 17$. Chen (1997) completed this problem by filling the missing pieces of the puzzle, thereby settling this for all primes $p$, $p \equiv 1 \pmod 4$. We now summarize the current state of knowledge about abelian groups that contain a Menon-Hadamard difference set:

**Theorem 11.7.** *Let H be an abelian group of order $2^{2a+2}$ satisfying $\exp(H) \le$ $2^{a+2}$. Let $b_1, \ldots, b_s$ be positive integers and $p_1, p_2, \ldots, p_t$ be odd primes (not necessarily distinct). Then the group $H \times Z_{3^{b_1}}^2 \times \ldots \times Z_{3^{b_s}}^2 \times Z_{p_1}^4 \times \ldots \times Z_{p_t}^4$ contains a Menon-Hadamard difference set.*

Recent interesting developments on $(v, k, \lambda)$ difference sets with g.c.d. $(v, k-\lambda)$ $> 1$ are follow-up work of the groundbreaking "unification" theorems of Davis and Jedwab. We refer the reader to Davis and Jedwab (1996, 1997) and Chen (1996, 1997) for further details.

We now turn to nonexistence results. All these results rely on character theoretic approach. Most of them require the so-called "self-conjugacy" assumption. Given positive integers $m$ and $w$, we call $m$ self-conjugate modulo $w$ if for each prime divisor $p$ of $m$, there exists an integer $j_p$ such that $p^{j_p} \equiv -1 \pmod{w_p}$ where $w_p$ is the largest divisor of $w$ relatively prime to $p$.

McFarland (1990) proved that under certain conditions the existence of a Menon-Hadamard difference set in an abelian group implies that of one in a subgroup:

**Theorem 11.8.** *Let $G = H \times K$ be an abelian group containing a Menon-Hadamard difference set, where H has even order coprime to $|K|$. Assume further that $|K|$ is self-conjugate modulo $(\exp H)$. Then H contains a Menon-Hadamard difference set, too.*

In theorem 11.8, if $K$ is restricted to a Sylow p-subgroup of order $p^{2b}$, where $p$ is an odd prime, additional results can be obtained. The exponent bound $\exp(K) \le$ $p^b$ follows from Turyn (1965). Arasu, Davis, Jedwab (1995) used the binary array viewpoint to show that $K = Z_{p^b}^2$ if $\exp(K) = p^b$:

**Theorem 11.9.** *Suppose that there exists a Menon-Hadamard difference set in an abelian group $H \times K \times Z_{p^b}$ where $|K| = p^b$ and p is an odd prime self-conjugate modulo $\exp(H)$. Then K is cyclic.*

Recently, Ma and Schmidt (1995a) and Davis and Jedwab (1997) strengthened theorem 9 by proving:

**Theorem 11.10.** *If a group of the form $H \times Z_{p^r}^2$ has a $(hp^{2r}, \sqrt{h}p^r(2\sqrt{h}p^r - 1), \sqrt{h}p^r(\sqrt{h}p^r - 1))$ Menon-Hadamard difference set, p a prime not dividing $|H| = h$ and $p^j \equiv -1 \pmod{\exp(H)}$ for some j, then $H \times Z_{p^t}^2$ has a $(hp^{2t}, \sqrt{h}p^t(2\sqrt{h}p^t - 1), \sqrt{h}p^t(\sqrt{h}p^t - 1))$ Menon-Hadamard difference set for every $0 \le t \le r$.*

**Remark** A special case of theorem 10 above is contained in earlier results of McFarland (1990a, b) and Chan et al. (1994):

**Theorem 11.11.** *Suppose there exists a Menon-Hadamard difference set in an abelian group $H \times Z_{p^b}^2$ of order $hp^{2b}$ where $p > 3$ is a prime self-conjugate mod $\exp(H)$. Then $(p + 1)|h$ and $h > (p + 1)^2$, and there exists a Menon-Hadamard difference set in $H \times Z_{p^c}^2$ for each nonnegative integer $c \le b$.*

Chan (1993) obtained a nonexistence result without self-conjugacy assumption:

**Theorem 11.12.** *Suppose there exists a Menon-Hadamard difference set in an abelian group $Z_2^2 \times Z_p^2 \times Q$ of order $4p^2q^{2b}$, where $p$ and $q$ are distinct odd primes for which $ord_p(q)$ is odd. Then $p \leq 2q^b(p-1)/ord_p(q)$.*

We finally mention a result of Ray-Chaudhuri, and Xiang (1997):

**Theorem 11.13.** *There are no Menon-Hadamard difference sets in abelian group $Z_2 \times Z_2 \times P$, where $|P| = p^{2a}$, $a$ is odd and $p$ is a prime, $p \equiv 1$ (mod 4).*

Theorem 11.13 is a partial generalization of the result of Mann and McFarland (1973).

## 12. Updating Recent Tables of Abelian Difference Sets

Admissible parameters $(v, k, \lambda)$ for abelian difference sets satisfying $k(k-1) = \lambda(v-1)$, conclusions of theorem 4.3 and $k \leq v/2$ have been compiled in tables by Lander (1983), Kopilovich (1989), and Lopez and Sanchez (1997).

All of the 25 missing entries in Lander's table can now be filled as follows:

| $(v, k, \lambda)$ | Group | Exist? | Reference |
|---|---|---|---|
| (27, 13, 6) | $Z_3 \times Z_9$ | NO | Arasu (1988) Wei (1990) Bozikov (1985) |
| (81, 16, 3) | $Z_9^2$ | NO | Arasu (1986) Bozikov (1985) |
| | $Z_3^2 \times Z_9$ | NO | Arasu (1986) Bozikov (1985) |
| (96, 20, 4) | $Z_4 \times Z_8 \times Z_3$ | NO | Arasu & Sehgal (1995) |
| | $Z_2^2 \times Z_8 \times Z_3$ | NO | Arasu *et al.* (1996) |
| | $Z_2 \times Z_4^2 \times Z_3$ | YES | Arasu & Sehgal (1995) |
| (64, 28, 12) | $Z_4 \times Z_{16}$ | YES | Davis (1991) |
| | $Z_8^2$ | YES | Davis (1991) |
| (375, 34, 3) | $Z_3 \times Z_5 \times Z_{25}$ | NO | Arasu (1988) Wei (1990) |
| | $Z_3 \times Z_5^3$ | NO | Iiams, Liebler & Smith (1996) |
| (704, 38, 2) | $Z_4^3 \times Z_{11}$ | NO | Arasu (1988) |
| | $Z_2^2 \times Z_4^2 \times Z_{11}$ | NO | Arasu (1988) |
| | $Z_2^4 \times Z_4 \times Z_{11}$ | NO | Arasu (1988) |
| | $Z_2^6 \times Z_{11}$ | NO | Arasu (1988) |
| (288, 42, 6) | $Z_4 \times Z_8 \times Z_3^2$ | NO | Iiams (*personal communication*) |
| | $Z_2^2 \times Z_8 \times Z_3^2$ | NO | Iiams (*personal communication*) |
| (100, 45, 20) | $Z_4 \times Z_5^2$ | NO | McFarland (1989) |
| (208, 46, 10) | $Z_4^2 \times Z_{13}$ | NO | Iiams (*personal communication*) |
| | $Z_2^2 \times Z_4 \times Z_{13}$ | NO | Iiams (*personal communication*) |
| | $Z_2^4 \times Z_{13}$ | NO | Iiams (*personal communication*) |
| (189, 48, 12) | $Z_3 \times Z_9 \times Z_7$ | NO | Arasu & Sehgal (1992) Arasu, McDonough & Sehgal (1993) |
| | $Z_3^3 \times Z_7$ | NO | Iiams (*personal communication*) |
| (176, 50, 14) | $Z_4^2 \times Z_{11}$ | NO | Iiams (*personal communication*) |
| | $Z_2^2 \times Z_4 \times Z_{11}$ | NO | Iiams (*personal communication*) |
| | $Z_2^4 \times Z_{11}$ | NO | Iiams (*personal communication*) |

Next, we update Kopilovich's table (1989).

| | | | |
|---|---|---|---|
| $(160, 54, 18)$ | $Z_2 \times Z_{16} \times Z_5$ | NO | Ma & Schmidt (1997) |
| | $Z_4 \times Z_8 \times Z_5$ | NO | Ma & Schmidt (1997) |
| $(400, 57, 8)$ | $Z_{16} \times Z_5^2$ | NO | Arasu (1997) |
| $(144, 66, 30)$ | $Z_2 \times Z_8 \times Z_3^2$ | YES | Arasu, Davis, Jedwab, Sehgal (1993) |
| $(783, 69, 6)$ | $Z_3^3 \times Z_{29}$ | | Arasu & Voss (1994) (*unpublished*) Schmidt (1997) |
| $(640, 72, 8)$ | $Z_2^4 \times Z_8 \times Z_5$ | NO | Ma & Schmidt (1995) |
| | $Z_2^3 \times Z_{16} \times Z_5$ | NO | Ma & Schmidt (1995) |
| | $Z_2^2 \times Z_4 \times Z_8 \times Z_5$ | NO | Ma & Schmidt (1995) |
| | $Z_4^2 \times Z_8 \times Z_5$ | NO | Ma & Schmidt (1995) |
| | $Z_8 \times Z_{16} \times Z_5$ | NO | Arasu & Sehgal (1995) Ma & Schmidt (1995) |
| | $Z_2 \times Z_4 \times Z_{16} \times Z_5$ | NO | Ma & Schmidt (1995) |
| | $Z_2 \times Z_8^2 \times Z_5$ | NO | Ma & Schmidt (1995) |
| $(333, 84, 21)$ | $Z_3^2 \times Z_{37}$ | NO | Arasu & Sehgal (1992) Arasu, Sehgal & McDonough (1993) |
| $(320, 88, 24)$ | $Z_2^6 \times Z_5$ | YES | Davis & Jedwab (1997) |
| | $Z_2^4 \times Z_4 \times Z_5$ | YES | Davis & Jedwab (1997) |
| | $Z_2^3 \times Z_8 \times Z_5$ | NO | Ma & Schmidt (1997a) |
| | $Z_2^2 \times Z_4^2 \times Z_5$ | YES | Davis & Jedwab (1997) |
| | $Z_2 \times Z_4 \times Z_8 \times Z_5$ | NO | Ma & Schmidt (1997a) |
| | $Z_8^2 \times Z_5$ | NO | Arasu & Sehgal (1995) |
| $(891, 90, 9)$ | $Z_3^2 \times Z_9 \times Z_{11}$ | NO | Arasu & Ma (1998) |
| | $Z_9^2 \times Z_{11}$ | NO | Arasu & Ma (1998) |

Finally, we mention that ten missing entries in the recent table of Lopez and Sanchez (1997) can be filled with answer 'NO' (See Arasu & Sehgal (1998)). We now list the open cases for $k \leq 150$:

| $(v, k, \lambda)$ | Group |
|---|---|
| $(640, 72, 8)$ | $Z_2 \times Z_4^3 \times Z_5$ |
| | $Z_2^3 \times Z_4^2 \times Z_5$ |
| $(320, 88, 24)$ | $Z_4^3 \times Z_5$ |
| $(1200, 110, 10)$ | $Z_2^2 \times Z_4 \times Z_3 \times Z_5^2$ |
| | $Z_4^2 \times Z_3 \times Z_5^2$ |
| | $Z_2 \times Z_8 \times Z_3 \times Z_5^2$ |
| $(261, 105, 42)$ | $Z_3^2 \times Z_{29}$ |
| $(429, 108, 27)$ | $Z_{429}$ |
| $(768, 118, 18)$ | $Z_4 \times Z_{64} \times Z_3$ |
| | $Z_8 \times Z_{32} \times Z_3$ |
| | $Z_{16}^2 \times Z_3$ |
| $(243, 121, 60)$ | $Z_9 \times Z_{27}$ |
| | $Z_3 \times Z_9^2$ |
| $(841, 120, 17)$ | $Z_{29}^2$ |
| $(715, 120, 20)$ | $Z_{715}$ |

| | |
|---|---|
| (364, 121, 40) | $Z_2^2 \times Z_7 \times Z_{13}$ |
| (351, 126, 45) | $Z_3 \times Z_9 \times Z_{13}$ |
| | $Z_{351}$ |
| (837, 133, 21) | $Z_3^3 \times Z_{31}$ |
| | $Z_3 \times Z_9 \times Z_{31}$ |
| | $Z_{837}$ |
| (419, 133, 42) | $Z_{419}$ |
| (361, 136, 51) | $Z_{19}^2$ |
| (465, 145, 45) | $Z_{465}$ |
| (1225, 136, 15) | $Z_5^2 \times Z_7^2$ |
| (448, 150, 50) | $Z_2 \times Z_{32} \times Z_7$ |
| | $Z_4 \times Z_{16} \times Z_7$ |
| | $Z_2^2 \times Z_{16} \times Z_7$ |
| | $Z_2 \times Z_4 \times Z_8 \times Z_7$ |
| (5440, 148, 4) | $Z_4^3 \times Z_5 \times Z_{17}$ |
| | $Z_8^2 \times Z_5 \times Z_{17}$ |
| | $Z_2 \times Z_4 \times Z_8 \times Z_5 \times Z_{17}$ |

## Acknowledgement

The authors are *very* grateful to Dr. Siu-Lun Ma for reading this manuscript carefully and providing valuable comments.

## References

Arasu, K.T., (81, 16, 3) Abelian difference sets do not exist. *J. Combin. Th.*, **A43**, 350–353, 1986.

Arasu, K.T., Abelian projective planes of square order. *Europ. J. Combin.*, **10**, 207–209, 1989.

Arasu, K.T., More missing entries in Lander's table could be filled. *Arch. Math.*, **51**, 188–192, 1988.

Arasu, K.T., On abelian difference sets. *Arch. Math.*, **48**, 491–494, 1987.

Arasu, K.T., Recent results on difference sets, In: *Coding Theory and Design Theory, Part II* (Ed. D. Ray-Chaudhuri) Springer, Heidelberg, 1–23, 1990.

Arasu, K.T., Singer groups of biplanes of order 25. *Arch. Math.*, **53**, 622–624, 1989.

Arasu, K.T., On the existence of (400, 57, 8) abelian difference sets, (1997), *manuscript*.

Arasu, K.T., Davis, J.A., Jedwab, J., Ma, S.L., and McFarland, R.L., Exponent bounds for a family of abelian difference sets. In: *Groups, Difference Sets, and the Monster.* (*Eds.* K.T. Arasu, J.F. Dillon, K. Harada, S.K. Sehgal and R.L. Solomon) DeGruyter Verlag, Berlin/New York, 129–143, 1996.

Arasu, K.T., Davis, J.A., Jedwab, J., and Sehgal, S.K., New constructions of Menon difference sets, *J. Combin. Th.*, **A64**, 329–336, 1993.

Arasu, K.T., Davis, J.A., and Jedwab, J., A nonexistence result for abelian Menon difference sets using perfect binary arrays, *Combinatorica*, **15**(3), 311–317, 1995.

Arasu, K.T., Davis, J.A., Jungnickel, D., and Pott, A., A note on intersection numbers of difference sets. *Europ. J. Combin.*, **11**, 95–98, 1990.

Arasu, K.T., and Ma, S.L., Abelian difference sets without self-conjugacy. *Design, Codes, and Cryptography, 1998 to appear.*

Arasu, K.T., and Mavron, V.C., Biplanes and singer groups, In: *Coding Theory, Design Theory, Group Theory*, (*Eds.* D. Jungnickel, S.A. Vanstone) Wiley, New York, 111–119, 1993.

Arasu, K.T., McDonough, T., and Sehgal, S.K., Sums of roots of unity, In: *Group Theory*, (*Eds.* R. Solomon and S.K. Sehgal) World Scientific, Singapore, 6–20, 1993.

Arasu, K.T., and Ray-Chaudhuri, D.K., Multiplier theorem for a difference list. *Ars Combin.*, **22**, 119–137, 1986.

Arasu, K.T., and Sehgal, S.K., Some new results on abelian difference sets, *Utilitas Math.*, **42**, 225–233, 1992.

Arasu, K.T., and Sehgal, S.K., Difference sets in abelian groups of $p$-rank two. *Designs, Codes, and Cryptography*, **5**, 5–12, 1995.

Arasu, K.T., and Sehgal, S.K., Non-existence of some difference sets. *JCMCC, 1998 to appear.*

Arasu, K.T., and Sehgal, S.K., Some new difference sets. *J. Combin. Th.*, **A69**, 170–172, 1995.

Arasu, K.T., and Xiang, Q., Multiplier theorems. *J. Comb. Des.*, **3**, 257–267, 1995.

Baumert, L.D., (1977), Cyclic difference sets. *Lect. Not. Math.*, Springer, New York, 182, 1971.

Baumert, L.D., and Fredricksen, H., The cyclotomic numbers of order 18 with applications to difference sets. *Math. Comp.*, **21**, 204–219, 1967.

Baumert, L.D., Mills, W.H., and Ward, R.L., Uniform cyclotomy. *J. Nr. Th.*, **14**, 67–82, 1982.

Beth, T., Jungnickel, D., and Lenz, H., *Design Theory*, Cambridge Univ. Press, Cambridge (2$^{nd}$ edition), 1998.

Bozikov, Z., Abelian Singer groups of certain symmetric block designs. *Radovi Mat.*, **1**, 247–253, 1985.

Broughton, W.J., A note on table 1 of "Barker sequences and difference sets". *L'Ens. Math.*, **50**, 105–107, 1994.

Brualdi, R.A., A note on multipliers of difference sets. *J Res. Nat Bur. Standards B*, **69**, 87–89, 1965.

Bruck, R.H., Difference sets in a finite group. *Trans. Amer. Math. Soc.*, **78**, 464–481, 1955.

Bruck, R.H., and Ryser, H.J., The non-existence of certain finite projective planes. *Canad. J. Math.*, **1**, 88–93, 1949.

Camion, P., and Mann, H.B., Antisymmetric difference sets. *J. Nr. Th.*, **4**, 266–268; 1972.

Chan, W.K., Necessary conditions for Menon difference sets, *Design, Codes, and Cryptography*, **3**, 147–154, 1993.

Chan, W.K., and Siu, M.K., Summary of perfects $s \times t$ arrays, $1 \leq s \leq t \leq 100$, *Electr. Lett.*, **27**, 709, 710, 1991.

Chan, W.K., and Siu, M.K., Correction to "summary of perfect $s \times t$ arrays, $1 \leq s \leq t \leq 100$", *Electr. Letter*, **27**, 1112, 1991.

Chan, W.K., Siu, M.K., and Ma, S.L., Nonexistence of certain perfect arrays, *Discrete Math.*, **125**, 107–113, 1994.

Chen, Y.Q., On the existence of abelian Hadamard difference sets and a new family of difference sets. *Finite Fields Appl.*, **3**, 234–256, 1997.

Chen Y.Q., Xiang, Q., and Sehgal, S.K., An exponent bound on skew Hadamard abelian difference sets. *Designs, Codes, and Cryptography*, **4**, 313–317, 1994.

Chowla, S., A property of biquadratic residues. *Proc Nat. Acad. Sci. India*, **A14**, 45–46, 1994.

Chowla, S., and Ryser, H.J., Combinatorial problems. *Canad. J. Math.*, **2**, 93–99, 1950.

Cohen, S.D., Generators in cyclic difference sets. *J. Combin. Th.*, **A51**, 227–236, 1989.

Colbourn, C.J., and Dinitz, J.H., *(Eds.)*, *The CRC Handbook of Combinatorial Designs*, CRC Press, Boca Raton, 1996.

Davis, J., A generalization of Kraemer's result on difference sets. *J. Combin. Th.*, **A59**, 187–192, 1992.

Davis, J., A result on Dillon's conjecture in difference sets. *J. Combin. Th.*, **A57**, 238–242, 1991.

Davis, J.A., Difference sets in abelian 2-groups, *J. Combin. Th.*, **A57**, 262–286, 1991.

Davis, J.A., and Jedwab, J., A Survey of Hadamard difference sets, In: *Groups. Difference Sets, and the Monster*, *(Eds.* K.T. Arasu, J. Dillon, K. Haradu, S.K. Sehgal and R. Solomon), deGruyter Verlag, Berlin/New York, 145–156, 1996.

Davis, J.A., and Jedwab, J., A unifying construction of difference sets. *J. Combin. Th.*, **A80**, 13–78, 1997.

Davis, J.A., and Jedwab, J., A summary of Menon difference sets. *Congr. Nr.*, **93**, 203–207, 1993.

Davis, J.A., and Jedwab, J., Some recent developments in difference sets. In: Combinatorical Designs and Applications *(Eds.* K. Quinn *et al.*,) Addison-Wesley, London, *(to appear)*.

Davis, J.A., and Jedwab, J., Nested Hadamard difference sets, *J. Stat. Planning & Inference*, **62**, 13–20, 1997.

Dillon, J.F., A Survey of difference sets in 2-groups. Presented at the Marshall Hall Memorial Conference, Burlington, VT, 1990a.

Dillon, J.F., Cyclic difference sets and primitive polynomials. In: *Finite Fields, Coding Theory and Advances in Communications and Computing (Eds:* G.L. Mullen and P.J.S. Shiue), Marcel Dekker, New York, 436–437, 1993.

Dillon, J.F., Difference sets in 2-groups. In: *Proc. NSA Math. Sci. Meet.*, Ft. George Meade, Maryland, *(Ed.* R.L. Ward) 165–172, 1987.

Dillon, J.F., Difference sets in 2-groups, In: Finite Geometries and Combinatorial Designs, Contemporary Mathematics. Vol. 111 *(Ed.* E.S. Kraemer), Birkhäuser, Boston, 65–72, 1990a.

Dillon, J.F., Elementary Hadamard difference sets. Ph.D. Thesis, University of Maryland, 1974.

Dillon, J.F., Elementary Hadamard difference sets. In: Proc. *Sixth Southeastern Conf. On Combinatorics, Graph Theory and Computing*, 237–249, 1975.

Dillon, J.F., Variations on a scheme of McFarland for noncyclic difference sets. *J. Combin. Th.*, **A40**, 9–21, 1985.

Eckmann, A., Eliahou, S., and Kervaire, M., Čompressible difference lists. *J. Statist. Planning & Inf.*, **62**(1), 35–38, 1997.

Eliahou, S., and Kervaire, M., A note on the equation $\theta\bar{\theta} = n + \lambda\Sigma$. *J. Statist. Planning & Inf.*, **62**(1), 21–34, 1997.

Eliahou, S., and Kervaire, M., Barker sequences and difference sets. *L'Ens. Math.*, **38**, 345–382, 1992.

van Eupen, M., and Tonchev, V.D., Linear codes and the existence of a reversible Hadamard difference set in $Z_2 \times Z_2 \times Z_5^4$, *J. Combin. Th.*, **A79**, 161–167, 1997.

Golomb, S.W., and Song, H.Y., A conjecture on the existence of cyclic Hadamard difference sets. *J. Statist. Planning Inf.*, **62**(1), 39–42, 1997.

Gordon, B., Mills, W.H., and Welsch, L.R., Some new difference sets. *Canad. J. Math.*, **14**, 614–625, 1962.

Gordon, D.M., The prime power conjecture is true for $n \leq 2,000,000$. *Electron. J. Comb.*, **1**, 1994.

Hall, M., Jr., Cyclic projective planes. *Duke J. Math.*, **14**, 1079–1090, 1947.

Hall, M., Jr., A survey of difference sets. *Proc. Amer. Math. Soc.*, **7**, 975–986, 1956.

Hall, M., Jr., *Combinatorial Theory*. $2^{nd}$ *Ed.* Wiley, New York, 1986.

Hall, M. Jr., and Ryser, H.J., Cyclic incidence matricess. *Canad. J. Math.*, **3**, 495–502, 1951.

Hayashi, H.S., Computer investigation of difference sets. *Math. Comp.*, **19**, 73–78, 1965.

Ho, C.Y., On bounds for groups of multipliers of planar difference sets. *J. Algebra*, **148**, 325–336, 1992.

Ho, C.Y., On multiplier groups of finite cyclic planes. *J. Algebra*, **122**, 250–259, 1989.

Ho, C.Y., Planar Singer groups with even order multiplier groups. In: *Finite Geometry and Combinatorics. (Eds.* A. Beutelspacher, *et al.,)* Cambridge University Press, Cambridge, 187–198, 1993.

Ho, C.Y., Projective planes with a regular collineation group and a question about powers of a prime. *J. Algebra*, **154**, 141–151, 1993.

Ho, C.Y., Singer groups, an approach from a group of multipliers of even order. *Proc. Amer. Math. Soc.*, **119**, 925–930, 1993.

Ho, C.Y., Some basic properties of planar Singer groups. *Geom. Ded.*, **55**, 59–70, 1995.

Ho, C.Y., and Pott, A., On multiplier groups of planar difference sets and a theorem of Kantor. *Proc. Amer. Math. Soc.*, **109**, 803–808, 1990.

Iiams, J., Liebler, R., and Smith, K.W., In: Groups, Difference sets and the monster, (Eds: K.T. Arasu et al.,), DeGruyter Verlag, Berlin/New York, 157–168, 1996.

Ireland, K., and Rosen, M., *A Classical Introduction to Modern Number Theory.* Springer, New York, 1982.

Jedwab, J., Generalized perfect arrays and Menon difference sets, *Designs, Codes, and Cryptography*, **2**, 19–68, 1992.

Jedwab, J., Non-existence of perfect binary arrays. *Electron. Lett.*, **27**, 1252–1254, 1991.

Jedwab, J., Non-existence of perfect binary arrays. *Electron. Lett.*, **29**, 99–101, 1993.

Jedwab, J., Mitchell, C., Piper, F., and Wild, P., Perfect binary arrays and difference sets. *Discr. Math.*, **125**, 241–254, 1994.

Johnsen, E.C., Skew-Hadamard abelian group difference sets. *J. Algebra*, **4**, 388–402, 1966.

Jungnickel, D., Difference sets, In: *Comtemporary Design Theory: A Collection of Surveys (Eds.* J.H. Dinitz and D.R. Stinson), Wiley, New York, 241–324, 1992*a*.

Jungnickel, D., On Lander's multiplier theorem for difference lists. *J. Comb. Inf. System Sci.*, **17**, 123–129, 1992*b*.

Jungnickel, D., and Pott, A., Difference sets: abelian. In: *The CRC Handbook of Combinatorial Designs (Eds.* C.J. Colbourn and J. Dinitz) CRC Press, Boca Raton, 297–312, 1996.

Jungnickel, D., and Pott, A., Two results on difference sets. *Coll. Math. Soc. János Bolyai*, **52**, 325–330, 1988.

Jungnickel, D., and Schmidt, B., Difference sets: an update, In: *Combinatorial Designs and Related Structure (Eds.* J.W.P. Hirschfeld, S.S. Magliveras, and M.J. de Resmini), *London Math. Soc. Lect. Not.*, **245**, 89–112, 1997.

Jungnickel, D., and Vedder, K., On the geometry of planar difference sets. *Europ. J. Combin.*, **5**, 143–148, 1984.

Kantor, W.M., Exponential numbers of two-weight codes, difference sets and symmetric designs. *Discrete Math.*, **46**, 95–98, 1984.

Kantor, W.M., Primitive permutation groups of odd degree, and an application to finite projective planes. *J. Alg.*, **106**, 15–45, 1987.

Kibler, R.E., A summary of non-cyclic difference sets, $k \leq 20$. *J. Combin. Th.*, **A25**, 62–67, 1978.

Kopilovich, L.E., Difference sets in non-cyclic abelian groups. *Kibernetika*, **2**, 20–23, 1989.

Kraemer, R.G., Proof of a conjecture on Hadamard 2-groups, *J. Combin. Th.*, **A63**, 1–10, 1993.

Lander, E.S., *Symmetric Designs: An Algebraic Approach*, Cambridge University Press, Cambridge, 1983.

Lander, E.S., Restrictions upon multipliers of an abelian difference set. *Arch. Math.*, **50**, 241–242, 1988.

Lehmer, E., On residue difference sets. *Canad. J. Math.*, **5**, 425–432, 1953.

Ma, S.L., and Schmidt, B., A sharp exponent bound for McFarland difference sets with $p = 2$. *J. Comb. Th.*, **A80**, 347–352, 1997a.

Ma, S.L., and Schmidt, B:, Difference sets corresponding to a class of symmetric designs. *Designs, Codes and Cryptography*, **10**, 223–236, 1997.

Ma, S.L., and Schmidt, B., The structure of abelian groups containing McFarland difference sets, *J. Comb. Th.*, **A70**, 313–322, 1995.

Ma, S.L., and Schmidt, B., On $(p^a, p, p^a, p^{a-1})$-relative difference sets. *Designs, Codes and Cryptography*, **6**, 57–71, 1995a.

Mann, H.B., Some theorems on difference sets. *Canad. J. Math.*, **4**, 222–226, 1952.

Mann, H.B., Balanced incomplete block designs and abelian difference sets. *Illinois J. Math.*, **8**, 252–261, 1964.

Mann, H.B., *Addition Theorems*, Wiley, New York, 1965.

Mann, H.B., Difference sets in elementary abelian groups. *Illinois J. Math.*, **9**, 212–219, 1965.

Mann, H.B., and McFarland, R.L., On Hadamard difference sets, In: *A Survey of Combinatorial Theory* (*Eds.* J.N. Srivastava *et al.*), North-Holland, Amsterdam, 333–334, 1973.

Mann, H.B., and Zaremba, S.K., On multipliers of difference sets. *Illinois J. Math.*, **13**, 378–382, 1969.

McFarland, R.L., Difference sets in abelian groups of order $4p^2$, *Mitt. Math. Sem., Giessen*, **192**, 1–70, 1989.

McFarland, R.L., A family of difference sets in non-cyclic abelian groups. *J. Comb. Th.*, **A15**, 1–10, 1973.

McFarland, R.L., On multipliers of abelian difference sets, *Ph. D. Thesis.*, Ohio State University, 1970.

McFarland, R.L., and Mann, H.B., On multipliers of abelian difference sets. *Canad. J. Math.*, **17**, 541–542, 1965.

McFarland, R.L., Necessary conditions for Hadamard difference sets, In: *Coding Theory and Design Theory* (*Ed.* D.K. Ray-Chaudhuri), Springer, New York, 257–272, 1990a.

McFarland, R.L., Subdifference sets of Hadamard difference sets, *J. Combin. Th.*, **A54**, 112–122, 1990b.

McFarland, R.L., and Rice, B.F., Translates and multipliers of abelian difference sets. *Proc. Amer. Math. Soc.*, **68**, 375–379, 1978.

Menon, P.K., Difference sets in abelian groups. *Proc. Amer. Math. Soc.,* **11**, 368–376, 1960.

Menon, P.K., On difference sets whose parameters satisfy a certain relation, *Proc. Amer. Math. Soc.,* **13**, 739–745, 1962.

Muzychuk, M., Difference sets with $n = 2p^m$. *J. algeb. Combin.,* **7**, 77–89, 1998.

Newman, M., Multipliers of difference sets. *Canad. J. Math.,* **15**, 121–124, 1963.

Ostrom, T.G., Concerning difference sets. *Canad. J. Math.,* **5**, 421–424, 1953.

Pott, A., Applications of the DFT to abelian difference sets. *Arch. Math.,* **51**, 283–288, 1988.

Pott, A., Differenzenmengen und Gruppenalgebren, *Ph.D. Thesis,* Univ. Giessen, 1988.

Pott, A., Finite geometry and character theory, *Lect. Not. Math.,* 1601, Springer-Verlag, Berlin, 1995.

Pott, A., New necessary conditions on the existence of abelian difference sets. *Combinatorica,* **12**, 89–93, 1992.

Pott, A., On abelian difference set codes. *Designs, Codes and Crypt.,* **2**, 263–271, 1992.

Pott, A., On multipliers theorems. In: Coding Theory and Design Theory, Part II (*Ed.* D. Ray-Chaudhuri). Springer, New York, 286–289, 1990.

Qiu, W., A method for studying the multiplier conjecture and a partial solution to it. *Ars. Comb.,* **39**, 5–23, 1995.

Qiu, W., A necessary condition on the existence of abelian difference sets. *Discr. Math.,* **137**, 383–386, 1995.

Qiu, W., Further results on the multiplier conjecture for the case $n = 2n_1$. *J. Comb. Math. Comb. Comp.,* **20**, 27–31, 1996.

Qiu, W., On the multiplier conjecture. *Acta. Math. Sinica, New Series,* **10**, 49–58, 1994.

Qiu, W., Proving the multiplier theorem using representation theory of groups. *Northeast Math. J.,* **9**, 169–172, 1993.

Qiu, W., The multiplier conjecture for elemenatary abelian groups. *J. Comb. Des.,* **2**, 117–129, 1994.

Qiu, W., The multiplier conjecture for the case $n = 4n_1$. *J. Comb. Des.,* **3**, 393–397, 1995.

Ray-Chaudhuri, D.K., and Xiang, Q., New necessary conditions for abelian Hadamard difference sets, *J. Stat. Planning & Inf.,* **62**, 69–80, 1997.

Rothaus, O.S., On "bent" functions. *J. Combin. Th.,* **A20**, 300–305, 1976.

Schmidt, B., Circulant Hadamard matrices: overcoming non-selfconjugacy. *Submitted.*

Schmidt, B., Cyclotomic integers of prescribed absolute value and the class group. *Submitted.*

Schmidt, B., "Defferenzmengen und relative Defferenzmengen", Augsburger Mathematisch-Naturwissenschaftliche Schriften Band 4, Augsburg, 1995.

Schmidt, B., Non-existence results on Chen and Davis-Jedwab difference sets. *Submitted.*

Schutzenberger, M.P., A non-existence theorem for an infinite family of symmetrical block designs. *Ann. Eugenics.,* **14**, 186–187, 1949.

Singer, J., A theorem in finite projective geometry and some applications to number theory. *Trans. Amer. math Soc.,* **43**, 377–385, 1938.

Smith, K.W., A table on non-abelian difference sets. In: *"CRC Handbook of Combinatorial Designs"* (*Eds.* C.J. Colbourn and J.H. Dinitz), CRC press, Boca Raton, 308–312, 1996.

Song, H.Y., and Golomb, S.W., On the existence of cyclic Hadamard difference sets. *IEEE Trans. Inf. Th.,* **40**, 1266–1268, 1994.

Spence, T., A family of difference sets in non-cyclic groups. *J. Combin. Th.,* **A22**, 103–106, 1977.

Stanton, R.G., and Sprott, D.A., A family of difference sets. *Canad. J. Math.*, **11**, 73–77, 1958.

Storer, J., *Cyclotomy and Difference Sets*. Markham, Chicago, 1967.

Takeuchi, K., A table of difference sets generating balanced incomplete block designs. *Rev. Inst. int. Statist.*, **30**, 361–366, 1962.

Turyn, R.J., Character sums and difference sets, *Pacific J. Math.*, **15**, 319–346, 1965.

Turyn, R.J., A special class of Willaimson matrices and difference sets, *J. Combin. Th.*, **A36**, 111–115, 1984.

Turyn, R.J., The multiplier theorem for difference sets. *Canad. J. Math.*, **16**, 386–388, 1964.

van Eupen, M., and Tonchev, V.D., Linear codes and the existence of reversible Hadamard difference sets in $Z_2 \times Z_2 \times Z_5^4$, *J. Combin. Th.*, **A79**, 161–167, 1997.

Vera Lopez, A., and Garcia Sanchez, M.A., On the existence of abelian difference sets with $100 < k \leq 150$. *J. Comb. Math. Com. Comp.*, **23**, 97–112, 1997.

Wei, R., Non-existence of some abelian difference sets. In: *Combinatorial Design and Applications* (*Eds.* W.D. Wallis, H. Shen, W. Wei, and L. Zhu). Marcel Dekker, New York, 159–164, 1990.

Whiteman, A.L., A family of difference sets. *Illinois J. Math.*, **6**, 107–121, 1962.

Wilbrink, H.A., A note on planar difference sets. *J. Combin. Th.*, **A38**, 94–95, 1985.

Wilson, R.M., and Xiang, Q., Constructions of Hadamard difference sets. *J. Combin. Th.*, **A77**, 148–160, 1997.

Xia, M.Y., Some infinite classes of special Willaimson matrices and difference sets, *J. Combin. Th.*, **A36**, 111–115, 1992.

Xiang, Q., Some results on multipliers and numerical multiplier groups of difference sets. *Graph. Combin.*, **10**, 293–304, 1994.

Xiang, Q., and Chen, Y.Q., On the size of the multiplier groups of cyclic difference sets. *J. Combin. Th.*, **A69**, 168–169, 1995.

Xiang, Q., and Chen, Y.Q., On Xia's construction of Hadamard difference sets. *Finite Field. Appl.*, **2**, 86–95, 1996.

Yamamoto, K., Decomposition fields of difference sets. *Pacific J. Math.*, **13**, 337–352, 1953.

Yamamoto, K., On Jacobi sums and difference sets. *J. Combin. Th.*, **3**, 146–181, 1967.

Yamamoto, K., On the application of half-norms to cyclic difference sets. In: *Combinatorial Mathematics and Its Applications* (*Eds.* R.C. Bose and T.A. Dowling). University of North Carolina Press, Chapel Hill, 1969.

Department of Mathematics and Statistics, Wright State University, Dayton, Ohio 45435, U.S.A.
Email: karasu@math.wright.edu

Department of Mathematics, Ohio State University, Columbus, Ohio 43210, U.S.A.
Email: sehgal@math.ohio-state.edu

# Unit Groups of Group Rings

*Ashwani K. Bhandari and I.B.S. Passi*

## Introduction

For a commutative ring $R$ with identity and an arbitrary group $G$, let $RG$ denote the group ring of $G$ over $R$ and $U(RG)$ its group of units. It is of interest, see the survey by Dennis (1977), to determine the necessary and sufficient conditions on $R$ and $G$ in order that $U(RG)$ has a specific group-theoretic property, e.g., solvability, nilpotence, etc.. Considerable work has been done, by various authors, on these questions over the last thirty years or so. Besides solvability and nilpotence, other group-theoretic problems for the unit groups of group rings like the characterization of residual nilpotence, residual solvability, being an FC-group, torsion elements forming a subgroup, as well as the behaviour of the upper central series of the unit groups have also been studied. S.K. Sehgal's book (1978) covers the main results obtained in this direction up to 1977 while the article by C. Polcino Milies (1981) gives some later developments (see also the books of Sehgal, 1989 and Karpilovsky, 1989). In this article our main aim is to survey the more recent developments. In §1 we review the case when $R$ is a field and in §2 the case of the integral group ring is considered.

## 1. Unit Groups of Group Algebras

Throughout this section the coefficient ring $R$ is a field, denoted by $K$.

Given a group $H$, its FC-subgroup $\Phi(H)$ is defined as

$$\Phi(H) = \{h \in H \,|\, h \text{ has only a finite number of conjugates in } H\}.$$

A group $H$ is called an $FC$-group if $H = \Phi(H)$. The problem of determining the conditions for $U(KG)$ to be an $FC$-group has been considered by Cliff and Sehgal (1978), Polcino Milies (1978$a$, $b$) and Sehgal and Zassenhaus (1977) while the conditions for the unit group to be a locally $FC$-group and related properties are examined by Bist (1992).

The nilpotent or $FC$-groups $G$ for which the torsion elements of $U(KG)$ form a subgroup have been studied by Pollino Milies (1981). The nilpotence of $U(KG)$

is studied by Bateman and Coleman (1968), Bhattacharya and Jain (1970), Fisher *et al.*, (1976), Khripta (1972) and Motose and Tominaga (1969). Most of these results are available in Sehgal's book (1978).

For a class $C$ of groups, let $\mathcal{RC}$ denote the residually-$C$ groups, i.e., the class of those groups $G$ which satisfy the property that for every $1 \neq x \in G$, there exists a normal subgroup $N_x$ of $G$ such that $x \notin N_x$ and $G/N_x \in C$. It is easy to see that a group $G$ is residually nilpotent (resp: residually solvable) if and only of $\cap_{n=1}^{\infty} \gamma_n(G) = \{1\}$ (resp: $\cap_{n=0}^{\infty} G^{(n)} = \{1\}$), where $\gamma_n(G)$ and $G^{(n)}$ denote respectively the $n^{\text{th}}$ term of the lower central and the derived series of $G$. Goncalves (1984) has studied the residual nilpotence of $U(KG)$ and proved that if $G$ is a finite group, then $U(KG)$ is residually nilpotent if and only if $U(KG)$ is nilpotent. For a field $K$ of characteristic $p > 0$, and a nontorsion nilpotent group $G$ having no $p$-element and having every torsion element of prime order, Goncalves has proved that $U(KG)$ is residually nilpotent if and only if the torsion subgroup of $G$ is central in $G$. A complete characterization of the residual nilpotence of $U(KG)$ is still not known.

Once $U(KG)$ is nilpotent, the next question is to determine its nilpotency class $\text{cl}(U(KG))$. This problem has been considered by Baginski (1987), Coleman and Passman (1970) and more extensively by A. Shalev (1990b, 1991, 1993) and Shalev and Mann (1990). It naturally relates to the *Lie nilpotency indices* of group algebras as follows:

Let $A$ be an associative ring. For $n \geq 1$, define $A^{[1]}$ to be $A$ and $A^{[n]}$, for $n > 1$, to be the two-sided ideal of $A$ generated by all left normed Lie-commutators $[x_1, \ldots, x_n]$, $x_i \in A$, where $[\alpha, \beta] = \alpha\beta - \beta\alpha, \alpha, \beta \in A$. Let $A^{(n)}, n \geq 1$, be defined inductively by setting $A^{(1)} = A$ and $A^{(n)} = [A^{(n-1)}, A]A, (n > 1)$, the two-sided ideal of $A$ generated by $[\alpha, \beta], \alpha \in A^{(n-1)}, \beta \in A$. Whereas $A^{(n)} \supseteq A^{[n]}$ always, Gupta and Levin (1983) have given an example of a group algebra $KG$ for which $kG^{(n)} \subsetneq KG^{[3]}$ for any $n \geq 3$. It is proved by Passi *et al.*, (1973) that a group algebra $KG$ is Lie nilpotent, i.e., $KG^{[n]} = 0$, for some $n > 1$, if and only if *either* the characteristic of $K$ is zero and $G$ is Abelian *or* the characteristic of $K$ is a prime $p$, $G$ is nilpotent and $G'$, the derived group of $G$, is a finite $p$-group. Consequently, it follows that if $KG^{[n]} = 0$, then $KG^{(m)} = 0$ for some $m$ (cf. Passi, 1979, Chapter VI, Theorem 1.4). For a Lie nilpotent group algebra $KG$, the lower and upper Lie nilpotency indices are defiend respectively by

$$t_L(KG) = \min\{n : KG^{[n]} = 0\}$$

and

$$t^L(KG) = \min\{m : KG^{(m)} = 0\}.$$

If the characteristic of $K$ is zero and $U(KG)$ is nilpotent, then clearly $\text{cl}(U(KG)) + 1 = t_L(KG) = t^L(KG) = 2$. In the case when the characteristic of $K$ is greater than zero, the upper Lie nilpotency index $t^L(KG)$ can be computed precisely from the Lie dimension subgroups $D_{(n),K}(G) := G \cap (1 + KG^{(n)})$, by an extension of Jenning's theory [Shalev (1991), p. 33]. The subgroups $D_{(n),K}(G)$ are computed by Passi and Sehgal (1975) (see also Passi, 1979). The related subgroups $D_{[n],K}(G) := G \cap (1 + KG^{[n]})$ were determined by Bhandari-Passi (1992) and Riley (1991) independently. Shalev (1990a) raised the problem of the computation of the lower Lie nilpotency index $t_L(KG)$, the motivation being a conjecture of Jennings (1955) which, in this context, states that if $G$ is a finite $p$-group, then $cl(U(KG)) + 1 = t_L(KG)$. Du (1992) has confirmed this conjecture. We have proved the following result (cf. Bhandari and Passi, 1992):

**Theorem 1.1.** *If $K$ is a field of characteristic $p > 3$ and $G$ a group such that $KG$ is Lie nilpotent, then*

$$t_L(KG) = t^L(KG) = 2 + (p-1) \sum_{m \geq 1} m d_{(m+1)},$$

*where, for $m \geq 2$, $p^{d_{(m)}} = [D_{(m),K}(G) : D_{(m+1),K}(G)]$.*

Combining the above results we thus have:

**Theorem 1.2.** *Let $K$ be a field of characteristic $p > 3$ and let $G$ be a finite $p$-group, then*

$$cl(U(KG)) = t_L(KG) - 1 = 1 + (p-1) \sum_{m \geq 1} m d_{(m+1)}.$$

The corresponding problem when the characteristic of $K$ is 2 or 3 is still open.

Shalev (1991, 1993) has extensively studied the series $\{d_{(m)}\}$ and proved several results about the nilpotency class of $U(KG)$; in particular, it has been shown that $cl(U(KG)) \leq p^n - 1$ always. One of the main results proved by Shalev in this direction is the following:

**Theorem 1.3. (Shalev, 1993)** *Suppose that the characteristic of $K$ is $p > 3$ and that $G$ is a finite $p$-group such that $G'$ has order $p^n$ and exponent $p^e$, where $e < n$. Let $a = \max\{e, n - e\}$, $b = \min\{e, n - e\}$. Then $cl(U(KG)) \leq p^a + 2p^b - 2$, and the equality occurs if and only if $G'$ has a cyclic subgroup of index $p$ and $G'$ is not powerfully embedded in $G$.*

The study of those groups $G$ for which $U(KG)$ is solvable is carried out by Bateman (1971), Bovdi and Khripta (1974 & 1977), Motose and Tominaga (1971), Motose and Ninomiya (1972) and Sehgal (1975). A synthesis of the results for finite groups has been given by Passman (1977).

The Lie derived series of $KG$ is defined inductively by setting

$$\delta^{[0]}(KG) = KG, \delta^{[n+1]}(KG) = [\delta^{[n]}(KG), \delta^{[n]}(KG)],$$

the $K$-subspace of $KG$ generated by all Lie products $[a, b] = ab - ba, a, b \in \delta^{[n]}(KG)$. From the results of Bovdi & Khripta (1977) and Passi et al., (1973) it follows that for torsion groups $G$, $U(KG)$ is solvable if and only if $KG$ is Lie solvable, i.e., for some some $n$, $\delta^{[n]}(KG) = 0$. Goncalves (1986) has given a counter-example to show that these two statements are not equivalent in general and he has also given an example of a field $F$ containing a field $K$ and of a group $G$ such that $U(KG)$ is solvable whereas $U(FG)$ is not.

Since a non-cyclic free group is not solvable, one would naturally like to understand the conditions for the presence, as also for the absence, of non-cyclic free subgroups in $U(KG)$. This problem is studied by Goncalves (1984 & 1985) and more recently by Goncalves and Passman (1996).

The conditions for the residual solvability of $U(KG)$ turn out to be different from those for solvability, even for finite groups $G$ (in contrast to the residual nilpotence of $U(KG)$). Bhandari and Passi (1987, 1989) have investigated the residual solvability of $U(KG)$, $G$ finite, when $K$ is either a $\wp$-adic field of characteristic zero, i.e., a finite extension of the field $Q_p$ of $p$-adic numbers, or an algebraic number field. For a field $K$ of characteristic zero, if $U(KG)$ is residually solvable, it turns out (cf. Bhandari and Passi, 1987) that $G$ has to be either an Abelian group or a Hamiltonian group $G = A \times E \times Q_8$, where $E$ is an elementary Abelian 2-group, $A$ is an Abelian group of odd exponent and $Q_8$ is the quaternion group. Furthermore, for such a group $G = A \times E \times Q_8$, $U(KG)$ is residually solvable if and only if the group of units of $H_{K_d}$ is residually solvable for every $d|e$, where $K_d$ denotes the $d^{\text{th}}$ cyclotomic extension of $K$, and for a field $L$, $H_L$ denotes the usual quaternion algebra over $L$. Consequently, for a finite group $G$ and a field $K$ of characteristic zero, the investigation of the residual solvability of $U(KG)$ rests on the following two facts:-

$(i)$ If $D$ is a central division algebra over a $\wp$-adic field $\hat{K}$ of characteristic zero, then the group $D^\times$ of units $D$ is residually solvable (see Bhandari and Passi, 1989, Theorem 2.1).

$(ii)$ For an algebraic number field $K$, if $H_K$ is a division albegra, then there exists a finite prime $\wp$ of $K$ such that for the completion $\hat{K}_\wp$ of $K$ at $\wp$, the quaternion algebra $H_{\hat{K}_\wp}$ is a division algebra (this is a result of Platonov and Rapincuk, see Bhandari and Passi, 1989).

It may be recalled that $H_K$ is a division algebra if and only if the quadratic form $X_1^2 + X_2^2 + X_3^2$ is anisotropic over $K$, i.e., the stufe $s(K)$ of $K$ (defined to be the least integer $n$ such that $-1 = \alpha_1^2 + \ldots + \alpha_n^2$, $\alpha_i \in K$, if it exists, and $\infty$ otherwise) is more than 2.

Combining the above facts and analysing the stuffs of cyclotomic fields it follows that $U(QG)$ is residually solvable if and only if $G$ is *either* Abelian *or* is of the type

$(*)$ a Hamiltonian group of order $2^n m$, $m$ odd, such that the multiplicative order of 2 modulo $m$ is odd.

We also have the following results (Bhandari & Passi, 1979, Theorem 2.3 and Corollary 3.6):

**Theorem 1.4.** *Let $\hat{K}$ be a $\wp$-adic field of characteristic zero and with residue class field $F_q, q = p^f, p$ prime. Let $G$ be a finte non-Abelian group. Then $U(KG)$ is residually solvable if and only if $p = 2, [\hat{K} : Q_2]$ is odd and $G$ is of type (\*).*

**Theorem 1.5.** *Let $K$ be an algebraic number field and $\mathfrak{o}$ its ring of integers and let $G$ be a finite non-Abelian group. Let $(2) = \prod_i \wp_i^{e_i}$, where $\wp_i$ are distinct prime ideals in $K$ and let $|\mathfrak{o}/\wp_i| = 2^{f_i}$. Then $U(KG)$ is residually solvable if and only if at least one of $e_i f_i$ is odd and $G$ is of type (\*).*

From the considerations similar to the ones above, it is easy to deduce that for a field of characteristic $p > 0$ and a finite group $G$ having no element of order $p, U(KG)$ is residually solvable if and only if $G$ is Abelian. In case $G$ is infinite or $K$ is of characteristic $p > 0$ and $G$ has elements of order $p$, the investigation of residual solvability of $U(KG)$ looks more intractable and has not so far been carried out.

The properties of solvability and nilpotence for a group $G$ are special cases of the requirement that the group satisfies a group identity, i.e., there exists a non-trivial word $w(x_1, x_2, \ldots, x_n)$ in the free group generated by $x_1, \ldots, x_n$ such that $w(u_1, \ldots, u_n) = 1$ for all $u_i \in G$. B. Hartley suggested the following:

**Conjecture 1.6.** *If $G$ is a torsion group and $U(KG)$ satisfies a group identity, then $KG$ satisfies a polynomial identity, i.e., there exists a non-zero polynomial $f(y_1, \ldots, y_n) \in K\{y_1, \ldots, y_n\}$, in non-commuting variables $y_1, \ldots, y_n$, such that $f(\alpha_1, \ldots, \alpha_n) = 0$ for all $\alpha_i \in KG$.*

Certain special cases of this conjecture are studied by Giambruno *et al.,* (1994). In Dokuchaev and Goncalves (1997) and Goncalves and Mandel (1991), the group algebras for which $U(KG)$ satisfies a semigroup identity are studied. Recently in Giambruno *et al.,* (1997) Hartley's conjecture has been confirmed for an infinite field $K$. The result has been further strengthened by Billig *et al.,* (1997) and Passman (1997) and it has been shown that:

**Theorem 1.7. (Passman, 1997)** *Let $K$ be an infinite field of characteristic $p > 0$ and $G$ a torsion group. Then the following are equivalent:-*

  (i) *$U(KG)$ satisfies a group identity.*

  (ii) *$G$ has a normal p-Abelian subgroup of finite index and $G'$ is a p-group of bounded period.*

  (iii) *$U(KG)$ satisfies the group identity $(x, y)^{p^k} = 1$, for some $k \geq 0$, where $(x, y) = x^{-1}y^{-1}xy$.*

**Theorem 1.8. (Billig et al., 1997)** *Let $KG$ be a group algebra of a torsion group over an infinite field $K$ of characteristic $p > 0$. Then the following are equivalent:*

  (i) *$U(KG)$ satisfies a group identity.*

*(ii) K G satisfies a non-matrix identity (i.e., a polynomial identity not satisfied by $M_2(K)$).*

*(iii) G contains a normal subgroup A such that $G/A$ and $A'$ are finite and $G'$ is a p-group of bounded exponent.*

*(iv) $KG^{(2)}(=[KG, KG]KG)$ is nil of bounded index.*

*(v) $(U(KG))'$ is a p-group of bounded exponent.*

## 2. Unit Groups of Integral Group Rings

Throughout this section $Z$ will denote the ring of integers, $Z_{(p)}$ its localization at the prime $p$ and $\hat{Z}_p$, the ring of $p$-adic integers.

The properties of nilpotence, solvability, being an FC-group and the torsion units forming a subgroup for $U(ZG)$ are studied in Goncalves (1985), Parmenter and Polcino Milies (1978), Polcino Milies (1976, 1981) and Sehgal and Zassenhaus (1977a & b). It turns out that for a finite group $G$, $U(ZG)$ has any of these properties if and only if $G$ is *either* Abelian *or* a Hamiltonian 2-group. One of the reasons is the presence, most of the time, of non-cyclic free subgroups in $U(ZG)$. Hartley and Pickel (1980) conjectured that this would be case unless every subgroup of $G$ is normal in $G$ and the set of torsion elements of $G$ is either Abelian or a Hamiltonian 2-group. In Sehgal (1978), this conjecture has been affirmed for solvable groups and in Hartley and Pickel (1980) it is established for solvable-by-finite groups. The occurrence of non-cyclic free subgroups in subnormal subgroups of $U(ZG)$ is considered by Goncalves (1989) and Hartley (1989). Classification of those finite groups $G$ for which $U(ZG)$ contains a subgroup of finite index, which is isomorphic to a direct product of non-Abelian free groups, is carried out in Jespers *et al.*, (1996) and Leal and del Rio (1997).

A unified approach to several group-theoretic properties for $U(ZG)$ is given by Bist (1994). The study of the unit group being solvable or nilpotent when the coefficient ring is $Z_{(p)}$, $\hat{Z}_p$ or the ring of integers of a totally real algebraic number field has been done in Goncalves (1985) & (1986), Merklen and Polcino Milies (1980) and Polcino Milies (1978).

Müsson and Weiss (1982) initiated the investigation of the residual nilpotence of $U(ZG)$. They proved that for a finite group $G$, $U(ZG)$ is residually nilpotent if and only if $G$ is nilpotent and $G'$ is a $p$-group. They also obtained partial results on this problem when $G$ is a finitely generated nilpotent or an FC-group. The residual nilpotence of $U(\hat{Z}_pG)$, for $G$ finite, is characterized by Goncalves (1985).

For a group $G$, the (transfinite) derived series is defined as follows:

$G^{(0)} = G$; for any ordinal $\alpha$, $G^{(\alpha)} = (G^{(\alpha-1)})'$, if $\alpha$ is not a limit ordinal and $G^{(\alpha)} = \cap_{\beta<\alpha}G^{(\beta)}$, otherwise. The group $G$ is said to be *generalized solvable* if $G^{(\lambda)} = \{1\}$ for some ordinal $\lambda$. Hartley (1976) has proved that if $G$ is a solvable group, then $U(ZG)$ is generalized solvable. This led to the belief (see Müsson and Weiss, 1982, p. 530) that for a finite solvable group, $U(ZG)$ is perhaps always residually solvable, i.e., $\cap_{n=0}^{\infty}(U(ZG))^{(n)} = \{1\}$. However, this has turned out not to be the case; a counter-example is given in Bhandari and Weiss (1989) (where the residually solvability of $U(Z_{(p)}G)$ has also been studied). We mention some of the main results in this direction.

**Theorem 2.1. (Bhandari and Weiss, 1989)** *Let $G$ be a finite group and let $\bar{G} = G/O_p(G)$, where $O_p(G)$ denotes the maximal normal $p$-subgroup of $G$. If $U(Z_{(p)} G)$ is residually solvable, then one of the following holds:-*

(i) *$\bar{G}$ is Abelian.*

(ii) *$p = 2$ and $\bar{G} = A \rtimes < y >$, where $A$ is an elementary Abelian 3-group on which $y$ acts by inversion and $y^2 = 1$.*

(iii) *$p = 3$ and $\bar{G}$ is a 2-group with $U(F_3\bar{G})$ is solvable.*

(iv) *$p$ is odd and $\bar{G} = Q_8 \times E \times A$ with $E^2 = 1$, $A$ Abelian of odd order $n$ which divides $2^r - 1$ for some odd $r$.*

*Conversely, if $G$ satisfies one of (i), (ii) or (iii), then $U(Z_{(p)}G)$ is residually solvable. Also, if $O_p(G) = 1$ and $G$ satisfies (iv), then $U(Z_{(p)}G)$ is residually solvable.*

**Example 2.2.** Following is an example of a finite solvable group $G$ for which $U(ZG)$ is not residually solvable, and yet the unit group contains no perfect subgroup.

Let $n \neq p^a$ or $2p^a$, for any prime $p$, and let $D_n$ be the dihedral group

$$D_n = \langle x, a : x^2 = 1 = a^n, xax^{-1} = a^{-1} \rangle.$$

Let $\eta = \zeta_n + \zeta_n^{-1}$, where $\zeta_n$ is the primitive $n^{\text{th}}$ root of unity. It turns out that $U(ZD_n)$ is not residually solvable whenever $Z[\eta]$ has units $u, v$ with $v = u^2 - 1$ (see Bhandari and Weiss, 1989, p. 2661 and Lemmas A and 1.2). In particular, if $n$ is also a multiple of 5, then $Z[\eta]$ contains $\epsilon = \zeta_5 + \zeta_5^{-1}$ for which $\epsilon^2 + \epsilon - 1 = 0$ and so $u = \epsilon$, $v = -\epsilon$ is a possible choice. Thus, for example, $U(ZD_n)$ is not residually solvable for $n = 15, 20$.

As in §1, the residual nilpotence and the residual solvability of $U(ZG)$ is naturally related to the residual Lie nilpotence and the residual Lie solvability of the augmentation ideal $\Delta_Z(G)$, the kernel of the augmentation map $\epsilon : ZG \rightarrow Z$. Indeed, defining $\delta_Z^{(0)}(\Delta_Z(G)) = \Delta_Z(G)$, $\delta_Z^{(k+1)}(\Delta_Z(G)) = [\delta_Z^{(k)}(\Delta_Z(G)), \delta_Z^{(k)}(\Delta_Z(G))] ZG$, it is easy to see that $(U(ZG))^{(n)} \subseteq \delta_Z^{(n)}(\Delta_Z(G))$. Hence $U(ZG)$ is residually solvable if $\cap_{k=0}^{\infty}\delta_Z^{(k)}(\Delta_Z(G)) = (0)$. Similarly, $U(ZG)$ is residually nilpotent if $\cap_{k=1}^{\infty}\Delta_Z^{(k)}(G) = (0)$ (or even if $\cap_{k=0}^{\infty}\Delta_Z^{[k]}(G) = (0)$). Thus, in Bhandari (1985), a sufficient condition for $U(ZG)$ to be residually solvable is derived. We may remark here that the necessary and sufficient conditions for

$\cap_{k=1}^{\infty} \Delta_Z^{(k)}(G) = (0)$, $\cap_{k=0}^{\infty} \Delta_Z^{[k]}(G) = (0)$, and for $G$ finite, $\cap_n \delta_Z^{(n)}(G) = (0)$ are obtained in Musson and Weiss (1982), Bhandari and Passi (1992) and Mitsuda (1986) respectively (see also Kiraly, 1997 for more general coefficient rings).

The complete characterization of those groups $G$ for which $U(ZG)$ is residually solvable is still an open question.

Finally, we mention some of the recent results about the centre and upper central series for $U(ZG)$. In Ritter and Sehgal (1990), the finite groups $G$ for which the centre of $U(ZG)$ consists only of $\{\pm g | g \in G\}$ are characterized. Let

$$(1) = Z_0(U) \leq Z_1(U) \leq Z_2(U) \leq \ldots \leq Z_n(U) \leq \ldots$$

be the upper central series of $U = U(ZG)$. In Arora *et al.*, (1993) it is proved that for $G$ finite, $Z_3(U) = Z_2(U)$. This result is extended to torsion groups by Li. The finite groups $G$ for which $U(ZG)$ is of central height precisely 2 are completely determined in Arora and Passi (1993).

## References

Arora, Satya R., Hales A.W. and Passi I.B.S., Jordan decomposition and hypercentral units in integral group rings, *Comm. Algebra*, **21**, 25–35, 1993.

Arora, Satya R. and Passi I.B.S., Central height of the unit group of an integral group ring, *Comm. Algebra*, **21**, 3673–3683, 1993.

Baginski C., Groups of units of modular group algebras, *Proc. Amer. Math. Soc.*, **101**, 619–624, 1987.

Bateman J.M., On the solvability of unit groups of group algebras, *Trans. Amer. Math. Soc.*, **157**, 73–86, 1971.

Bateman J.M. and Coleman D.B., Group algebras with nilpotent unit groups, *Proc. Amer. Math. Soc.*, **19**, 448–449, 1968.

Bhandari A.K., Some remarks on the unit groups of integral group rings, *Arch. Math.*, **44**, 319–322, 1985.

Bhandari A.K. and Passi I.B.S., Residual solvability of the unit groups of rational group algebras, *Arch. Math.*, **48**, 213–216, 1987.

Bhandari A.K. and Passi I.B.S., Residual solvability of the unit groups of group algebras, In: *Group Theory*, Proceedings of the 1987 Singapore Conference, Walter de Gruyter, Berlin-New York, 267–274, 1989.

Bhandari A.K. and Passi I.B.S., Residually Lie nilpotent group rings, *Arch. Math.*, **58**, 1–6, 1992.

Bhandari A.K. and Passi I.B.S., Lie nilpotency indices of group algebras, *Bull. London Math. Soc.*, **24**, 68–70, 1992.

Bhandari A.K. and Weiss A., Residual solvability of unit groups of local group rings, *Comm. Algebra*, **17**, 2635–2662, 1989.

Bhattacharya P.B. and Jain S.K., A note on the adjoint group of a ring, *Arch. Math.*, **21**, 366–368, 1970.

Billig Y. Riley D. and Tasić V., Non-matrix varieties and nil-generated algebras whose units satisfy a group identity, *J. Algebra*, **190**, 241–252, 1997.

Bist V., Group of units of group algebras, *Comm. Algebra*, **20**, 1747–1761, 1992.

Bist V., Unit groups of integral group rings, *Proc. Amer. Math. Soc.*, **120**, 13–17, 1994.

Bovdi A.A. and Khripta I.I., Finite dimensional group algebras having solvable unit groups, In: *Trans. Science Conf. Uzgorod State University*, 227–233, 1974.

Bovdi A.A. and Khripta I.I., Group algebras of periodic groups with solvable multiplicative groups, *Math. Notes Acad. Sci. USSR*, **22**, 725–731, 1977.

Cliff G.H. and Sehgal S.K., Group rings whose units form an FC-group, *Math. Z.*, **161**, 163–168, 1978.

Coleman D.B. and Passman D.S., Units in modular group rings, *Proc. Amer. Math. Soc.*, **25**, 510–512, 1970.

Dennis R.K., The structure of the unit groups of group rings, In: *Ring Theory II*, Pro. Sec. Conf. Univ. Oklahoma, Norman, Oklahoma, 1975, 103–130, Marcel Dekker, New York, 1977.

Dokuchaev M. and Goncalves J.Z., Semigroup identities on units of integral group rings, *Glasgow Math. J.*, **39**, 1–6, 1997.

Du X.K., The centre of a radical ring, *Canad. Math. Bull.*, **35**, 174–179, 1992.

Fisher J. Parmenter M.M. and Sehgal S.K., Group rings with solvable $n$-Engel unit groups, *Proc. Amer. Math. Soc.*, **59**, 195–200, 1976.

Giambruno, A. Jespers E. and Valenti A., Group identities on units of rings, *Arch. Math.*, **63**, 291–296, 1994.

Giambruno, A. Sehgal S.K. and Valenti A., Group algebras whose units satisfy a group identity, *Proc. Amer. Math. Soc.*, **125**, 629–634, 1997.

Goncalves J.Z., Free subgroups in subnormal subgroups and residual nilpotence of the group of units of group rings, *Canad. Math. Bull.*, **27**, 365–370, 1984.

Goncalves J.Z., Free subgroups of units in group rings, *Canad. Math. Bull.*, **27**, 309–312, 1984.

Goncalves J.Z., Free subgroups in the group of units of group rings II, *J. Nr. Th.*, **21**, 121–127, 1985.

Goncalves J.Z., Integral group rings whose group of units is solvable—an elementary proof, *Bol. Soc. Bras. Mat.*, **16**, 1–9, 1985.

Goncalves J.Z., Group rings with solvable unit groups, *Comm. Algebra*, **14**, 1–20, 1986.

Goncalves J.Z., Free subgroups and the residual nilpotence of the group of units of modular and $p$-adic group rings, *Canad. Math. Bull.*, **29**, 321–328, 1986.

Goncalves J.Z. and Mandel A., Semigroup identities on units of group algebras, *Arch. Math.*, **57**, 539–545, 1991.

Goncalves J.Z., and Passman D.S., Construction of free subgroups in the group of units of modular group algebras, *Comm. Algebra*, **24**, 4211–4215, 1996.

Goncalves, J.Z., Ritter J. and Sehgal S.K., Subnormal subgroups in $U(ZG)$, *Proc. Amer. Math. Soc.*, **103**, 375–382, 1989.

Gupta N.D. and Levin F., On the Lie ideals of a ring, *J. Algebra*, **81**, 225–231, 1983.

Hartley B., A conjecture of Bachmuth and Muchizuki on automorphisms of soluble groups, *Canad. J. Math.*, **28**, 1302–1310, 1976.

Hartley B., Free groups in normal subgroups of unit groups and arithmetic groups, *Contemporary Math.*, **93**, 173–178, 1989.

Hartley B. and Pickel P.F., Free subgroups in the unit groups of integral group rings, *Canad. J. Math.*, **32**, 1342–1352, 1980.

Jennings S.A., Radical rings with nilpotent associated groups, *Trans. Royal Soc. Canada*, **49**, Ser. III, 31–38, 1955.

Jespers, E. Leal G. and del Rio A., Product of free groups in the unit groups of integral group rings, *J. Algebra*, **180**, 22–40, 1996.

Karpilovsky G., *Unit Groups of Group Rings*, Longman, Essex, 1989.

Khripta I.I., The nilpotence of the multiplicative group of a group ring, *Math. Notes*, **11**, 119–124, 1972.

Király B., Residual Lie nilpotence of the augmentation ideal, *Acta Academiae Paedagogicae Agriensis*, **24**, 75–88, 1997.

Leal G. and del Rio A., Products of free groups in the unit groups of integral group ring II, *J. Algebra*, **191**, 240–251, 1997.

Li Y., The hypercentre and the $n$-centre of the unit group of an integral group ring, *Canad. J. Math.*, **50**, 401–411, 1998.

Merklen H. and Polcino Milies C., Group rings over $Z_{(p)}$ with FC unit groups, *Canad. J. Math.*, **32**, 1266–1269, 1980.

Mitsuda T., Group rings whose augmentation ideals are residually Lie solvable, *Proc. Japan Acad, Ser.*, **A 62**, 264–266, 1986.

Motose K. and Tominaga H., Group rings with nilpotent unit groups, *Math. J. Okayama Univ.*, **14**, 43–46, 1969.

Motose K. and Tominaga H., Group rings with solvable unit groups, *Math. J. Okayama Univ.*, **15**, 37–40, 1971.

Motose K. and Ninomiya Y., On the solvability of the unit groups of group rings, *Math J. Okayama Univ.*, **15**, 209–214, 1972.

Musson I. and Weiss A., Integral group rings with residually nilpotent unit groups, *Arch. Math.*, **38**, 514–530, 1982.

Parmenter M.M. and Polcino Milies C., Group rings whose units form a nilpotent or FC group, *Proc. Amer. Math. Soc.*, **68**, 247–248, 1978.

Passi I.B.S., *Group Rings and Their Augmentation Ideals, LNM*, **715**, Springer-Verlag, 1979.

Passi, I.B.S., Passman D.S. and Sehgal S.K., Lie solvable group rings, *Canad. J. Math.*, **25**, 748–757, 1973.

Passi I.B.S. and Sehgal S.K., Lie dimension subgroups, *Comm. Algebra*, **3**, 59–73, 1975.

Passman D.S., Observations on group rings, *Comm. Algebra*, **5**, 1119–1162, 1977.

Passman D.S., Group algebras whose units satisfy a group identity II, *Proc. Amer. Math. Soc.*, **125**, 657–662, 1997.

Polcino Milies C., Integral group rings with nilpotent unit groups, *Canad. J. Math.*, **28**, 954–960, 1976.

Polcino Milies C., $p$-adic group rings with nilpotent unit groups, *J. pure appl. Algebra*, **12**, 147–151, 1978.

Polcino Milies C., Group rings whose units form an FC-group, *Arch. Math.*, **30**, 380–384, 1978*a*.

Polcino Milies C., Group rings whose units form an FC-group: corrigendum, *Arch. Math.*, **31**, 528, 1978*b*.

Polcino Milies C., Group rings whose torsion units form a subgroup I, *Proc. Amer. Math. Soc.*, **81**, 172–174, 1981.

Polcino Milies C., Group rings whose units torsion form subgroup II, *Comm. Algebra*, **9**, 699–712, 1981.

Polcino Milies C., Units of group rings: A short survey, *Groups - St. Andrews, London Math. Soc. Lect. Not. Sr.*, **71**, 281–297, 1981.

Riley D., Restricted Lie dimension subgroups, *Comm. Algebra*, **19**, 1493–1499, 1991.

Ritter J. and Sehgal S.K., Integral group rings with trivial central units, *Proc. Amer. Math. Soc.*, **108**, 327–329, 1990.

Sehgal S.K., Nilpotent elements in group rings, *Manuscripta Math.*, **15**, 65–80, 1975.

Sehgal S.K., *Topics in Group Rings*, Marcel Dekker, New York and Basel, 1978.

Sehgal S.K., *Units in Integral Group Rings*, Longman, Essex, 1989.

Sehgal S.K. and Zassenhaus H., Group rings whose units form an FC-group, *Math. Z.*, **153**, 29–35, 1977a.

Sehgal S.K. and Zassenhaus H., Integral group rings with nilpotent unit groups, *Comm. Algebra*, **5**, 101–111, 1977b.

Shalev A., On some conjectures concerning units in *p*-group algebras, Proceedings of the second International Group Theory Conference (Bressanone, 1989), *Ren. Circ. mat. Polermo (2) Supp.*, **23**, 279–288, 1990a.

Shalev A., The nilpotency class of the unit group of a modular group algebra I, *Israel J. Math.*, **70**, 257–266, 1990b.

Shalev A. and Mann A., The nilpotency class of the unit group of a modular group algebra II, *Israel J. Math.*, **70**, 267–277, 1990.

Shalev A., The nilpotency class of the unit group of a modular group algebra III, *Arch. Math.*, **60**, 136–145, 1993.

Shalev A., Lie dimension subgroups, Lie nilpotency indices, and the exponent of the group of normalized units, *J. London Math. Soc.*, **43**, 23–36, 1991.

Centre for Advanced Study in Mathematics, Panjab University, Chandigarh 160014, India

# Projective Modules Over Polynomial Rings

*S.M. Bhatwadekar*

## 1. Introduction

In 1976, Quillen [Q] and Suslin [Su 1] proved the following conjecture of Serre:-

**Conjecture: (Serre)** *Every finitely generated projective module over a polynomial ring $k[T_1, \ldots, T_n]$ over a field $k$ is free.*

Actually Serre did not state it as a conjecture; in his famous paper 'Faisceaux algebriques coherents' (1955), he made a statement which amounts to saying that one did not know whether the implication contained in the above conjecture was true. But to the mathematical world this statement is known as "Serre's Conjecture". From the time Serre made this conjecture, there had been many attempts at the conjecture and several important partial results were obtained. An excellent account of these attempts and historical developments, which after almost twenty years culminated in the final solution of this conjecture, can be found in T.Y. Lam's book [L].

Apart from settling Serre's conjecture, Quillen's and Suslin's proofs have had a profound effect on the subsequent developments of the study of projective modules, especially projective modules over polynomial rings. This is because of two reasons: their proofs were elementary, hence accessible, yet contained some very powerful techniques. To cite one such, we quote a generalisation of Horrocks' result [H] to the non local situation:

**Monic Inversion Theorem (Horrocks/Quillen-Suslin)** *Let $R$ be a commutative noetherian ring and let $P$ be a finitely generated projective $R[X]$-module such that $P_f$ is $R[X]_f$- free for some monic polynomial $f \in R[X]$. Then $P$ is free.*

As a result, many problems which were stated earlier but inaccessible were solved completely. Moreover, new problems were posed and some of them were solved, thus continuing the growth of the subject.

This survey article deals with such developments that took place after Quillen-Suslin solution, though it does not encompass all aspects of these developments. I have only selected two topics viz.

(1) Existence of unimodular elements in projective modules of large rank over polynomial extensions and cancellative property of such modules (section 3).

(2) The Bass - Quillen conjecture (section 4).

A glaring omission is the related topic of 'complete intersections'. Interested readers can refer to the Springer lecture notes of Mandal [Ma 2] entitled 'Projective modules and complete intersections' or ([L], Appendix) and [Mu 3].

In section 2 we have set up notation, stated a few basic definitions and proved some simple results. A reader who is to some extent familiar with the subject can safely skip this section.

I conclude this introduction by thanking my friend and collaborator Amit Roy for his valuable suggestions which led to considerable improvement of the text of this article. An apology is in order to those authors whose relevant work I might not have mentioned inadvertently.

## 2. Preliminaries

Throughout this article **all rings are commutative noetherian with unity and all modules are finitely generated**.

For standard material in Commutative Algebra one may use [AM], [Go] or [Mt].

In this section we assemble a few definitions, set up notation and state some simple results for later use. We first set up notation which will be used throughout the article and start this section with a definition of a projective module over a commutative ring $R$.

$R^n$: free $R$-module of rank $n$.

$\text{Spec}(R)$: the set of all prime ideals of $R$.

$V(I)$: the set of all prime ideals of $R$ containing an ideal $I$ of $R$.

$D(f)$: the set of all prime ideals of $R$ not containing an element $f \in R$.

$GL_n(R)$: the group of all $n \times n$ invertible matrices over $R$.

$SL_n(R)$: the group of all $n \times n$ matrices over $R$ of determinant 1.

**Definition 2.1.** An $R$-module $P$ is said to be *projective* if it satisfies one of the following equivalent conditions:-

(i) Given $R$-modules $M$, $N$ and an $R$-linear surjective map $\alpha : M \twoheadrightarrow N$, the canonical map from $\text{Hom}_R(P, M)$ to $\text{Hom}_R(P, N)$ sending $\theta$ to $\alpha\theta$ is surjective.

(ii) Given an $R$-module $M$ and a surjective $R$-linear map $\alpha : M \twoheadrightarrow P$ there exists an $R$-linear map $\beta : P \to M$ such that $\alpha\beta = 1_P$.

(iii) There exists an $R$-module $Q$ such that $P \oplus Q \simeq R^n$ for some positive integer $n$, i.e., $P \oplus Q$ is free.

**Remark 2.2.** Let $P, Q$ be projective $R$-modules. Then from $((2.1),$ (iii)) it follows that $\text{Hom}_R(P, Q)$ is also a projective $R$-module. In particular $P^* = \text{Hom}_R(P, R)$ is projective. Moreover, if $R \to S$ is a homomorphism of rings then $P \otimes_R S$ is a projective $S$-module.

**Definition 2.3.** Let $R \hookrightarrow S$ be a ring extension. A projective $S$-module $Q$ is said to be *extended* from $R$ if there exists a projective $R$-module $P$ such that $Q \simeq P \otimes_R S$.

**Definition 2.4. (Zariski Topology)** The Zariski topology on $\text{Spec}(R)$ is the topology for which all closed sets are of the form $V(I)$, where $I$ is an ideal of $R$ or equivalently the basic open sets are of the form $D(f)$, $f \in R$.

**Lemma 2.5.** *Let $I$ be an ideal of $R$ contained in the Jacobson radical of $R$. Let $P, Q$ be projective $R$-modules such that projective $R/I$-modules $P/IP$ and $Q/IQ$ are isomorphic. Then $P$ and $Q$ are isomorphic as $R$-modules.*

**Proof.** Let $\bar{\alpha} : P/IP \simeq Q/IQ$ be an isomorphism. Then, since $P$ is projective, $\bar{\alpha}$ can be lifted to an $R$-linear homomorphism $\alpha : P \to Q$. We will show now that $\alpha$ is an isomorphism.

Since $\bar{\alpha}$ is surjective and $\alpha$ is a lift, it follows that $Q = \alpha(P) + IQ$. Therefore, as $I$ is contained in the Jacobson radical of $R$ (and all modules are finitely generated by assumption), $Q = \alpha(P)$ by Nakayama lemma. Thus $\alpha$ is surjective.

Since $Q$ is projective there exists an $R$-linear map $\beta : Q \to P$ such that $\alpha\beta = 1_Q$. Let $\bar{\beta} : Q/IQ \to P/IP$ be the map induced by $\beta$. Then we have $\bar{\alpha}\bar{\beta} = 1_{Q/IQ}$. As $\bar{\alpha}$ is an isomorphism, we get that $\bar{\beta}$ is also an isomorphism and in particular $\bar{\beta}$ is surjective. Therefore $P = \beta(Q) + IP$. Hence, as before, we see that $\beta$ is surjective. Now injectivity of $\alpha$ follows from the fact that $\alpha\beta = 1_Q$. ∎

**Corollary 2.6.** *Let $R$ be a local ring. Then every projective $R$-module is free.*

**Proof.** Let $\mathfrak{m}$ be the maximal ideal of $R$ and let $k = R/\mathfrak{m}$ be the residue field of $R$. Let $P$ be a projective $R$-module and let $n = \dim_k P/\mathfrak{m}P$. Now applying (2.5) to the projective modules $P$ and $R^n$ we see that $P \simeq R^n$. ∎

Let $P$ be a projective $R$-module. In view of (2.6) we can define the rank function $\text{rank}_P$ as follows:

**Definition 2.7.** $\text{rank}_P : \text{Spec}(R) \to \mathbf{Z}$ is the function defined by $\text{rank}_P(\mathfrak{q}) = $ rank of the free $R_\mathfrak{q}$-module $P \otimes_R R_\mathfrak{q}$. If $\text{rank}_P$ is a constant function taking the value $n$ then we define rank of $P$ to be $n$ and denote it by $\text{rank}(P)$.

**Remark 2.8.** It is easy to see that $\text{rank}_P$ is a continuous function (with the discrete topology on $\mathbf{Z}$ and Zariski topology on $\text{Spec}(R)$). Moreover, $\text{rank}_P$ is a constant function for every projective $R$-module $P$ if and only if $R$ has no nontrivial idempotent elements.

Let $e_{ij}$, $i \neq j$ denote the $n \times n$ matrix with 1 in the $(i, j)$ coordinate and zeros elsewhere and $E_{ij}(a) = I_n + ae_{ij}$, $a \in R$. $E_{ij}(a)$ is called an *elementary matrix*. We denote by $E_n(R)$ the subgroup of $SL_n(R)$ generated by elementary matrices.

We can identify $GL_n(R)$ with the group $Aut_R(R^n)$ of automorphisms of $R^n$ by fixing a basis of $R^n$ and hence we can talk of the determinant of an automorphism of $R^n$. It is easy to see that the determinant of an automorphism of $R^n$ is independent of a choice of a basis. We can also define the determinant of an automorphism of a projective module (of constant rank) as follows. Let $P$ be a projective $R$-module of rank $n$. Let $\wedge^n(P)$ denote the $n^{\text{th}}$ exterior power of $P$. Then $\wedge^n(P)$ is a projective $R$-module of rank 1. An $R$-linear endomorphism $\alpha$ of $P$ gives rise, in a natural way, to an endomorphism $\wedge^n(\alpha)$ of $\wedge^n(P)$. Since rank $\wedge^n(P) = 1$, we have $\text{End}_R(\wedge^n(P)) = R$ and hence $\wedge^n(\alpha) \in R$. It can be proved that $\alpha$ is an automorphism if and only if $\wedge^n(\alpha)$ is an invertible element of $R$.

**Definition 2.9.** Let $P$, $\alpha$ be as in the above paragraph. We define the determinant of $\alpha$ to be $\wedge^n(\alpha)$ and denote it by $\det(\alpha)$. We denote the group of automorphisms of $P$ of determinant 1 by $SL(P)$.

**Definition 2.10.** Given a projective $R$-module $P$ and an element $p \in P$ we define $\mathcal{O}_P(p) = \{\alpha(p) \mid \alpha \in P^*\}$. We say that $p$ is *unimodular* if $\mathcal{O}_p(p) = R$. The set of all unimodular elements of $P$ is denoted by $\text{Um}(P)$. If $P = R^n$ then we write $\text{Um}_n(R)$ for $\text{Um}(R^n)$.

**Remark 2.11.** It is easy to see that $\mathcal{O}_p(p)$ is an ideal of $R$ and $p$ is unimodular if and only if there exists $\alpha \in P^*$ such that $\alpha(p) = 1$. An element $(a_1, \ldots, a_n) \in R^n$ is unimodular if and only if there exist elements $b_1, \ldots, b_n \in R$ such that $\sum_{i=1}^{n} a_i b_i = 1$. If $(a_1, \ldots, a_n) \in R^n$ is unimodular then we say the row $[a_1, \ldots, a_n]$ is a *unimodular row*.

If $I$ is an ideal of $R$ then (as $P$ is projective) $\mathcal{O}_{\bar{P}}(\bar{p}) = (\mathcal{O}_P(p) + I)/I$ where $\bar{P} = P/IP$ and $\bar{p}$ is the image of $p$ in $P/IP$. In particular, if $I$ is contained in the Jacobson radical of $R$ then $p$ is a unimodular element if and only if $\bar{p}$ is a unimodular element of $\bar{P}$.

The unimodularity of an element $p \in P$ is also equivalent to the conditions: $Rp \simeq R$ and $Rp$ is a direct summand of $P$. In this case there exists an isomorphism $\theta : (P/Rp) \oplus R \xrightarrow{\sim} P$ such that $\theta(0, 1) = p$. Conversely, given an isomorphism $\sigma : Q \oplus \dot{R} \xrightarrow{\sim} P$, if $p = \sigma(0, 1)$ then $p$ is a unimodular element of $P$ and $Q \simeq P/Rp$. In particular giving a unimodular row $[a_1, \ldots, a_n]$ is equivalent to giving a projective $R$-module $Q$ with the property $Q \oplus R \simeq R^n$.

In the above situation we say that $Q$ is obtained from $[a_1, \ldots, a_n]$.

Let $P$ a projective $R$-module. Given an element $\varphi \in P^*$ and an element $p \in P$, we define an endomorphism $\varphi_p$ as the composite

$$P \xrightarrow{\varphi} R \xrightarrow{p} P.$$

If $\varphi(p) = 0$, then $\varphi_p^2 = 0$ and $1 + \varphi_p$ is a unipotent automorphism of $P$ and hence is an element of $SL(P)$.

**Definition 2.12.** By a *transvection*, we mean an automorphism of $P$ of the form $1 + \varphi_p$ where $\varphi(p) = 0$ and either $\varphi$ is unimodular in $P^*$ or $p$ is unimodular in $P$. We denote by $E(P)$ the subgroup of $SL(P)$ generated by all transvections of $P$.

**Remark 2.13.** $E(P)$ is a normal subgroup of $Aut(P)$. Moreover, an elementary matrix gives rise to a transvection of $R^n$ when we identify $GL_n(R)$ with $Aut(R^n)$. Hence, $E_n(R)$ can be regarded as a subgroup of $E(R^n)$. If $n \geq 3$ then, by a result of Suslin ([Su 2], Corollary 1.4), $E_n(R) = E(R^n)$.

**Definition 2.14.** Projective $R$-modules $P$ and $Q$ are said to be *stably isomorphic* if there exists a projective $R$-module $Q'$ such that $P \oplus Q' \simeq Q \oplus Q'$. Equivalently, $P$ and $Q$ are said to be stably isomorphic if $P \oplus R^n \simeq Q \oplus R^n$ for some positive integer $n$. A projective $R$-module $P$ is said to be *stably free* if it is stably isomorphic to a free $R$-module. If $P$ and $Q$ are stably isomorphic then rank functions $\text{rank}_P$ and $\text{rank}_Q$ are same.

**Definition 2.15.** A projective $R$-module $P$ is said to be *cancellative* if every projective module $Q$ which is stably isomorphic to $P$ is isomorphic to $P$.

**Proposition 2.16.** *Every projective $R$-module of rank 1 is cancellative.*

**Proof.** Let $L$ be a projective $R$-module of rank 1. Suppose that $L$ is stably isomorphic to a projective $R$-module $L'$. Then rank $(L') = 1$ and there exists a positive integer $n$ such that $L \oplus R^n \simeq L' \oplus R^n$. Therefore $L \simeq \wedge^{n+1}(L \oplus R^n) \simeq \wedge^{n+1}(L' \oplus R^n) \simeq L'$.

**Proposition 2.17.** *For a projective $R$-module $P$ the following are equivalent:*

(i) *For any projective $R$-module $Q$, if $P \oplus R \simeq Q \oplus R$, then $P \simeq Q$.*

(ii) *Given a unimodular element $(p, a) \in P \oplus R$, there exists an automorphism $\Delta$ of $P \oplus R$ such that $\Delta(p, a) = (0, 1)$.*

**Proof.** $(i) \Rightarrow (ii)$.

Since $(p, a)$ is a unimodular element of $P \oplus R$, there exists an element $\alpha \in (P \oplus R)^*$ such that $\alpha(p, a) = 1$. Let $Q = ker(\alpha)$. Then we get the following short exact sequence of $R$-modules:

$$0 \to Q \to P \oplus R \xrightarrow{\alpha} R \to 0.$$

Let $\beta : R \to P \oplus R$ be an $R$-linear map such that $\beta(1) = (p, a)$. Then $\alpha\beta = 1_R$. Hence, the cyclic submodule $R(p, a)$ of $P \oplus R$ is isomorphic to $R$ and $P \oplus R = Q \oplus R(p, a)$. Therefore, by assumption, there exists an isomorphism $\sigma : Q \xrightarrow{\sim} P$.

Let $\Delta : Q \oplus R(p, a)(= P \oplus R) \to P \oplus R$ be an endomorphism of $P \oplus R$ defined by $\Delta(q, 0) = (\sigma(q), 0)$ for $q \in Q$ and $\Delta(p, a) = (0, 1)$. Then, as $\sigma$ is an isomorphism and $R(p, a) \simeq R$, it follows that $\Delta$ is an automorphism of $P \oplus R$ which sends $(p, a)$ to $(0, 1)$.

$(ii) \Rightarrow (i)$.

Let $\Psi : Q \oplus R \xrightarrow{\sim} P \oplus R$ be an isomorphism and let $\Psi(0, 1) = (p, a)$. Then, as $\Psi$ is an isomorphism, $(p, a)$ is a unimodular element of $P \oplus R$. Therefore, by assumption, there exists an automorphism $\Delta$ of $P \oplus R$ such that $\Delta(p, a) = (0, 1)$. Hence the isomorphism $\Delta\Psi : Q \oplus R \xrightarrow{\sim} P \oplus R$ sends the element $(0, 1)$ of $Q \oplus R$ to $(0, 1)$ of $P \oplus R$. Now we are through by noting the fact that $Q \simeq (Q \oplus R)/R(0, 1)$ and $P \simeq (P \oplus R)/R(0, 1)$.                                          ∎

As a consequence of this proposition we have the following result:

**Corollary 2.18.** *The following statements are equivalent.*

(i) *For any projective R-module Q, if $R^n \simeq Q \oplus R$, then $R^{n-1} \simeq Q$.*

(ii) *Every unimodular row $[a_1, \ldots, a_n]$ over R is completable i.e., $[a_1, \ldots, a_n]$ is a first row of an invertible matrix over R.*

## 3. Projective Modules Over Polynomial and Laurent Polynomial Extensions of An Arbitrary Ring

In this section we state results which show that projective modules of large rank over polynomial and Laurent polynomial rings have unimodular elements and such projective modules are cancellative. Before giving the precise statements, we give the genesis of these results and some historical background.

One way to solve Serre's conjecture would have been to show that every projective module over a polynomial ring $A$ over a field has a unimodular element. For suppose we have a projective $A$-module $P$. If $P$ has a unimodular element, then $P$ decomposes as a direct sum $Q \oplus A$ for some $Q$. Now $Q$ in turn has a similar decomposition if it has a unimodular element. This process can be continued as far as possible hoping to eventually prove that $P$ is free.

Therefore it was natural to look for conditions under which a projective module over a *general* ring $R$ would have a unimodular element.

In this context, motivated by a topological result, Serre [S 2] proved in 1957 the following fundamental result.

**Theorem 3.1. (Serre)** *Let R be a ring of finite Krull dimension d and let P be a projective R-module of rank $\geq d + 1$. Then P has a unimodular element.*

First note that, in view of (2.16), the above theorem of Serre says that if dimension of $R$ is 1 then every projective $R$-module (of constant rank) is cancellative. The second point to notice is that a Dedekind domain which is not a UFD will furnish an example which will show that Serre's theorem is best possible for general rings. We shall shortly see (Example 3.3) that (3.1) is also the best possible result even if $R$ is a UFD.

Meanwhile let $P$ be a projective $R$-module which can be decomposed as $Q \oplus R^l$. Then a natural question would be to ask whether such a $Q$ is unique (up to isomorphism). This is the so-called cancellation question. Addressing this we have the following fundamental theorem due to Bass [B 2] proved in 1964.

**Theorem 3.2. (Bass)** *Let $R$ be a ring of finite Krull dimension $d$ and let $P$ be a projective $R$-module of rank $\geq d + 1$. Then $P$ is cancellative.*

Now we present an example to show that (3.1) and (3.2) are the best possible in the general setup.

**Example 3.3.** Let $R$ denote the field of real numbers and let $R = R[X, Y, Z]/(X^2 + Y^2 + Z^2 - 1)$ denote the coordinate ring of the real 2-sphere. Then $R$ is a UFD of dimension 2. Let $x, y, z$ denote the images of $X, Y, Z$ respectively in $R$.

Let $Q$ be the projective $R$-module given by the unimodular row $[x, y, z]$. Then rank $(Q) = 2$ and $Q \oplus R \simeq R^3 = R^2 \oplus R$. It is known (using topology) that the unimodular row $[x, y, z]$ over $R$ is not completable. Hence, $Q$ is not free proving that $Q$ is not cancellative.

Next we show that $Q$ does not have a unimodular element. If it were so then $Q \simeq L \oplus R$ for some projective $R$-module $L$ of rank one. But then $L \oplus (R \oplus R) \simeq Q \oplus R \simeq R^3 = R \oplus (R \oplus R)$ implying $L \simeq R$ by (2.16). This in turn would mean that $Q$ is free. Contradiction.

Bass proved (3.2) by proving the following stronger result.

**Theorem 3.4. (Bass)** *Let $R$ be a ring of finite Krull dimension $d$ and let $P$ be a projective $R$-module of rank $\geq d + 1$. Let $(p, a) \in P \oplus R$ be a unimodular element. Then there exists an element $q \in P$ such that $p + aq$ is unimodular element of $P$.*

We will show now how one can deduce (3.2) from (3.4).

**Proof (of (3.2) using (3.4)).** To prove (3.2), it is enough to show that if $Q$ is a projective $R$-module such that $P \oplus R \simeq Q \oplus R$ then $P \simeq Q$. Therefore, by (2.17), the theorem is proved if we show that given a unimodular element $(p, a) \in P \oplus R$ there exists an automorphism $\Delta$ of $P \oplus R$ such that $\Delta(p, a) = (0, 1)$.

Let $(p, a)$ be a unimodular element. Then, by (3.4), there exists an element $q \in P$ such that $p + aq$ is unimodular. Therefore, there exists $\alpha \in P^*$ such that $\alpha(p + aq) = 1$. Let $\beta = (1 - a)\alpha$. We now regard $\beta$ as an element of $(P \oplus R)^*$ by putting $\beta(0, 1) = 0$. Let $\varphi \in (P \oplus R)^*$ be such that $\varphi(0, 1) = 1$ and $\varphi(p', 0) = 0$ for every $p' \in P$. Let $p_1 = p + aq$. Now we construct transvections $\Delta_i, i = 1, 2, 3$ (see (2.12) for definition) of $P \oplus R$ as follows:

$$\Delta_1 = 1 + \varphi_{(q, 0)},$$
$$\Delta_2 = 1 + \beta_{(0,1)},$$
$$\Delta_3 = 1 + \varphi_{(-p_1, 0)}.$$

Let $\Delta = \Delta_3 \Delta_2 \Delta_1$. Then $\Delta(p, a) = (0, 1)$.   ■

Note that, by construction, $\Delta \in E(P \oplus R)$, i.e., $\Delta$ is a product of transvections. In practice transvections are automorphisms which get used very often because of their simple and explicit description. Another very important property of transvections is their liftability in the following sense: Let $Q$ be a projective $R$-module and let $I$ be an ideal of $R$. Then every transvection of $Q/IQ$ (this of course assumes that $Q/IQ$ has a unimodular element) can be lifted to a (unipotent) automorphism of $Q$ (see ([BRy 1], Proposition 4.1)).

The next result, which is a generalisation of (3.4), is due to Eisenbud and Evans. The nice thing about this result is that it yields proofs of (3.1) and (3.4) in one stroke. A proof of this result follows from the remark following Theorem A of [EE 1] or from ([P], p.1420, Eisenbud-Evans Theorem).

**Theorem 3.5.** *Let $P$ be a projective $R$-module of rank $n$. Let $(p, a) \in (P \oplus R)$. Then there exists an element $q \in P$ such that $\mathrm{ht}\,(I R_a) \geq n$, where $I = \mathcal{O}_P(p+aq)$.*

As we have seen above, the questions of existence of unimodular elements and cancellation are important. We have also seen that the theorems of Serre and Bass take us as far as it is possible to go over arbitrary rings. So it is natural to investigate whether these theorems can be improved upon in the case of special rings. More specifically, in the spirit of Serre's conjecture, at least for polynomial rings. And, by extension, to Laurent polynomial rings. Let us see what happened in this direction after Serre's theorem.

In 1958, Seshadri [Ss 1] showed that every projective module over $R[X]$ is free when $R$ is a PID, thus settling 'Serre conjecture' for two variables. Subsequently, it was shown in ([Ss 2], [S 3], [B 1]) that when $R$ is a Dedekind domain then every projective module over $R[X]$ is extended from $R$. This result was generalised by Endo ([En], Theorem 4.7, Theorem 5.4) by proving that if $R$ is a seminormal domain of Krull dimension 1 with finite normalisation then every projective $R[X]$-module is extended from $R$ and furthermore if $R$ is semilocal then every projective module over $R[X, Y]$ is free. Around the same time, Murthy ([Mu 1], Theorem 3) showed that if $R$ is a 1-dimensional domain with finite normalisation then evey projective $R[X]$-module of rank $\geq 2$ has a unimodular element. Note that this result of Murthy, in view of (2.16), shows that every projective module over $R[X]$, $\dim(R) = 1$ is cancellative. Later on Bass-Murthy ([BM], Theorem 9.1) extended this result of Murthy by showing that if $R$ is a domain of dimension 1 and with finite normalisation then every projective module over $R[X, X^{-1}]$ of rank $\geq 2$ has a unimodular element. Moreover, if $R$ is semilocal, then they showed that similar result holds for $R[X, X^{-1}, Y, Y^{-1}]$.

These results eventually led to the two questions stated below which will be the subject of our discussion in the rest of this section.

Question(I) was stated by Bass ([B 4], Question $(XIV)_n$) along with many other important $K$-theoretic questions.

**Question (I).** *Let $R$ be a noetherian commutative ring of Krull dimension $d$ and let $A = R[X_1, \ldots, X_n]$ be a polynomial ring in $n$ variables over $R$. Let $P$ be a* *finitely generated projective A-module. If rank $(P) \geq d + 1$, are the following statements true?*

*(i) P has a unimodular element.*

*(ii) P is cancellative.*

If $R$ is a field then, as projective $A$-modules are stably free (see [S 2]), both statements $(i)$ and $(ii)$ are equivalent to Serre's conjecture. Therefore, one can regard Question(I) as a generalisation of Serre's conjecture. This question in nutshell asks whether the behaviour, with respect to questions of existence of unimodular elements and of cancellation, of projective modules over polynomial extensions depends only on the Krull dimension of the base ring and *not* on the number of polynomial variables.

The following question can be regarded as a natural generalisation of Question(I).

**Question** **(II).** *Let R be as in Question(I) and let $A = R[X_1, \ldots, X_n, Y_1^{\pm 1}, \ldots, Y_m^{\pm 1}]$ be a Laurent polynomial ring over R. Let P be a finitely generated projective A-module of rank $\geq d + 1$. Then*

*(i) Does P have a unimodular element?*

*(ii) Is P cancellative?*

Both questions now have been answered affirmatively (see [BRy 1], [BLR], [Ra 1], [Li 2]). Prior to complete solutions of these questions, apart from results mentioned above, many partial results were proved. We now record these results in chronological order.

*Question* *(I):* Suslin ([Su 2], Theorem 2.6) proved that $E(A^r)$ act transitively on the set $\text{Um}(A^r) = \text{Um}_r(A)$ if $r \geq \max(3, d + 2)$. In particular every free $A$-module of rank $\geq d + 1$ is cancellative.

Swan ([Sw 1], Theorem 1.1) proved that every projective module over $A = R[X_1, \ldots, X_n, Y_1^{\pm 1}, \ldots, Y_m^{\pm 1}]$ of rank $\geq d + 1$ which is stably extended from $R$ (i.e there exists a projective $R$-module $Q$ such that $P$ is stably isomorphic to $Q \otimes_R A$) is extended from $R$, thus giving an affirmative answer to both questions (Questions(I) and (II)) when $P$ is stably extended from $R$.

When $d = 1$, by ([MP]), every projective $A$-module $P$ of rank $r$ is stably isomorphic to $\wedge^r(P) \oplus A^{r-1}$ and hence the statements (i) and (ii) are equivalent. In this case (i.e., $d = 1$) Kang ([K], Theorem 4.3), under the assumption $R_{\text{red}} (= R/\text{nilradical})$ has finite normalisation, showed that if rank$(P) = r \geq 3$ then $P \simeq \wedge^r(P) \oplus A^{r-1}$ and hence $P$ has a unimodular element and it is cancellative. Kàng proved this result even when $A = R[X_1, \ldots, X_n, Y_1^{\pm 1}, \ldots, Y_m^{\pm 1}]$. Moreover he ([K], Theorem 5.3) proved similar result (only when $A$ is a polynomial algebra over $R$) for rank 2 projective modules under the additional assumption that $1/2 \in R$.

Roy [Ry 1] generalised ([K], Theorem 4.3) by removing the assumption of finite normalisation and Greither [Gr] generalised ([K], Theorem 5.3) by removing the assumption $1/2 \in R$ (though still needing the assumption on finite normalisation).

For $n = 1$ (and arbitrary $d$), the question was also conjectured by Eisenbud and Evans [EE 2] and settled affirmatively by Plumstead [P], who in turn conjectured the truth of this question for arbitrary $n$.

*Question (II):* Suslin ([Su 2], Corollary 7.4) and Swan ([Sw 1], Theorem 1.1) independently proved that if $R$ is a field or a PID then all projective modules over $R[X_1, \ldots, X_n, Y_1^{\pm 1}, \ldots, Y_m^{\pm 1}]$ are free thus settling this question affirmatively for $d = 0$. We have mentioned above results of Swan and Kang regarding this question.

When $n = 0, m = 1$, Mandal ([Ma 2], Theorem 2.1 and Corollary 2.4) gave an affirmative answer to this question.

This was the state of affairs till Bhatwadekar-Roy ([BRy 1], Theorem 3.1) proved the validity of the statement (i) of Question(I) (i.e., polynomial case) in full generality. Their proof assumes the theorems of Quillen and Suslin ([Q], [Su 1]) viz. all projective modules over $k[X_1, \ldots, X_n]$, where $k$ is a field, are free. Subsequently, by developing a nice criterion for the existence of unimodular elements, Lindel ([Li 2], Theorem 1.12) gave another (and very elegant) proof which does not assume these theorems. In fact his proof gives another proof (there are plenty of such proofs) of Serre's conjecture. In this section we will give his proof.

Bhatwadekar-Lindel-Rao ([BLR], Theorem 4.1) gave an affirmative answer to (i) of Question(II) (i.e., the Laurent case). Thus the 'unimodular problem' was completely settled.

Regarding the 'cancellation problem' over polynomial extensions ((ii), Question(I)), there were partial results (see [BRy 1], [BRy 2], [Bh 1]) which settled the cases (1) $A$ is a polynomial ring in *two* variables over a *normal* domain $R$, (2) $\dim(R) = 2, n \geq 2$ until Rao ([Ra 1], Theorem 2.5) proved the validity of ((ii), Question(I)) in full generality. Note that, under the assumptions of Question(I) or (II), showing that "projective $A$-modules of rank $\geq d + 1$ are cancellative" is equivalent to proving the following statement:

(iii) *If* rank$(P) \geq d + 1$ *then the group* Aut$(P \oplus A)$ *of automorphisms of* $P \oplus A$ *acts transitively on the set* Um$(P \oplus A)$ *of unimodular elements.*

Subsequently, Lindel ([Li 2], Theorem 2.6) showed that; if $A = R[X_1, \ldots, X_n, Y_1^{\pm 1}, \ldots, Y_m^{\pm 1}]$ and $P$ is a projective $A$-module of rank $\geq \max(2, \dim(R) + 1)$ Then $E(P \oplus A)$ acts transitively on Um$(P \oplus A)$ thus giving an affirmative answer to (ii) of Question(II).

Now we will give Lindel's proof of the 'unimodular theorem'. We first state some auxiliary results needed for the proof.

**Lemma 3.6.** *Let $B[T]$ be a polynomial ring in one variable over a commutative ring $B$. Let $s, b$ be elements of $B$ and let $I = (f_1(T), \ldots, f_r(T))$ be an ideal of*

$B[T]$ such that $sb \in I$ and $s(1 + Tf(T)) \in I$ for some element $f(T) \in A$. Then the ideal $\tilde{I}$ of $B[T]$ generated by $f_1(bT), \ldots, f_r(bT)$ contains $s$.

**Proof.** Since $s(1+Tf(T))$, $sb \in I$ there exist elements $g_i(T)$, $h_i(T) \in B[T]$, $1 \le i \le r$ such that $\sum_{i=1}^r f_i(T)g_i(T) = s(1 + Tf(T))$ and $\sum_{i=1}^r f_i(T)h_i(T) = sb$. These equalities, as $s, b \in B$, show that $\sum_{i=1}^r f_i(bT)g_i(bT) = s(1 + bTf(bT))$ and $\sum_{i=1}^r f_i(bT)h_i(bT) = sb$. ∎

Since the elements $b$ and $1 + bTf(bT)$ are comaximal in $B[T]$ we are through.

**Lemma 3.7.** *Let $I$ be an ideal in $B[T]$ containing a monic polynomial (i.e., a polynomial whose leading coefficient is 1). Suppose $s \in B$ is an element such that $I + sB[T] = B[T]$. Then there exists an element $c \in B$ such that $1 + sc \in I$.*

**Proof.** Let $J = I \cap B$. Then, as $I$ contains a monic polynomial, $B[T]/I$ is an integral extension of the subring $B/J$. Therefore, as the image of $s$ in $B[T]/I$ is invertible, the image of $s$ in $B/J$ is also invertible. Hence there exists $c \in B$ such that $1 + sc \in J \subset I$. ∎

The following observation is due to Suslin (see [L], Chap.III, Section 3)

**Proposition 3.8.** *Let $D$ be a ring of finite Krull dimension and let $I$ be an ideal of $D[Z_1, \ldots, Z_m]$ of height $\ge \dim(D) + 1$. Then there exist positive integers $r_1, \ldots, r_{m-1}$ and a (Nagata) transformation of variables:*

$$T_i = Z_i + Z_m^{r_i}, 1 \le i \le m - 1, T_m = Z_m$$

*such that $I$ contains a monic polynomial in $T_m$ with coefficients in $B = D[T_1, \ldots, T_{m-1}]$.*

We now state a criterion due to Lindel ([Li 2], Proposition 1.6) for obtaining a unimodular element in a projective module over a polynomial ring.

**Proposition 3.9.** *Let $P$ be a projective module over $B[T]$ of rank $r$. Let $s \in B$ be a non-zero-divisor such that $P_s$ is free. Suppose there exists $p \in P$ such that*

*(1) $\mathcal{O}_p(p)$ contains an element of the type $1 + Tf(T)$.*

*(2) There exists $c \in B$ such that $1 + sc \in \mathcal{O}_P(p)$.*

*Then there exists a unimodular element $q \in P$ such that $q - p \in sTP$.*

**Proof.** Since $P_s$ is free of rank $r$ and $s$ is a non-zero-divisor, it is easy to see that there exist $e_1, \ldots, e_r \in P$ such that $F = \sum_{i=1}^r B[T]e_i$ is a free $B[T]$-module of rank $r$ and $s^l P \subset F$ for some positive integer $l$. Replacing $s$ by $s^l$, if necessary, we assume that $sP \subset F \subset P$. Hence $sp \in F$.

Let $sp = f_1(T)e_1 + \cdots + f_r(T)e_r$ and let $b = 1 - s^2c^2$. Since $s\mathcal{O}_P(p) = \mathcal{O}_P(sp) \subset \mathcal{O}_F(sp)$, it follows, from assumptions (1) and (2), that $s(1+Tf(T))$, $sb \in \mathcal{O}_F(sp) = (f_1(T), \ldots, f_r(T))$. Now consider the element $m = f_1(bT)e_1 + \cdots +$

$f_r(bT)e_r$ of $F$. Then, by (3.6), $s \in \mathcal{O}_F(m)$. Moreover, since $f_i(bT) - f_i(T) \in s^2TB[T]$, we have $m = sp + s^2Tm'$ for some $m' \in F$.

Let $q = p + sTm'$. Then $q - p \in sTP$ and $sq = m \in F$. Hence, as $s \in \mathcal{O}_F(m)$. and $P_s = F_s$, it follows that $q \in \text{Um}(P_s)$. This shows that $s^t \in \mathcal{O}_P(q)$ for some positive integer $t$. Moreover, as $q - p \in sP$ and $1 + sc \in \mathcal{O}_P(p)$, there exists $g(T) \in B[T]$ such that $1 + sg(T) \in \mathcal{O}_P(q)$. These two facts together show that $\mathcal{O}_P(q) = B[T]$.                                                           ∎

Now we are ready to state and prove the 'unimodular theorem' of Bhatwadekar-Roy/Lindel. The proof given below is due to Lindel. This proof also gives another proof of 'Serre's conjecture'.

**Theorem 3.10.** *Let $R$ be a ring of Krull dimension $d$ and let $A = R[X_1, \ldots, X_n]$. Then every projective $A$-module of rank $\geq d + 1$ has a unimodular element.*

**Proof.** We prove the result by induction on $\dim(A) = d + n$.

Let $N$ denote the nilradical of $R$ and let $I = NA$. Then $I$ is contained in the Jacobson radical of $A$. Therefore, by (2.11), it is enough to prove the result for the ring $A/I = (R/N)[X_1, \ldots, X_n]$. Hence, without loss of generality, we can assume that $R$ is a reduced ring. Moreover we can also assume that $\text{Spec}(R)$ is connected (i.e., $R$ does not have any nontrivial idempotent). Now we proceed keeping these reductions in mind.

If $d + n = 0$ then $A$ is a field and the assertion is clear. So we assume that $d + n > 0$. If $n = 0$ then this is Serre's unimodular element theorem (3.1). If $d = 0, n = 1$ then, as $A$ is a PID and all projective modules over a PID are free, we are through. So we assume that $n > 0, d + n > 1$.

Let $P$ be a projective $A$-module of rank $r \geq d + 1$. If $r = 1$ then $d = 0$ and hence, as $A$ is a UFD, $P$ is free. Therefore we now assume that $r > 1$.

Let $D = R$ if $d > 0$ and let $D = R[X_1]$ if $d = 0$. Then $\dim(D) > 0, A = D[Z_1, \ldots, Z_m], m > 0$ and $P$ is a projective $A$-module with rank $(P) = r \geq \dim(D) + 1$.

Let $S$ denote the set of all non-zero-divisors of $D$. Then $\dim(D_S) = 0, A_S = D_S[Z_1, \ldots, Z_m]$ and $\dim(A_S) < \dim(A)$. Therefore, by induction hypothesis, $P_S$ not only has a unimodular element but in fact $P_S$ is a *free* $A_S$-module of rank $r \geq 2$ (here is the clever use of induction which avoids the Quillen-Suslin Theorem). Let $s \in S$ be such that $P_s$ is $A_s$-free.

Now consider the projective module $P/sZ_mP$ over the ring $A/sZ_mA = D[Z_m]/(sZ_m)[Z_1, \ldots, Z_{m-1}]$. Since $\dim(D[Z_m]/(sZ_m)) = \dim(D) < r = \text{rank}(P/sZ_mP)$ and $\dim(A/(sZ_m)A) < \dim(A)$, by the induction hypothesis, there exists $p' \in P$ whose image in $P/sZ_mP$ is unimodular. Therefore, the element $(p', sZ_m)$ is unimodular in $P \oplus A$. Hence, by (3.5), there exists $p_1 \in P$ such that $\text{ht}\,\mathcal{O}_P(p) \geq r$ where $p = p' + sZ_mp_1$. Therefore, by (3.8), by changing variables suitably, we can write $A = B[Z_m]$ and assume that $\mathcal{O}_P(p)$ contains a monic polynomial in $Z_m$ with coefficients in $B$. Note that, by construction, $(p, sZ_m) \in P \oplus A$ is unimodular. Hence, there exists $f(Z_m) \in A$ such that

$1 + Z_m f(Z_M) \in \mathcal{O}_P(p)$. Moreover, as $\mathcal{O}_P(p)$ contains monic polynomial, by (3.7), there exists $c \in B$ such that $1 + sc \in \mathcal{O}_P(p)$. Therefore, by (3.9), $P$ contains a unimodular element $q$ such that $q - p \in s Z_m P$. ∎

**Remark 3.11.** The above proof gives the stronger result viz., the canonical map $\mathrm{Um}(P)$ to $\mathrm{Um}(P/X_n P)$ is surjective.

As mentioned in the introduction, one of the main ingredients of the Quillen-Suslin proof ([Q], [Su 1]) of Serre's Conjecture is the following result:

**Theorem 3.12.** *Let $P$ be a projective module over $R[X]$ such that $P_f$ is free over $R[X, 1/f]$ for some monic polynomial $f(X) \in R[X]$. Then $P$ is free. More generally, if $Q$ is a projective $R[X]$-module which is extended from $R$ and $P_f \simeq Q_f$ then $P \simeq Q$.*

Therefore, one is led to wonder whether the behaviour of a projective module over $R[X]$ is completely determined by its behaviour under inversion of a monic polynomial. Hence we ask the following questions. Note that affirmative answers to these questions will also give an affirmative answer to Question (I) mentioned in the begining of this section.

**Question (III).** *Let $P$ be a projective $R[X]$-module and let $f = f(X)$ be a monic polynomial. If $P_f$ has a unimodular element, then does $P$ have a unimodular element?*

When $R$ is local, this question has been answered affirmatively by Roitman ([Ro 1], Lemma 10). Other known cases are: (1) $R$ is a normal domain of finite Krull dimension, $\mathrm{rank}(P) = \dim(R)$ and $P/XP$ has a unimodular element ([BLR], Theorem 5.2). (2) $R$ is an affine normal domain over an algebraically closed field and $\mathrm{rank}(P) = \dim(R)$ ([Bh 2], Remark 3.8).

**Question (IV).** *Let $P, f(X)$ be as in Question (III). If $P_f$ is cancellative, is $P$ cancellative?*

An affirmative answer to the following question will give an affirmative answer to Question (IV).

**Question (V).** *Let $P, Q$ be projective modules over $R[X]$ such that $P_f \simeq Q_f$ for some monic $f \in R[X]$. Then is $P$ isomorphic to $Q$?*

When $Q$ is extended from $R$ then the answer is affirmative (due to Quillen and Suslin). When $R$ is local and $P$ has a rank 1 direct summand then Roy [Ry 2] has given an affirmative answer to this question.

**Remark 3.13.** If $P, Q$ are projective $R[X]$-modules such that $P_f \simeq Q_f$ for some monic $f(X)$ then it is easy to see that $P$ and $Q$ are stably isomorphic (e.g., see [MP]).

## 4. Projective Modules Over Polynomial Extensions of A Regular Ring

In this section we deal with the following conjecture of Bass and Quillen ([B 4], Question (IX)) and ([Q]):

**Bass-Quillen Conjecture.** *Let $R$ be a regular ring. Then every projective $R[X_1, \ldots, X_n]$-module is extended from $R$.*

This statement is not true for arbitrary rings. For instance, if $k$ is a field and $R$ denotes the $k$-subalgebra $k[t^2, t^3]$ of the polynomial ring $k[t]$, then the ideal $(1 - t^2 X^2, 1 - t^3 X^3)$ of $R[X]$ furnishes an example of a projective $R[X]$-module (of rank 1) which is not extended (hence not stably extended) from $R$. This example is due to Schanuel (see ([B 1], Proposition 2.1 (b))).

However, by a theorem of Grothendieck (see [B 3], p.635), if $R$ is regular, then every projective module over $R[X_1, \ldots, X_n]$ is stably extended from $R$. This might have led Bass to pose this conjecture as a question (1972). Since an affirmative solution to this problem would imply Serre's conjecture and at that time the status of Serre's conjecture was not clear, he cautioned to approach it by possibly seeking a counterexample. But after giving an affirmative answer to Serre's conjecture, Quillen posed this problem in its present conjectural form (possibly hoping to have an affirmative answer). Hence this conjecture is known as the "Bass-Quillen Conjecture". For brevity, here onwards we shall refer to this conjecture as the BQ conjecture.

The following theorem of Quillen [Q] shows that it is enough to prove the BQ conjecture when $R$ is a regular *local* ring. In this theorem $R$ denotes an arbitrary ring.

**Theorem 4.1. (Quillen Localization Theorem)** *If $P$ is a projective $R[X_1, \ldots, X_n]$-module, where $R$ is an arbitrary ring, such that the $R_{\mathfrak{m}}[X_1, \ldots, X_n]$-module $P_{\mathfrak{m}}$ is free for every maximal ideal $\mathfrak{m}$ of $R$, then $P$ is extended from $R$.*

Thus the BQ conjecture is reduced to proving the following equivalent conjecture (we still refer to it as the BQ conjecture):

**BQ Conjecture.** *Let $R$ be a regular local ring. Then every projective $R[X_1, \ldots, X_n]$-module is free.*

The Quillen-Suslin Theorem shows the validity of this conjecture when $\dim(R) \leq 1$. Horrocks [H] proved this conjecture for $n = 1$ when $R$ is a regular local ring of dimension 2 containing a coefficient field. Later on, Murthy ([Mu 2]) proved it for any regular local ring of dimension 2 and for $n = 1$. Therefore, thanks to the 'monic inversion theorem' of Quillen and Suslin and (4.1), the conjecture can be shown to be true for $\dim(R) = 2, n \geq 1$ (see ([L], Theorem 3.3, p. 138)). So we assume for the discussion that $\dim(R) \geq 3$.

This conjecture has an affirmative answer when $R$ contains a field. Mohan Kumar [Mk] and Lindel-Lutkebohmert [LL] independently proved it in the case when $R$ is a power series ring over a field. Later on, Lindel ([Li 1], Theorem) settled the case when $R$ is a geometric regular local ring. A result of Popescu (see (4.2) below) shows that to prove the conjecture in general (provided $R$ contains a field), it is enough to prove it in the geometric case.

We now indicate briefly how results of Lindel ([Li 1], Theorem) and Popescu (4.2) culminated in the final answer.

Since $R$ is regular and local, by the above mentioned result of Grothendieck (see ([B 3], p. 635)), every projective $R[X_1, \dots, X_n]$-module is stably free. Therefore, to prove the conjecture, it is enough to show that if $P$ is a projective $R[X_1, \dots, X_n]$-module such that $P \oplus R[X_1, \dots, X_n]$ is free, then $P$ is free. By (2.11), Such a projective module $P$ will be given by a unimodular row $[f_1, \dots, f_r]$ where $r = \text{rank}(P) + 1$. By (2.18), $P$ is free if and only the row $[f_1, \dots, f_r]$ is *completable*. Thus the following statement is another equivalent formulation of BQ conjecture:

*Let $R$ be a regular local ring. Then every unimodular row $[f_1, \dots, f_r]$ over $R[X_1, \dots, X_n]$ is completable.*

The following theorem of Popescu (see ([Sw 2], Theorem 1.1)) for a proof) and the above equivalent formulation show that to prove the BQ conjecture for a regular local ring $R$ containing a *field* it is enough to consider the special case viz. $R$ is the localisation of a regular affine domain over a prime field $k$ at some prime ideal. We call such a local ring a regular *k-spot*.

**Theorem 4.2.** *Let $R$ be a regular local ring containing prime field $k$ (i.e., $k = \mathbf{Q}$ if char $k = 0$ and $k = \mathbf{Z}/p$ if char $k = p > 0$). Let $B$ be an affine $k$-subalgebra of $R$. Then there exists a regular affine domain $C$ over $k$ and $k$-algebra homomorphisms $\pi : B \to C, \psi : C \to R$ such that the composite map $\psi\pi : B \to R$ is the inclusion map of $B$ into $R$.*

As an application of this theorem we have the following result:

**Proposition 4.3.** *Let $R$ be a regular local ring containing prime field $k$. Let $[f_1, \dots, f_r]$ be a unimodular row over $R[X_1, \dots, X_n]$. Then there exists a regular $k$-spot $S$ together with a $k$-algebra homomorphism $\theta : S[X_1, \dots, X_n] \to R[X_1, \dots, X_n]$ and a unimodular row $[g_1, \dots, g_r]$ over $S[X_1, \dots, X_n]$ such that $\theta(g_i) = f_i, 1 \le i \le r$.*

**Proof.** Since the row $[f_1, \dots, f_r]$ is unimodular, there exist elements $h_1, \dots, h_r \in R[X_1, \dots, X_n]$ such that $\sum_{i=1}^r f_i h_i = 1$. Let $B$ be an affine $k$-subalgebra of $R$ generated by coefficients of $f_i, h_i, 1 \le i \le r$. Then $f_i, g_i \in B[X_1, \dots, X_n]$ and hence $[f_1, \dots, f_r]$ is a unimodular row over $B[X_1, \dots, X_n]$.

Let $C, \pi, \psi$ be as in (4.2). Let m denote the maximal ideal of $R$ and let q $= \psi^{-1}(\text{m})$. Then q is a prime ideal of $C$ and hence $S = C_q$ is a regular $k$-spot as $C$ is a regular affine domain over $k$. It is easy to see that, as $R$ is local, the $k$-algebra homomorphism $\psi : C \to R$ induces a $k$-algebra homomorphism $\theta : S \to R$ such that $\psi$ factors through $\theta$. We regard $\theta$ as a map from $S[X_1, \dots, X_n]$ to $R[X_1, \dots, X_n]$ by sending $X_i$ to $X_i$.

Let $\pi(f_i) = g_i$. Then, as $[f_1, \ldots, f_r]$ is unimodular over $B[X_1, \ldots, X_n]$, $[g_1, \ldots, g_r]$ is unimodular over $C[X_1, \ldots, X_n]$ and hence over $S[X_1, \ldots, X_n]$. Now we are through as $\pi\psi : B \to R$ is the inclusion map of $B$ into $R$. ∎

From (4.3) we see that, if the BQ conjecture is true for any regular $k$-spot ($k$: prime field) then it is true for any regular local ring $R$ containing a field. Lindel ([Li 1], Theorem) had settled the case of regular $k$-spot. Thus the BQ conjecture is completely settled when $R$ contains a field.

We now indicate Lindel's proof. We start with a definition of an *analytic isomorphism*.

**Definition 4.4.** Let $A$ be an arbitrary domain, $B$ be a subdomain of $A$ and $s$ be an element of $B$. We say that $A$ is *analytically isomorphic* to $B$ along $s$ if $B/sB = A/sA$.

Before proceeding further, we give some examples of analytic isomorphisms. The simplest example is to take two comaximal elements $s, t$ in a domain $B$ and put $A = B_t$ so that $B/sB = A/sA$. To obtain non-trivial examples, we need first the following definition of Weierstrass polynomial.

**Definition 4.5.** Let $R$ be a local ring with maximal ideal m. A monic polynomial $f(T) \in R[T]$ of degree $n$ is called a *Weierstrass* polynomial (with respect to $T$) if it is of the form

$$f(T) = \sum_{i=0}^{n-1} a_i T^i + T^n, a_i \in \mathfrak{m} \text{ for } 0 \le i \le n - 1.$$

**Example 4.6.** Let $(R, \mathfrak{m})$ be a local domain and let $f(T) \in R[T]$ be a Weierstrass polynomial. Let $\mathfrak{M} = (\mathfrak{m}, T)$ be a maximal ideal of $R[T]$ and $S = R[T]_{\mathfrak{M}}$. Then $S$ is analytically isomorphic to $R[T]$ along $f$. Moreover, if $R$ is m-adically complete, then $R[[T]]$ is analytically isomorphic to $R[T]$ along $f$.

In the case of an analytic isomorphism we have the following "Descent Lemma". This lemma was first discovered by Mohan Kumar [Mk] and Lindel-Lutkebohmert [LL] independently while proving BQ conjecture when $R$ is a power series ring over a field. Subsequently this lemma has been used by many others (see [BRa], [Li 1]). For its proof readers can refer to ([L], 5.11) or [LL] or [Mk].

**Lemma 4.7. (Descent Lemma)** *Let $B$ be a domain and let $A$ be an overdomain of $B$ such that $A$ is analytically isomorphic to $B$ along an element $s \in B$. Let $P$ be a projective $A$-module such that $P_s \simeq Q' \otimes_{B_s} A_s$ for some projective $B_s$-module $Q'$. Then there exists a projective $B$-module $Q$ such that $P \simeq Q \otimes_B A$ and $Q_s \simeq Q'$.*

The following proposition, which is a variant of a result of Lindel ([Li 1], Proposition 2), gives another example of an analytic isomorphism and is very crucial for his proof of the BQ conjecture. For a proof of this proposition reader can refer to ([N], Theorem 2.8) or ([O], Proposition, p. 114).

**Proposition 4.8.** *Let $k$ be a perfect field and let $(R, \mathfrak{m})$ be a regular $k$-spot of dimension $d$. Let $a \in \mathfrak{m}$ be a non-zero element. Then there exists a field extension $K$ of $k$ contained in $R$ and elements $Z_1, \ldots, Z_d \in R$ such that*

(i) *$Z_1, \ldots, Z_d$ are algebraically independent over $K$.*

(ii) *$K[Z_1, \ldots, Z_d] \cap \mathfrak{m} = (f(Z_1), Z_2, \ldots, Z_d)$ where $f(Z_1) \in K[Z_1]$ is an irreducible polynomial.*

(iii) *$R$ is analytically isomorphic over $R'$ along $s$ for some $s \in aR \cap R'$ where $R'$ is a localisation of $K[Z_1, \ldots, Z_d]$ at the maximal ideal $(f(Z_1), Z_2, \ldots, Z_d)$.*

The following result, which is in some sense a converse of the 'localisation theorem' of Quillen (4.1), is due to Roitman [Ro 1].

**Proposition 4.9.** *Let $R$ be a ring such that every projective $R[X_1, \ldots, X_n]$-module of rank $r$ is extended from $R$. Let $S$ be a multiplicatively closed subset of $R$. Then every projective $R_S[X_1, \ldots, X_n]$-module of rank $r$ is extended from $R_S$.*

Now we are ready to give Lindel's proof of the BQ conjecture for a regular $k$-spot ([Li 1], Theorem).

**Theorem 4.10.** *Let $k$ be a perfect field and let $R$ be a regular $k$-spot. Then every projective $R[X_1, \ldots, X_n]$-module is free.*

**Proof.** Let $P$ be a projective $R[X_1, \ldots, X_n]$ module. Let $L$ denote the quotient field of $R$. By the Quillen-Suslin theorem, $P \otimes L[X_1, \ldots, X_n]$ is free and hence there exists a non-zero element $a \in R$ such that the projective $R_a[X_1, \ldots, X_n]$-module $P_a$ is free. If $a$ is invertible then there is nothing to prove. So we assume that $a \in \mathfrak{m}$ where $\mathfrak{m}$ denotes the maximal ideal of $R$.

By (4.8), $R$ contains a field extension $K$ of $k$ and a local ring S of a *polynomial extension* $K[Z_1, \ldots, Z_d]$; $d = \dim(R)$, of $K$ such that $R$ is analytically isomorphic to $S$ along $s$ for some $s \in aR \cap S$. Note that $s$ is a multiple of $a$ and hence $P_s$ is free over $R_s[X_1, \ldots, X_n]$. Therefore, by (4.7), there exists a projective $S[X_1, \ldots, X_n]$-module $Q$ such that $P \simeq Q \otimes R[X_1, \ldots, X_n]$. By (4.9) and the Quillen-Suslin theorem, $Q$ and hence $P$ is free. ∎

Now we cite another class of rings for which the BQ conjecture is true. If $R$ is a regular local ring of characteristic 0 whose residue field $k$ is of characteristic $p > 0$, we say that $R$ is *unramified* if $p \notin \mathfrak{m}^2$ where $\mathfrak{m}$ denotes the maximal ideal of $R$. Popescu [Po] has proved the following result when $R$ is unramified.

**Theorem 4.11.** *Let $R$ be an unramified regular local ring. Then all projective modules over $R[X_1, \ldots, X_n]$ are free.*

The following question of Suslin ([Su 3], Problem 4) is related to the BQ conjecture (see equivalent formulation of the BQ conjecture).

**Question 4.12.** *Let $R$ be an arbitrary local ring such that $n!$ is invertible in $R$. Then, is every unimodular row $[f_1, \ldots, f_{n+1}]$ over $R[X]$ completable? In particular, if $R$ contains the field $\mathbf{Q}$ of rationals, then is every stably free projective $R[X]$-module free?*

This problem is still open. But there are some partial results which we quote below. If $n = 1$ then the answer is obviously affirmative (in fact for any unimodular row of length two over any ring). So we consider only the case that $n \geq 2$. In this case, if $n \geq \dim(R) + 1$, then Suslin ([Su 2], Theorem 7.2) proved that in fact $[f_1, \ldots, f_{n+1}]$ is a first row of an elementary matrix. He proved this result even without the assumption on the invertibility of $n!$. If $\dim(R) = 1$, then above mentioned result of Suslin shows that every unimodular row over $R[X]$ is completable. If $\dim(R) = 2$ and $1/2 \in R$, then Murthy (see ([Ra 2], Lemma 2.9)) showed that every unimodular row over $R[X]$ is completable. In ([Ro 2], Corollary 6), Roitman has shown that if $R$ is a local ring of characteristic $p$, where $p$ is an odd prime $> d = \dim(R)$, then every stably free $R[X]$-module of rank $\geq d/2+1$ is free. Rao ([Ra 2], Theorem 2.4) has shown that if $R$ is a local ring of dimension $d$ and $d!$ is invertible in $R$, then every unimodular row $[f_1, \ldots, f_{d+1}]$ over $R[X]$ is completable, thus extending ([Su 2], Theorem 7.2) one step further. Moreover, using this result, subsequently Rao ([Ra 3], Corollary 3.3) showed that if $R$ is a local ring of dimension 3 and $1/6 \in R$, then every stably free $R[X]$-module is free.

The following question is an analogue of the BQ conjecture for Laurent polynomial extensions.

**Question 4.13.** *Let $R$ be a regular ring. Is every projective $R[X_1, \ldots, X_n, Y_1^{\pm 1}, \ldots, Y_m^{\pm 1}]$-module extended from $R$?*

If $R$ is regular, then it is well known (see ([B 3], Corollary 4.3, p. 643)) that every $R[X_1, \ldots, X_n, Y_1^{\pm 1}, \ldots, Y_m^{\pm 1}]$-module is stably extended from $R$. If $R$ is of finite Krull dimension, then, by ([Sw 2], Theorem 1.1), every projective $R[X_1, \ldots, X_n, Y_1^{\pm 1}, \ldots, Y_m^{\pm 1}]$-module of rank $\geq \dim(R)+1$ is extended from $R$. The following example ([BRa], Example (2), p. 809) shows that this result is best possible (see [Sw 1] for another example).

**Example 4.14.** Let $R$ denote the field of real numbers and let $R = \mathbf{R}[X, Y, Z]/(X^2 + Y^2 + Z^2 - 1)$ denote the coordinate ring of the real 2-sphere. Let $x, y, z$ denote the images of $X, Y, Z$ respectively in $R$. Recall (see (3.3)) that the unimodular row $[x, y, z]$ over $R$ is not completable.

Let $P$ be a projective module over $R[T, T^{-1}]$ given by the unimodular row $[(1 + x) + (1 - x)T, y, z]$. Then $P$ is not extended from $R$. This is because the projective $R$-module $P/(T-1)P$ corresponds to the unimodular row $[2, y, z]$ over $R$ and hence is free, where-as the projective $R$-module $P/(T+1)P$ correspondence to the unimodular row $[2x, y, z]$ over $R$ and hence is not free.

However, using (4.2) and Corollary 3.9 of [BRa], we have the following result:

**Theorem 4.15.** *Let $R$ be a regular local ring containing the field $Q$ of rationals. Then every $R[X_1, \ldots, X_n, Y_1^{\pm 1}, \ldots, Y_m^{\pm 1}]$-module is free.*

This theorem and the example above together show that the Quillen localisation theorem (4.1) is not true for Laurent polynomial extensions.

Quillen, after solving the 'Serre conjecture', not only posed the original question of Bass in a conjectural form (BQ conjecture) but posed the following question as a way to solve it. This question is known as 'Quillen question'.

**Question 4.16.** *(Quillen question) Let $R$ be a regular local ring with maximal ideal $\mathfrak{m}$ and let $f$ be a regular parameter (i.e., $f \in \mathfrak{m} \backslash \mathfrak{m}^2$). Then every projective $R_f$-module is free.*

A priori there does not seem to be any connection between this question and the BQ conjecture. But soon we will show how this question in fact is a stronger form of the BQ conjecture.

Let $S$ be a ring and let $S(T)$ denote a localisation of $S[T]$ obtained by inverting all monic polynomials. Now we quote a result of Horrocks [H]. Horrocks proved this result using a geometric argument. Roy gave an elegant algebraic proof in [Ry 2].

**Theorem 4.17.** *Let $S$ be a local ring and let $P$ be a projective $S[T]$-module. If the projective $S(T)$-module $P \otimes_{S[T]} S(T)$ is free, then $P$ is free.*

Now assume that $S$ is a regular local ring with maximal ideal $\mathfrak{n}$. Let $R$ denote a local ring obtained from $S[T^{-1}]$ by localising at the maximal ideal $\mathfrak{m} = (\mathfrak{n}, T^{-1})$ and let $f = T^{-1}$. Then, since $S$ is regular, $R$ is a regular local ring and $f$ is a regular parameter of $R$. Moreover, $R_f = S(T)$. Thus $S(T)$ is a special case of a ring of the type $R_f$. In view of (4.17) it is clear how an affirmative solution to the 'Quillen question' would answer the BQ conjecture affirmatively. But, as it happened, the 'BQ conjecture' was proved in many cases without solving the 'Quillen question'. Still one feels that this question merits some interest.

When $R$ is a regular local ring of dimension 3 containing a coefficient field, Horrocks [H] answered this question affirmatively. Subsequently Gabber [G] generalised this result for any regular local ring of dimension 3. When $R$ is a power series ring over a field then an affirmative answer is due to Mohan Kumar [Mk]. Using (4.2) and results of Bhatwadekar - Rao ([BRa], Theorem 2.5 and Corollary 3.4) we have the following:

**Theorem 4.18.** *Let $R$ be a regular local ring containing a field. Let $f \in R$ be a regular parameter. Then every projective $R_f$-module is free. Moreover, if $R$ contains $Q$ then every projective $R_f[X_1, \ldots, X_n, Y_1^{\pm 1}, \ldots, Y_m^{\pm 1}]$-module is free.*

Now we conclude this section with the following interesting result which says that the 'BQ conjecture' is equivalent to the 'Quillen question' when $R_f$ is of the type $S(T)$. This result can be deduced from ([BRa], Theorem 2.2).

**Theorem 4.19.** *Let $S$ be a regular local ring. Then every projective $S[T]$-module is free if and only if every projective $S(T)$-module is free.*

# References

[AM]    M. F. Atiyah and I. G. Macdonald, *Introduction to Commutative Algebra*, Addison-Wesley, 1969.

[B 1]   H. Bass, Torsion-free and projective modules, *Trans. Amer. Math. Soc.* **bf 102**, 319–327, 1962.

[B 2]   H. Bass, $K$-theory and stable algebra, *I.H.E.S.* **22**, 5–60, 1964.

[B 3]   H. Bass, *Algebraic K-Theory*, Benjamin, New York, 1968.

[B 4]   H. Bass, Some problems in "classical" algebraic $K$-theory, *Algebraic K-Theory, II, Lect. Not. Math.* **342,** Springer-Verlag, 1972.

[BM]    H. Bass and M. P. Murthy, Grothendieck groups and Picard groups of abelian group rings, *Ann. Math.* **86,** 16–73, 1967.

[Bh 1]  S. M. Bhatwadekar, A note on projective modules over polynomial rings, *Math. Zeit.* **194,** 285–291, 1987.

[Bh 2]  S. M. Bhatwadekar, Inversion of monic polynomials and existence of unimodular elements, *Math. Zeit.* **200,** 1989.

[BLR]   S. M. Bhatwadekar, H. Lindel and R. A. Rao, The Bass Murthy question: Serre dimension of Laurent polynomial extensions, *Invent. Math.* **81,** 189–203, 1985.

[BRa]   S. M. Bhatwadekar and R. A. Rao, On a question of Quillen, *Trans. Amer. Soc.* **279,** 801–810, 1983.

[BRy 1] S. M. Bhatwadekar and A. Roy, Some theorems about projective modules over polynomial rings, *J. Alg.* **86,** 150–158, 1984.

[BRy 2] S. M. Bhatwadekar and A. Roy, Some cancellation theorems about projective modules over polynomial rings, *J. Alg.* **111,** 166–176, 1987.

[EE 1]  D. Eisenbud and E. G. Evans, Generating modules efficiently: theorems from algebraic $K$-Theory, *J. Alg.* **27,** 278–305, 1973.

[EE 2]  D. Eisenbud and E. G. Evans, Three conjectures about modules over polynomial rings, conference on commutative algebra, *Lect. Not. Math.* **311,** Springer-Verlag, 1973.

[En]    S. Endo, Projective modules over polynomial rings, *J. Math. Soc. Japan* **15,** 339–352, 1963.

[G]     O. Gabber, Groupe de Brauer, *Seminaire, Les Plans-sur-Bex, Lect. Not. Math.* **844,** 129–209, Springer-Verlag, 1980.

[Go]    N. S. Gopalakrishnan, *Commutative Algebra*, Oxonian Press, 1984.

[Gr]    C. Greither, Serre's problem for one-dimensional ground ring, *J. Alg.* **75,** 290–295, 1982.

[H]     G. Horrocks, Projective modules over an extension of a local ring, *Proc. London Math. Soc.* **14,** 714–718, 1964.

[K]     M.-C. Kang, Projective modules over some polynomial rings, *J. Alg.* **59,** 65–76, 1979.

[L]     T. Y. Lam, Serre's conjecture, *Lect. Not. Math.* **635,** Springer-Verlag, 1978.

[Li 1]  H. Lindel, On a question of Bass-Quillen and Suslin concerning projective modules over polynomial rings, *Invent. Math.* **65,** 319–323, 1981.

[Li 2]  H. Lindel, Unimodular elements in projective modules, *J. Alg.* **172,** 301–319, 1995.

[LL]    H. Lindel and W. Lutkebohmert, Projective moduln uber polynomialen erweiterungen von potenzreihenalgebren, *Arch. Math.* **28,** 51–54, 1977.

[Ma 1]   S. Mandal, Basic elements and cancellation over Laurent polynomial rings, *J. Alg.* **79,** 251–257, 1982.

[Ma 2]   S. Mandal, Projective modules and complete intersections, *Lect. Not. Math.* **1672,** Springer, 1997.

[Mk]     N. Mohan Kumar, On a question of Bass-Quillen, *J. Indian Math. Soc.* **43,** 13–18, 1979.

[Mt]     H. Matsumura, *Commutative Ring Theory,* Cambridge University Press, 1986.

[Mu 1]   M. P. Murthy, Projective modules over a class of polynomial rings, *Math. Zeit.* **88,** 184–189, 1965.

[Mu 2]   M. P. Murthy, Projective $A[X]$-modules, *J. London Math. Soc.* **41,** 453–456, 1966.

[Mu 3]   M. P. Murthy, Complete intersections, *Proc. Con. Commutative Alg. Queen's University,* 197–211, 1975.

[MP]     M. P. Murthy and C. Pedrini, $K_0$ and $K_1$ of polynomial rings, *Algebraic K-Theory II, Lect. Not. Math.* **342,** 109–121, Springer-Verlag, 1973.

[N]      B. S. Nashier, Efficient generation of ideals in polynomial rings, *J. Alg.* **85,** 287–302, 1983.

[O]      M. Ojanguren, Quadratic forms over regular local rings, *J. Indian Math. Soc.* **44,** 109–116, 1980.

[P]      B. Plumstead, The conjectures of Eisenbud and Evans, *Amer. J. Math.* **105,** 1417–1433, 1983.

[Po]     D. Popescu, Polynomial rings and their projective modules, *Nagoya Math. J.* **113,** 121–128, 1989.

[Q]      D. Quillen, Projective modules over polynomial rings, *Invent. Math.* **36,** 167–171, 1976.

[Ra 1]   R. A. Rao, A question of H. Bass on the cancellative nature of large projective modules over polynomial rings, *Amer. J. Math.* **110,** 641–657, 1988.

[Ra 2]   R. A. Rao, The Bass-Quillen conjecture in dimension three but characteristic $\neq 2, 3$ via a question of Suslin, *Invent. Math.* **93,** 609–618, 1988.

[Ra 3]   R. A. Rao, On completing unimodular polynomial vectors of length three, *Trans. Amer. Math. Soc.* **325,** 231–239, 1991.

[Ro 1]   M. Roitman, On projective modules over polynomial rings, *J. Alg.* **58,** 51–63, 1979.

[Ro 2]   M. Roitman, On stably extended projective modules over polynomial rings, *Proc. Amer. Math. Soc.* **97,** 585–589, 1986.

[Ry 1]   A. Roy, Application of patching diagrams to some questions about projective modules, *J. pure app. Alg.* **24,** 313–319, 1982.

[Ry 2]   A. Roy, Remarks on a result of Roitman, *J. Indian Math. Soc.* **44,** 117–120, 1980.

[S 1]    J. P. Serre, Faisceaux algebriques coherents, *Ann. Math.* **61,** 191–278, 1955.

[S 2]    J. P. Serre, Modules projectifs et espaces fibres a fibre vectorielle, *Sem. Dubreil-Pisot* **23,** 1957/58.

[S 3]    J. P. Serre, Sur les modules projectifs, *Sem. Dubreil-Pisot* **2,** 1960/61.

[Ss 1]   C. S. Seshadri, Triviality of vector bundles over the affine space $K^2$, *Proc. Nat'l Acad. Sci. U.S.A.* **44,** 456–458, 1958.

[Ss 2]   C. S. Seshadri, Algebraic vector bundles over the product of an affine curve and the affine line, *Proc. Amer. Math. Soc.* **10,** 670–673, 1959.

[Su 1]   A. A. Suslin, Projective modules over a polynomial ring are free, *Sov. Math. Dokl.* **17,** 1160–1164, 1976.

[Su 2]    A. A. Suslin, On the structure of the special linear group over polynomial rings, *Math. USSR-Izv.* **11,** 221–238, 1977.

[Su 3]    A. A. Suslin, On stably free modules, *Math. USSR Sbornik* **31,** 479–491, 1977.

[Sw 1]    R. G. Swan, Projective modules over Laurent polynomial rings, *Trans. Amer. Math. Soc.* **237,** 111–120, 1978.

[Sw 2]    R. G. Swan, Neron-Popescu Desingularization (*Preprint*).

School of Mathematics, Tata Institute of Fundamental Research, Homi Bhabha Road,
Mumbai-400 005, India. E-mail: smb@math.tifr.res.in

# Around Automorphisms of Relatively Free Groups

## C.K. Gupta

This article is intended to be a survey on *some* topics within the framework of *automorphisms* of free groups and relatively free groups of certain soluble varieties. The bibliography at the end is neither claimed to be exhaustive, nor it is necessarily connected with a reference in the text. I include it as I see it revolves *around* the concepts emerging from the investigation of automorphisms of free groups. The interested reader may find it useful to browse over the list occasionally.

## Introduction

Let $V$ $(\geq F')$ be a fully invariant subgroup of a free group $F = \langle x_1, \ldots, x_n \rangle$ of rank $n \geq 2$. A system $w(V) = (w_1 V, \ldots, w_m V), m \leq n$, of cosets of $F/V$ is said to be *primitive* in $F/V$ if it can be extended to a basis of $F/V$; and if $w(V)$ is a primitive system in $F/V$ then $w(V)$ is said to *lift* to a primitive system of the free group $F$ if there exists $v = (v_1, \ldots, v_m), v_i \in V$, such that the system $w(v) = (w_1 v_1, \ldots, w_m v_m)$ is primitive in $F$. Primitivity of a system $w(V) = (w_1 V, \ldots, w_n V)$, with $n = \text{rank } (F)$, defines an automorphism $\alpha : F/V \to F/V$ mapping $x_i V \to w_i V, i = 1, \ldots, n$. Clearly, every automorphism, $\underline{\alpha} : F \to F$ induces an automorphism $\alpha : F/V \to F/V$ $(\alpha \in \text{Aut } (F/V))$. An automorphism $\alpha \in \text{Aut } (F/V)$ is said to be *tame* if it is induced by an automorphism $\underline{\alpha} : F \to F$ $(\underline{\alpha} \in \text{Aut } (F))$ of the free group; or equivalently, if the corresponding primitive system $w(V) = (w_1 V, \ldots, w_n V)$ of cosets in $F/V$ lifts to a primitive system $w(v) = (w_1 v_1, \ldots, w_n v_n)$ in $F$. If $\alpha \in \text{Aut } (F/V)$ is not tame then $\alpha$ may also be called *wild* or simply *non-tame*. Thus, for instance, if $F$ is free of rank $n \geq 4$, then every automorphism of $F/F''$ is tame (Bachmuth & Mochizuki, 1985) and consequently, every primitive system mod $F''$ lifts to a primitive system of $F$. However, in the case when $n = 3$, while the primitive system $(x[y, z, x, x], y, z)$ mod $F''$ does not lift to a primitve system of $F$ (Chein, 1968); the primitive element $x[y, z, x, x]F''$ is the same as $x[x, [y, z, x]^{-1}]F''$ which in turn lifts to a part of the basis $\{x[x, [y, z, x]^{-1}], y[y, [y, z, x]^{-1}], z[z, [y, z, x]^{-1}]\}$ of $F$. More generally, since finitely generated relatively free groups of almost all varieties defined by outer commutator words have non-tame automorphisms (Gupta & Levin, 1991), lifting primitivity of free polynilpotent groups is naturally one of the most important problems.

## 1. Automorphisms of Free Groups

Let $F = F_n = \langle x_1, x_2, \ldots, x_n \rangle$, $n \geq 2$, be a free group of rank $n$ and let Aut $(F)$ denote the group of all automorphisms of $F$. Then we begin with a significant result of Nielsen.

**Theorem (Nielsen, 1924).** Aut $(F)$ *can be generated by the following four elementary automorphisms:*

$$
\begin{aligned}
\sigma &= \{x_1 \to x_2, x_2 \to x_3, \ldots, x_n \to x_1\}; \\
\tau &= \{x_1 \to x_2, x_2 \to x_1, x_i \to x_i, i \neq 1, 2\}; \\
\lambda &= \{x_1 \to x_1^{-1}, x_i \to x_i, i \neq 1\}; \\
\mu &= \{x_1 \to x_1 x_2, x_i \to x_i, i \neq 1\}.
\end{aligned}
$$

*[Note that when $n = 2$, $\sigma = \tau$ and so Aut $(F)$ is 3-generator. If $n \geq 4$, then Aut $(F)$ can, in fact, be generated by a set of two automorphisms (B. H. Neumann, 1932). For a presentation of Aut $(F)$ we refer to Magnus, Karass and Solitar (1965).]*

Let $F'$ denote the commutator subgroup of $F$ and consider the natural homomorphism

$$
\alpha : \text{Aut } (F) \to \text{Aut } (F/F').
$$

The kernel of this homomorphism clearly consists of all those automorphisms of $F$ which are identity modulo $F'$. These are the so-called IA-*automorphisms* of $F$. We denote by IA-Aut $(F)$ the subgroup of all IA-automorphisms of $F$. Elements of IA-Aut $(F)$ may be defined as

$$
\alpha = \{x_1 \to x_1 d_1, x_2 \to x_2\, d_2, \ldots, x_n \to x_n d_n\},
$$

where $d_i \in F'$ are such that $\{x_1 d_1, x_2 d_2, \ldots, x_n d_n\}$ is a basis of $F$.

An inner automorphism of $F$ is clearly an IA-automorphism. We denote by Inner-Aut $(F)$ the subgroup of Aut $(F)$ consisting of all inner automorphisms of $F$. The centre of $F$ is trivial, so Inner-Aut $(F) \cong F$ and the following inclusions of normal subgroups of Aut $(F)$ are now clear:

$$
F \cong \text{Inner-Aut } (F) \leq \text{IA-Aut } (F) \leq \text{Aut } (F).
$$

[Nielsen (1924) proved that if $F$ is free of rank 2 then IA-Aut $(F) = $ Inner-Aut $(F)$, see Lyndon and Schupp (1977) for a proof.]

Also, it is clear that Aut $(F_n/[F_n, F_n]) \cong GL(n, \mathbb{Z})$, the group of $n \times n$ invertible matrices over the integers.

Consider $w = (w_1, \ldots, w_m)$, $1 \leq m \leq n$, to be a system of elements of $F$ ($= F_n$). Whether or not $\bar{w}$ is primitive has been of primary importance. It is easily seen that if $\{f_1, f_2, \ldots, f_n\}$ is a basis of the free group $F$ then the map

$$
\alpha : \{x_1 \to f_1, x_2 \to f_2, \ldots, x_n \to f_n\}
$$

defines an automorphism $\alpha \in$ Aut $(F)$; and conversely each set $\{\alpha x_1, \alpha x_2, \ldots, \alpha x_n\}$ with $\alpha \in$ Aut $(F)$ defines a basis of $F$. Since an automorphism maps a basis to a basis, a given word $w$ is primitive in $F$ if and only if $\alpha(w) = x_1$ for some $\alpha \in$ Aut $(F)$. For a given word $w$ in $F$, testing its primitivity is in general a very difficult problem. This problem was resolved by Whitehead (1936) through topological arguments using a very large but finite set of the so-called Whitehead (elementary) automorphisms.

**Theorem (Whitehead, 1936).** *There is an algorithm to decide whether or not a given pair of words $u$, $v$ in $F$ are equivalent under an automorphism of $F$. More generally, there is an algorithm to decide if a given system $\mathbf{w} = (f_1, f_2, \ldots, f_m)$, $m \leq n$, of words in $F$ is primitive.*
*[Rapaport (1958) gave an algebraic proof of Whitehead's theorem. See also McCool (1974) for a presentation of* Aut $(F)$ *in terms of Whitehead automorphisms.]*

When $m = n$, algorithmic decidability of a system of $n$ elements in the free group $F$ of rank $n$ reduces to decidability of a given endomorphism of $F$ to be an automorphism. This, in turn, can be translated to a problem of invertibility of a given matrix over the free group ring $\mathbf{Z}F$. The following criterion is due to Joan Birman.

**Theorem (Birman, 1974).** *A given system $\mathbf{w} = (w_1, w_2, \ldots, w_n)$ is primitive in the free group $F$ of rank $n$ if and only if the $n \times n$ Jacobian matrix $J(\mathbf{w}) = (\partial w_i / \partial x_j)$ of the Fox-derivatives of the system is invertible over $\mathbf{Z}F$.*
*[If $w - 1 = \sum_i (x_i - 1) u_i$ then $u_i = \partial w / \partial x_i$ is the (right) Fox derivative of $w$, w.r.t. $x_i$.] Umirbaev (1994) gave a similar primitivity criterion for a system $\mathbf{w} = (w_1, w_2, \ldots, w_m)$ with $m \leq n$, thus extending Birman's criterion.*

## 2. Automorphisms of Relatively Free Groups

Consider a variety $V$ of groups defined by an outer commutator word $w$. Then we have the following important result.

**Theorem (Gupta-Levin, 1991).** *Except for the varieties: $M$ of metabelian groups, $A$ of abelian groups and $N_2$ of nilpotent groups of class at most 2, the relatively free groups of all varieties $V$ defined by outer commutator words have non-tame automorphisms.*
*[The reader is refered to (N. Gupta-Shpilrain, 1993), for a variation of this result for the varieties of polynilpotent groups.]*

We summarize some recent work on lifting primitivity of certain polynilpotent groups.
(i) **Lifting Primitivity of Free Metabelian Nilpotent Groups** (the case $V = \gamma_{c+1}(F)F''$)

Let $w(V) = (w_1 V, \ldots, w_m V)$, $m \leq n$, be primitive in $F/V$. Then lifting $w(V)$ to a primitive system in $F$ is not always possible. For example if $F = \langle x, y, z \rangle$, the system $(x[x, y, x], y, z)$ is primitive mod $V = \gamma_4(F)F''(= \gamma_4(F))$ but the extended system $w = (x[x, y, x]u, yv, zw)$ is not primitive in $F$ for any choice of $u, v$ in $V$ and $w$ in $F'$. This can be seen using Bachmuth's criteria by verifying that the Jacobian matrix $J(w)$ of the system is not invertible. However, when $n \geq 4$ and $V = \gamma_{c+1}(F)F''$, we can take advantage of the Bachmuth & Mochizuki's result which reduces the problem of lifting primitivity mod $\gamma_{c+1}(F)F''$ to that of mod $F''$. Thus we can restrict to free metabelian nilpotent-of-class-c groups $M_{n,c}$ and we need to study only the lifting of primitivity mod $\gamma_{c+1}(M_n)$ to the free metabelian group $M = M_n = \langle x_1, \ldots, x_n \rangle$, $n \geq 4$.

**Theorem (Gupta-Gupta-Romankov, 1992).** *For $n \geq 4$ and $m \leq n - 2$, every primitive system $g = (g_1, \ldots, g_m)$ mod $\gamma_{c+1}(M_n)$ can be lifted to a primitive system in $M_n$, and hence lifts to $F = F_n$.*

**Remarks**

(1) The restriction $m \leq n - 2$ in the above theorem can not be improved. To see this, choose $g_1 = x_1[x_1, x_3, x_3]$, $g_i = x_i$, $i \neq 1, 3$. Then for any choice of $g_3 = x_3 u$, $u \in M'_n$, and any choice of elements $w_i \in \gamma_4(M_n)$, $i = 1, \ldots, n$, the Jacobian matrix $J(g)$ of the system $g = \{g_1 w_1, \ldots, g_n w_n\}$ can be seen to be non-invertible.

(2) When rank of $F$ is 3, the metabelian approach does not apply as the metabelian group $M = M_3 = \langle x, y, z \rangle$ admits wild automorphisms (Chein 1968). As a result, the direct proof that every primitive element of $M_{3,c}$, $c \geq 3$, can be lifted to a primitve element of $F_3$ turns out to be quite technical.

(3) Since, every IA-automorphism of $M_2$ is inner (Bachmuth 1966), $g = x_1 u$ can be lifted to a primitive element of $M_2$ if and only if $u$ is of the form $[x_1, v]$. Thus, for $c \geq 3$, not every primitive element in $M_{2,c}$ can be lifted to a primitive element in $M_2$.

(4) It is easily seen that every endomorphism $\{x \rightarrow x[y, z]^{p(x,y,z)}, y \rightarrow y, z \rightarrow z\}$ of $M_3$ is an automorphism of $M_3$. So, for each $p(x, y, z) \in ZM_3$, the element $x[y, z]^{p(x,y,z)}$ which is primitive in $M_3$ can be lifted to a primitive element of $F_3$ (Gupta-Gupta-Romankov, 1992). A natural question is: Can every primitive element in $M_3$ be lifted to a primitive element in $F_3$? Remarkably, in contrast to the Chein elements, Romankov (1993) has proved the existence of an element which is primitive mod $F''$ but can not be lifted to a primitive element in $F$. This is only an existence proof and it will be of interest to have an explicit description of such elements.

**Problem** *Give a description of elements in $F_3$ which are primitive mod $F''$ but which cannot be lifted to any primitive element of $F_3$.*

**(5)** Evans (1994) has made a remarkable application of the existence of Romankov's element by establishing the existence of an epimorphism $\mu$ : $F_3 \to M_2$ with no primitive element of $F_3$ in Ker $\mu$. He achieves this by first proving that if a primitve element $w$ of $M_n$ is in the normal closure of a single element $\pi \in M_n$ then $w$ must be conjugate of $\pi^{\pm 1}$, which is itself a result of independent interest. Thus, since $\pi$ is primitive in $M_3$, the group $\langle x, y, z; \pi, F'' \rangle$ is a presentation of $M_2$.

**(ii) Lifting Primitivity of Free Nilpotent Groups** (the case $V = \gamma_{c+1}(F)$)
Let $w = (w_1, \ldots, w_m), m \leq n$, be primitive mod $V = \gamma_{c+1}(F_n)$. Here, we do not have the facility of working modulo $F''$, so certain further restrictions on $m$ may be necessary. The central result in this direction is,

**Theorem (Gupta & Gupta, 1992).** *For $m \leq n + 1 - c$, every primitive system $w = (w_1, \ldots, w_m), m \leq n$, mod $\gamma_{c+1}(F_n)$ can be lifted to a primitive system $w(v)$ in $F_n$.*

**Problem** *For $c \geq 4, n = c - 1$, it would be of interest to know whether every primitive element mod $\gamma_{c+1}(F)$ can be lifted to a primitive element of $F$. The simplest case of the problem is to decide whether or not, for $n = 3, c = 4$, the element $x_1[x_1, x_2, x_2, x_3]$ can be lifted to a basis of $F$.*

**(iii) Lifting Primitivity of Relatively Free $N_c$ A-groups**
Let $V$ be a fully invariant subgroup of $F$ ($= F_n$) such that $F'' \geq V \geq \gamma_{c+1}(F')$ for some $c \geq 2$.

**Theorem (Bryant & Gupta, 1993).** *For $n \geq m + 2c, n \geq 4$, every primitive system $w = (w_1, \ldots, w_m)$ mod $V$ can be lifted to a primitive system $w(v)$ of $F$.*

Let $G = F/[F', F', F']$, $F = \langle x, y, u, v \rangle$, and let $T$ denote the tame automorphisms of $G$. Then we have the following result,

**Theorem (Gupta and Levin, 1989).** Aut $(G) = gp \{T, \delta_0, \delta_1, \delta_2, \ldots\}$, *where* $\delta_k = \{x \to x[[x, y]^{x^k}, [u, v]], y \to y, u \to u, v \to v\}$ *is a non-tame automorphism of G for each $k \geq 0$.*
*[The non-tameness of $\delta_k$ is proved by showing that the Jacobian matrix $J(w)$ of the system $w = (x[[x, y]^{x^k}, [u, v]], y, u, v)$ over the free group ring $\mathbf{Z}F$ is not-invertible. This is achieved by building a homomorphism of the group $GL(4, \mathbf{Z}F)$ into $GL(2, \mathbf{Z}[t])$ which maps $J(w)$ to a non-invertible element of $GL(2, \mathbf{Z}[t])$. A quick proof of this fact is now possible (see, Application 3 in next section).]*

## 3. A Criterion for Non-Tameness and Some Applications

Recall that the $k$-th left partial derivative $\partial_k$ is defined linearly on the free group ring $\mathbf{Z}(F)$ ($= \mathbf{Z}(F_n)$) by: $\partial_k(x_k) = 1$; $\partial_k(x_i) = 0, i \neq k$; $\partial_k(uv) = \partial_k(u) + u\partial_k(v), u, v \in F$. In particular, for any $w \in \gamma_m(F)$, the partial derivative $\partial_k(w)$

lies in $\Delta^{m-1}(F)$, and hence, modulo $\Delta^m(F)$, it can be represented as a polynomial $f(X_1, \ldots, X_n)$ in the non-commuting variables $X_i = x_i - 1$, $i = 1, \ldots, n$. For any $S_i, T_i \in \{X_1, \ldots, X_n\}$ we define an equivalence relation "$\approx$" on monomials by: $S_1, \ldots, S_k \approx T_1, \ldots, T_k$ if one is a cyclic permutation of the other. Finally, a polynomial $f(X_1, \ldots, X_n)$ is called *balanced* if $f(X_1, \ldots, X_n) \approx 0$, or equivalently, *the sum of the co-efficients of its cyclically equivalent terms is zero*. Then through a technical analysis of the invertibility of the Jacobian matrix associated with a basis of $F$ we have the following useful test for an endomorphism of $F$ to be an automorphism.

**Criterion (Bryant, Gupta, Levin and Mochizuki, 1990).**

Let $w = w(x_1, \ldots, x_n) \in \gamma_m(F_n)$ for some $m \geq 2$ and let $\alpha$ be an endomorphism of $F_n$ defined by: $\alpha(x_1) \equiv x_1 w, \alpha(x_i) \equiv x_i \pmod{\gamma_{m+1}(F_n)}$, $i = 2, \ldots, n$. Let $\partial_1(w) \equiv f(X_1, \ldots, X_n) \pmod{\Delta^m(F)}$. If $\alpha$ defines an automorphism of $F_n$ then $f(X_1, \ldots, X_n)$ must be balanced.

**Application 1.** The following automorphism of free class-3 group $F_{n,3}$ of rank $n \geq 2$ is wild: $\alpha = \{x \rightarrow x [[x, y, x], y \rightarrow y, \ldots, z \rightarrow z\}$.
**Proof. (cf. Andreadakis, 1968).** We have,

$$\partial_x([x, y, x]) \equiv 2(y - 1)(x - 1) - (x - 1)(y - 1)(\Delta^3(F)).$$

Then $f = 2(y - 1)(x - 1) - (x - 1)(y - 1)$ is not balanced, and the proof follows.
**Application 2.** The following automorphism of free centre-by-metabelian group of rank $n \geq 4$ is wild: $\alpha = \{x \rightarrow x[[x, y], [u, v]], y \rightarrow y, \ldots, z \rightarrow z\}$. This answers a question of Stöhr (1987).
**Proof.** Since, $[F'', F] \leq \gamma_5(F)$, it suffices to prove that $\alpha$ is not a tame automorphism of the free class 4 group $F_{n,4}$. Indeed, $\partial_x([x, y], [u, v]]) \equiv (y - 1)([u, v] - 1) \pmod{\Delta^5(F)}$ and $f = (y - 1)(u - 1)(v - 1) - (y - 1)(v - 1)(u - 1)$ is clearly not balanced.
**Application 3. (Gupta-Levin, 1989).** For each $k \geq 1$ the following automorphism of free class-2 by abelian group of rank 4 is wild:

$$\alpha_k = \{x \rightarrow x[[x, y, x^k], [u, v]], y \rightarrow y, u \rightarrow u, v \rightarrow v\}.$$

**Proof.** Since, $[F'', F'] \leq \gamma_6(F)$, it suffices to prove that $\alpha$ is not a tame automorphism of free class 5 group $F_{4,5}$.

We have, $\partial_x([[x, y, x^k], [u, v]])$

$$\equiv (y - 1)(x^k - 1)([u, v] - 1) - (1 + x + \cdots + x^{k-1})([x, y] - 1)$$
$$([u, v] - 1)\pmod{\Delta^5(F)}$$
$$\equiv k(y - 1)(x - 1)([u, v] - 1) - k([x, y] - 1)([u, v] - 1)\pmod{\Delta^5(F)}.$$

Then $f = k(y - 1)(x - 1)([u, v] - 1) - k([x, y] - 1)([u, v] - 1)$ and the sum of the co-efficients of terms cyclically equivalent to $(x - 1)(y - 1)(u - 1)(v - 1)$ is $(-1)k$ which is non-zero.

## 4. Automorphisms of Relatively Free Groups of Countably Inifinte Rank

Consider the automorphism $\theta = \{x_1 \to x_1[x_1, x_2, x_1], x_i \to x_i, i \neq 1\}$ of the $n$-generator free nilpotent group $F_{n,3}$ of class 3. Then we know that $\theta$ is a non-tame automorphism for all $n \geq 2$. Gawron and Macedonska (1988) proved that every automorphism of a free class-3 group $G = F_{\omega,3} = \langle x_1, x_2, \ldots \rangle$ of countable infinite rank is tame. In particular, the extended automorphism $\theta = \{x_1 \to x_1[x_1, x_2, x_1], x_i \to x_i, i \neq 1\}$ of the group $F_{\omega,3} = \langle x_1, x_2, \ldots \rangle$ of class 3 is tame.

This result has now been extended to arbitrary nilpotency class-$c$.

**Theorem (Bryant & Macedonska, 1989).** *Every automorphism of a free nilpotent group of countable infinite rank is tame.*

This raises the following more general problem.

**Problem** *Is* Aut $(F/V)$ *always tame for relatively free countable infinite rank groups defined by outer-commutator words? In particular, are all automorphisms of free polynilpotent groups of countable infinite rank tame?*

For free metabelian groups of countable infinite rank, Bryant and Groves (1992) have shown that the above question has affirmative answer. If $V$ is the fully invariant subgroup of the variety defined by a finite simple group, then Bryant and Groves have shown that not all automorphisms lift. Finally, Gupta & Bryant (1993) have proved that if $V$ is a fully invariant subgroup of $F = F_\omega$ such that $F'' \geq V \geq \gamma_{c+1}(F')$, then Aut $(F/V)$ is tame.

## 5. Primitivity in the Free Groups of the Variety $A_m A_n$

For each $m, n \geq 0$, let $G_{m,n} = F_r(A_m A_n)$ be the free group of rank $r$ of the variety $A_m A_n$, where $A_n$, $n \geq 0$, is the variety of all abelian groups of exponent dividing $n$ ($A = A_0$ is the variety of all abelian groups) and let $A_n$ be the free group of the variety $A_n$ with a basis $a_1, \ldots, a_r$. The product variety $A_m A_n$) is the variety of all extensions of groups in $A_m$ by groups in $A_n$. With regard to primitivity, some of what is known in this direction is the following:

(*i*) A system $(v_1, \ldots, v_r)$ of words in $F$ is a primitive system of the group $G_{m,n}$ if and only if the $r \times r$ Jacobian matrix $J(v) = (\partial v_i / \partial x_j)$ over the group ring $Z_m A_n$ is invertible (Gupta & Timoshenko 1996);

(ii) A system $(v_1, \ldots, v_l)$ of words, $l \leq r$, can be included in a basis of the group $G_{m,0}$ if and only if the ideal generated by the $l \times l$ minors of the Jacobian matrix $(\partial v_i / \partial x_j)(l \times r)$ in the ring $Z_m A_0$ coincides with the ring $Z_m A_0$, and $\{v_1, \ldots, v_l\}$ is a primitive system mod $F'$ (Gupta-Timoshenko 1996).

In particular, the above results yield the previously well-known results for the free metabelian group $G_{0,0}$ (Bachmuth, 1965, Timoshenko, 1988, Gupta-Gupta-Noskov, 1994 and others).

Recall that an automorphism of a group $G = F/V(F)$ is tame if it is induced by an automorphism of the free group $F$. Bachmuth and Mochizuki (1989) proved that the group Aut $(G_{m,0})$ is not finitely generated if the ring $Z_m$ is not semisimple and the rank of the group $G_{m,0}$ exceeds 1. It follows that the group $G_{m,0}$ contains non-tame automorphisms if $Z_m$ is not semisimple. In this connection Papistas (1996) proved the following result: Let $m \geq 2$ be a positive integer but not a prime. Then IA-Aut $(G_{m,0})$ contains non-tame automorphisms for all $r \geq 2$, where $r$ is the rank of the group $G(m, 0)$. The following results are more recent:

**Theorem (Gupta-Timoshenko, 1998).** *Let $v_1, \ldots, v_l$ be elements of the free group $F$ of rank $r$, $l \leq r$, and assume that one of the following three conditions is not satisfied: $r - l = 1$, $m = 0$, $n > 0$. Consider the $l \times r$ matrix $J(v) = (\partial v_i / \partial x_j)$ of Fox derivatives over the ring $Z_m A_n$. Then the images of the elements $v_1, \ldots, v_l$ in the group $G_{m,n} \doteqdot F_r(A_m A_n)$ can be included in a basis of $G_{m,n}$ iff*

(i) *The ideal generated by the $l \times l$ minors of the matrix $J(v)$ in the ring $Z_m A_n$ coincides with $Z_m A_n$;*

(ii) *$(v_1, \ldots, v_l)$ is a primitive system mod $F'F^n$.*

**Theorem (Gupta-Timoshenko, 1998).** *Let $r \geq 4$, $s = p^n$ ($p$ a prime), $n \geq 2$, $F_r$ be the free group of rank $r$. For $1 \leq m \leq r - 2$ every primitive system of $m$ elements of the free group $F_r(A_s A_0)$ of the variety $A_s A_0$ of rank $r$ can be lifted to a primitive system of the free group $F_r$. For $m \geq r - 1$ and each $t > 0$, there exists primitive system of $m$ elements of the group $F_r(A_t A_0)$ which can not be lifted to a primitive system of $F_r$.*

## 6. Splitting Epimorphisms

If $\pi_1, \pi_2 : G \to H$ are epimorphisms of a group $G$ onto a group $H$, then there is a so-called *splitting epimorphism*

$$\pi_1 \times \pi_2 : G \to H \times H \text{ defined by } g \to (g\pi_1 \times g\pi_2).$$

Two splitting epimorphisms $\pi_1 \times \pi_2$ and $\pi_1' \times \pi_2'$ are said to be *equivalent* if

$$\pi_1' = \alpha\pi_1\beta \text{ and } \pi_2' = \alpha\pi_2\beta_2,$$

for some $\alpha \in$ Aut $(G)$ and $\beta_1, \beta_2 \in$ Aut $(H)$.

The central question here is the classification of the splitting epimorphisms upto their natural equivalence. This problem has its origin in low dimension topology. For instance, when

$$G = \Gamma_n = \langle x_1, \ldots, x_n, y_1, \ldots, y_n; [x_1, y_1] \ldots [x_n, y_n] = 1 \rangle, n \geq 2,$$

is the surface group of genus $n$, and

$$H(x) = F_n = \langle x_1, \ldots, x_n; \emptyset \rangle, H(y) = F_n = \langle y_1, \ldots, y_n; \emptyset \rangle$$

are free subgroups of $G$ then the famous Poincarè Conjecture admits the following equivalent group theoretic statement:

*The Poincarè conjecture is true if and only if for each $n \geq 2$, up to natural equivalence, there is one and only splitting epimorphism $\Gamma_n \to F_n \times F_n$.*

The problem remains difficult even in the special case when $G$ is assumed to be itself a free group

$$F_{2n} = \langle x_1, \ldots, x_n, y_1, \ldots, y_n; \emptyset \rangle, n \geq 2.$$

The following conjecture is due to Grigorchuk & Khurchanov (1993).

**GK-Conjecture:** *Up to natural equivalence, there is one and only one epimorphism $F_{2n} \to F_n(x) \times F_n(y)$.*

It is remarkable that the $GK$-conjecture, while formally being quite similar to the Poincarè Conjecture, in fact implies the well-known Andrews-Curtis conjecture which states:

If a balanced presentation $\langle x_1, \ldots, x_n; u_1, \ldots, u_n \rangle$ defines the trivial group then the $n$-tuple $(u_1, \ldots, u_n)$ can be transformed to the standard $n$-tuple $(x_1, \ldots, x_n)$ by a sequence of elementary transformations of the following four types:

$$\alpha(i, j) : (\ldots, w_i, \ldots, w_j, \ldots) \to (\ldots, w_j, \ldots, w_i, \ldots)$$
$$\beta(i) : (w_1, \ldots, w_i, \ldots, w_n) \to (w_1, \ldots, w_i^{-1}, \ldots, w_n)$$
$$\gamma(i, j) : (\ldots, w_i, \ldots, w_j, \ldots) \to (\ldots, w_i, \ldots, w_j w_i, \ldots)$$
$$\delta_f(i) : (w_1, \ldots, w_i, \ldots, w_n) \to (w_1, \ldots, w_i^f, \ldots, w_n), f \in F.$$

The relatively free equivalent of the $GK$-conjecture remains unresolved even for the metabelian groups. We refer the reader to Gupta-Noskov (1997) for some reduction theorems and partial results towards the classification of the Splitting epimorphisms: $M_{2n} \to M_n \times M_n$ of free metabelian groups.

## References

Andreadakis, S. On the automorphisms of free groups and free nilpotent groups, *Proc. London Math. Soc.* (3) **15**, 239–269, 1965.

Andreadakis, S. Generators for Aut $G$, $G$ free nilpotent, *Arch. Math.* **42**, 296–300, 1984.

Andreadakis, S. and Gupta, C.K. Automorphisms of free metabelian nilpotent groups, *Algebra i Logika*, **29**, 746–751, 1990.

Bachmuth, S. Automorphisms of free metabelian groups, *Trans. Amer. Math. Soc.* **118**, 93–104, 1965.

Bachmuth, S. Induced automorphisms of free groups and free metabelian groups, *Trans. Amer. Math. Soc.* **122**, 1–17, 1966.

Bachmuth, S., Baumslag, G., Dyer, J. and Mochizuki, H.Y. Automorphism groups of two-generator metabelian groups, *J. London Math. Soc.* **36**, 393–406, 1987.

Bachmuth, S. and Mochizuki, H.Y. IA-automorphisms of free metabelian group of rank 3 *J. Alg.* **55**, 106–115, 1979.

Bachmuth, S. and Mochizuki, H.Y. Aut $(F) \rightarrow$ Aut($F/F''$) is surjective for free group $F$ of rank $\geq 4$, *Trans. Amer. Math. Soc.* **292**, 81–101, 1985.

Bachmuth, S., Mochizuki, H.Y. The tame range of automorphism groups and $GL_n$, *Proc. Singapore Group Theory Conf.* 241–251, de Gruyter, New York, 1987/89.

Bachmuth, S. Foremanek, E. and Mochizuki, H.Y. IA-automorphisms of certain two generator torsion free groups, *J. Alg.* **40**, 19–30, 1967.

Baumslag, G. Automorphism groups of residually finite groups, *J. London Math. Soc.* **38**, 117–118, 1963.

Birman, Joan S. An inverse function theorem for free groups, *Proc. Amer. Math. Soc.* **41**, 634–638, 1974.

Bryant, R.M. and Groves, J.R.J. Automorphisms of free metabelian groups of infinite rank, *Comm. Alg.* **20**, 783–814, 1992.

Bryant, R.M. and Gupta, C.K. Automorphism groups of free nilpotent groups, *Arch. Math.* **52**, 313–320, 1989.

Bryant, R.M. and Gupta, C.K. Automorphisms of free nilpotent-by-abelian groups, *Math. Proc. Camb. phil. Soc.* **114**, 143–147, 1993.

Bryant, R.M., Gupta, C.K., Levin, F. and Mochizuki, H.Y. Non-tame automorphisms of free nilpotent groups, *Comm. Alg.* **18**, 3619–3631, 1990.

Bryant, Roger M. and Olga Macedonska, Automorphisms of relatively free nilpotent groups of infinite rank, *J. Alg.* **121**, 388-398, 1989.

Caranti, A. and Scoppola, C.M. Endomorphisms of two-generator metabelian groups that induce the identity modulo the derived subgroup, *Arch. Math.* **56**, 218–227, 1991.

Caranti, A. and Scoppola, C.M. Two-generator metabelian groups that have many IA-automorphisms, *(1989 Preprint)*.

Chein Orin, IA Automorphisms of free and free metabelian groups, *Comm. pure appl. Math.* **21**, 605–629, 1968.

Drensky, V. and Gupta, C.K. Automorphisms of free nilpotent Lie algebras, *Canad. J. Math.* **42**, 259–279, 1990.

Drensky, V. and Gupta, C.K. New automorphisms of generic matrix algebras and polynomial algebras, *J. Alg.* **194**, 408–414, 1997.

Evans, M.J. Presentations of the free metabelian group of rank 2, *Canad. Math. Bull.* **37**, 468–472, 1994.

Formanek, E. and Procesi, C. The automorphism group of free group is not linear, *J. Algebra* **149**, 494–499, 1992.

Fox, R.H. Free differential caculus 1. Derivations in the free group ring, *Ann. of Math.* **57**, 547–560, 1953.

Gawron, Piotr Wlodzimierz and Olga Macedonska, All automorphisms of the 3-nilpotent free group of countably infinite rank can be lifted, *J. Alg.* **118**, 120–128, 1988.

Goryaga, A.V. Generators of the automorphism group of a free nilpotent group, *Alg. and Logic*, **15**, 289–292 [English Translation], 1976.

Grigorchuk, R.I. and Kurchanko, P.F. Certain questions of group theory related to Geometry,

In: *Encyclopaedia of Mathematical Sciences, Algebra VII, Combinatorial Group Theory, Applications to Geometry*, Springer, 173–231, 1993.

Grossman, E.K. On the residual finiteness of certain mapping class groups, *J. London Math. Soc.* **9**, 160–164, 1974.

Gupta, C.K. IA-automorphisms of two generator metabelian groups, *Arch. Math.* **37**, 106–112, 1981.

Gupta, C.K. and Gupta, N.D. Lifting primitivity of free nilpotent groups, *Proc. Amer. Math. Soc.* **114**, 617–621, 1992.

Gupta, C.K., Gupta, N.D. and Noskov, G.A. Some applications of Artamonov-Quillen-Suslin theorems to metabelian inner-rank and primitivity, *Canad. J. Math.* **46**, 298–307, 1994.

Gupta, C.K., Gupta, N.D. and Roman'kov, V.A. Primitivity in free groups and free metabelian groups, *Canad. J. Math.* **44**, 516–523, 1992.

Gupta, C.K. and Frank Levin, Automorphisms of free class-2 by abelian groups, *Bull. Austral. Math. Soc.* **40**, 207–214, 1989.

Gupta, C.K. and Frank Levin, Tame range of automorphism groups of free polynilpotent groups, *Comm. Alg.* **19**, 2497–2500, 1991.

Gupta, C.K. and Noskov, G.A. Splitting epimorphisms of free metabelian groups, *Internat. J. Algebra and Computation* **7**, 697–711, 1997.

Gupta, C.K. and Roman'kov, V.A. Finite separability of tameness and primitivity in certain relatively free groups, *Comm. Alg.* **23**, 4101–4108, 1995.

Gupta, C.K. and Shpilrain, V. Lifting automorphisms - a survey, *London Math. Soc. Lecture Notes Ser.* **211**, 249–263, 1995 [Groups 93, Galway/St. Andrews].

Gupta, C.K. and Timoshenko, E.I. Primitivity in the free groups of the variety $A_m A_n$, *Comm. Alg.* **24**, 2859–2876, 1996.

Gupta, C.K. and Timoshenko, E.I. Automorphic and endomorphic reducibility and primitive endomorphisms of free metabeilan groups, *Comm. Alg.* **25**, 3057–3070, 1997.

Gupta, C.K. and Timoshenko, E.I. Primitive system in the variety $A_m A_n$: the criteria and lifting (*to appear*).

Gupta, Narain. Free Group Rings, *Contemporary Math.* **66**, Amer. Math. Soc. 1987.

Gupta, N.D. and Shpilrain, V. Nielsen's commutator test for two-generator groups, *Math. Proc. Camb. phil. Soc.* **114**, 295–301, 1993.

Imrich, W. and Turner, E.C. Endomorphisms of free groups and their fixed points, *Math. Proc. Camb. phil. Soc.* **105**, 421–422, 1989.

Ivanov, S.V. On endomorphisms of free groups that preserve primitivity, (*Preprint*).

Jaco, W. Heegard splittings and splitting homomorphisms, *Trans. Amer. Math. Soc.* **144**, 365–379, 1969.

Krasnikov, A.F. Generators of the group $F/[N, N]$, *Alg. Logika*, **17**, 167–173, 1978.

Lyndon, Roger C. and Schupp, Paul E. Combinatorial group theory, *Ergebnisse Math. Grenzgeb.* **89**, Springer-Verlag, 1977.

Magnus, W. Über $n$-dimensionale Gitter transformationen. *Acta Math.* **64**, 353–367, 1934.

Magnus, W. Karrass, A. and Solitar, D. *Combinatorial Group Theory*, Interscience Publ., New York, 1966.

Magnus, W. and Tretkoff, C. Representations of automorphism groups of free groups, *Word Problems II, Studies in Logic and Foundations of Mathematics*, **95**, 255–260, North Holland, Oxford, 1980.

McCool, J. A presentation for the automorphism group of finite rank, *J. London Math. Soc.* **8**(2) 259–266, 1974.

Meskin, S. Periodic automorphisms of the two-generator free group, *Proc. Second int. Group Theory Conf. Canberra 1973, Springer LN.* **372**, 494–498, 1974.

Mikhalev, A.A. and Zolotykh, A.A. Automorphisms and primitive elements in free Lie algebras, *Comm. Alg.* **22**, 5889–5901, 1994.

Neumann, B.H. Die Automorphismengruppe der freien Gruppen, *Math. Ann.* **107**, 367–376, 1932.

Nielsen, J. Die Isomorphismender allegmeinen unendlichen Gruppen mit zwei Erzugenden, *Math. Ann.* **78**, 385–397, 1918.

Nielsen, J. Die isomorphismengruppe der freien Gruppen, *Math. Ann.* **91**, 169–209, 1924.

Papistas, A.I. Automorphisms of metabelian groups, *Canad. Math. Bull.* **41**, 98–104, 1988.

Papistas, A.I. Free nilpotent groups of rank 2, *Comm. Alg.* **25**, 2141–2145, 1997.

Rapaport, E.S. On free groups and their automorphisms, *Acta Math.* **99**, 139–163, 1958.

Roman'kov, V.A. Primitive elements of free groups of rank 3, *Math. Sb.* **182**, 1074–1085, 1991.

Roman'kov, V.A. Automorphisms of groups, *Acta appl. Math.* **29**, 241–280, 1992.

Roman'kov, V.A. The automorphism groups of free metabelian groups, [Questions on pure and applied algebra] *Proc. Computer Centre, USSR Academy of Sciences, Novosibirsk*, 35–81 [*Russian*], 1985.

Shpilrain, V.E. Automorphisms of F/R' groups, *Int. J. Alg. Comput.* **1**, 177–184, 1991.

Shpilrain, V.E. On monomorphisms of free groups, *Arch. Math.* **64**, 465–470, 1995.

Elena Stöhr, On automorphisms of free centre-by-metabelian groups, *Arch. Math.* **48**, 376–380, 1987.

Timoshenko, E.I. Algorithmic problems for metabelian groups, *Alg. Logic* **12**, 132–137, [*Russian Edition: Alg. Logika* **12**, 232–240], 1973.

Timoshenko, E.I. On embedding of given elements into a basis of free metabelian groups, *Sibirsk. Math. Zh*, Novosibirsk, 1988.

Turner, E.C. Test words for automorphisms of free groups, *Bull. London Math. Soc.* **28**, 255–263, 1996.

Umirbaev, U.U. On primitive elements of free groups, *Russian Math. Surv.* **49**, 184–185, 1994.

Whitehead, J.H.C. On certain set of elements in a free group, *Proc. London Math. Soc.* **41**, 48–56, 1936.

Whitehead, J.H.C. On equivalent sets of elements in a free group, *Ann. Math.* **37**, 782–800, 1936.

Department of Mathematics, University of Manitoba, Winnipeg R3T 2N2, Canada.
Email: cgupta@cc.umanitoba.ca

# Jordan Decomposition

*A.W. Hales and I.B.S. Passi*

## 1. Introduction

The Jordan canonical form for matrices over algebraically closed fields is standard fare in many linear algebra courses. The Jordan decomposition (into semisimple and nilpotent parts) for matrices over perfect fields is perhaps less well known, though very useful in many areas and closely related to the canonical form. This Jordan decomposition extends readily to elements of group algebras over perfect fields. During the past decade or so there has been activity in extending the decomposition to group rings (and matrices) over integral domains. In this article, we give a survey of this recent work (Arora *et al.*, 1993 & 1998; Hales *et al.*, 1990 & 1991) as well as some background on the classical results.

We begin with an introductory section on Jordan canonical form and on both additive and multiplicative Jordan decomposition. This gives sketches of proofs of most of the basic results. The next section shows how the decompositions extend to group algebras. Section four considers additive and multiplicative Jordan decompositions for matrices over integral domains. Section five, the main part of this survey, deals with group rings over integral domains of characteristic zero.

## 2. Background

Let $M$ be a square $n$ by $n$ matrix over an algebraically closed field $F$. Then there exists an invertible $n$ by $n$ matrix $C$ over $F$ such that the matrix $A = C^{-1}MC$ is in *Jordan canonical form*, i.e. we have that $A$ is in block diagonal form

$$
A = \begin{pmatrix}
B_1 & 0 & . & . & . & 0 \\
0 & B_2 & 0 & . & . & 0 \\
. & . & . & . & . & . \\
. & . & . & . & . & . \\
. & . & . & . & . & . \\
0 & . & . & . & 0 & B_k
\end{pmatrix}
$$

where the block $B_i$ is an $n_i$ by $n_i$ upper triangular matrix of the form

$$
B_i = \begin{pmatrix}
\alpha_i & 1 & 0 & . & . & 0 \\
0 & \alpha_i & 1 & 0 & . & 0 \\
. & & & & & . \\
. & & & & & . \\
. & & & & & . \\
0 & . & . & 0 & \alpha_i & 1 \\
0 & . & . & . & 0 & \alpha_i
\end{pmatrix}.
$$

Note that $\alpha_i$ may equal $\alpha_j$ for $i \neq j$. The alphas are the eigenvalues of $A$ (and of $M$), the characteristic polynomial of $A$ (and of $M$) is

$$
\prod_{i=1}^{k} (x - \alpha_i)^{n_i},
$$

and the minimal polynomial of $A$ and $M$ is

$$
\prod_{i}' (x - \alpha_i)^{n_i'},
$$

where the primed product runs over the set of distinct $\alpha_i$ and $n_i'$ denotes the maximum of all $n_i$ associated with a given $\alpha_i$.

We may clearly write each block $B_i$ above as the sum of a diagonal (scalar) matrix with $\alpha_i$'s on the diagonal and a strictly upper triangular matrix whose $n_i^{th}$ power is 0. Piecing these together we can write $A$ as the sum of a diagonal matrix $D$ and a strictly upper triangular matrix $N$ whose $m^{th}$ power is zero, where $m$ is the maximum of the $n_i$. From the equation $A = D + N$ we obtain the equation $M = CDC^{-1} + CNC^{-1}$, i.e., $M$ is the sum of a diagonalizable matrix and a nilpotent matrix. Furthermore $D$ and $N$ clearly commute, and hence so do the matrices $M_s = CDC^{-1}$ and $M_n = CNC^{-1}$ whose sum is $M$.

To summarize, we have shown that any square matrix $M$ over $F$ can be written as the sum of commuting matrices $M_s$ and $M_n$, where $M_s$ is *semisimple* and $M_n$ is nilpotent. Here semisimple means diagonalizable (over $F$) or, equivalently, having a minimal polynomial with no repeated roots. This representation is, in fact, unique. To see this note first that $M_s$ and $M_n$ commute not only with each other but also with $M$. We can strengthen this and show that each of $M_s$ and $M_n$ is a polynomial in $M$ with coefficients in $F$ (and, if desired, zero constant term). To see this apply the Chinese Remainder Theorem to the pairwise coprime polynomials $(x - \alpha_i)^{n_i'}$ as $\alpha_i$ ranges over the distinct $\alpha_i$ to obtain a polynomial $p(x)$ such that $p(x)$ is congruent to $\alpha_i$ modulo $(x - \alpha_i)^{n_i'}$ for each $i$. It then follows by direct calculation with the matrices $B_i$ that $p(A) = D$. Hence we have $p(M) = M_s$ and $q(M) = M_n$ where $q(x) = x - p(x)$.

Now suppose we also have $M = M_s' + M_n'$ where $M_s'$ is semisimple and $M_n'$ is nilpotent and the two commute. Then they commute with $M$ and hence with each of $M_s$ and $M_n$. But the difference of two commuting nilpotents is clearly nilpotent, and the difference of two commuting semisimples is semisimple (since they can be simultaneously diagonalized). Hence, in the equation $M_s - M_s' = M_n' - M_n$ we have a semisimple equal to a nilpotent. This means each must be zero, so the two representations coincide. We are thus justified in speaking of $M_s$ and $M_n$ as the *semisimple component and nilpotent component* of $M$ respectively. We refer to the equation $M = M_s + M_n$ as the *(additive) Jordan decomposition of M*.

Consider now the case where $M$ is invertible. Then using the uniqueness of the additive decomposition it is straightforward to show that $M_s$ is also invertible and that $(M_s)^{-1} = (M^{-1})_s$. Therefore, we can write $M = M_s + M_n = M_s(I + M_s^{-1}M_n) = M_sM_u$ where we take $M_u = I + M_s^{-1}M_n$. Hence, any invertible matrix over $F$ can be written as the product of a semisimple matrix $M_s$ and a unipotent matrix $M_u$, where $M_s$ and $M_u$ commute. (Here unipotent means identity plus nilpotent, i.e., all eigenvalues equal 1.) Furthermore, since the minimal polynomial of $M_s$ has nonzero constant term, we can divide by $M_s$ to write $M_s^{-1}$ as a polynomial in $M_s$. This implies that $M_s^{-1}$ is a polynomial in $M$, and hence so is $M_u$. From the fact that both components of the multiplicative decomposition $M = M_sM_u$ are polynomials in $M$ we can show, as in the additive case above, that this decomposition is unique. We refer to this as the *(multiplicative) Jordan decomposition of M*.

If we relax the algebraically closed condition and only assume that the base field $F$ is perfect, the above results all still hold. In other words, if $M$ is a square $n$ by $n$ matrix over $F$, then the matrices $M_s$, $M_n$ and $M_u$ (all defined over the algebraic closure $\bar{F}$) have entries lying in $F$ and in fact can be written as polynomials in $M$ over $F$. To see this let $f(x)$ be the product of the distinct irreducible factors of the minimal polynomial of $M$ over $F$. Using the perfectness of $F$ this $f(x)$ has no repeated roots in $\bar{F}$ and is just the product of all $(x-\alpha_i)$ as $\alpha_i$ ranges over the distinct eigenvalues of $M$ in $\bar{F}$. Let $f'(x)$ be the derivative of $f(x)$. Then we can write 1 as a linear combination of $f(x)$ and $f'(x)$, $1 = r(x)f(x)+s(x)f'(x)$, where $r(x)$ and $s(x)$ have coefficients in $F$. Consider the polynomial $x - f(x)s(x)$. Direct calculation shows that evaluating this when $x$ equals one of the Jordan blocks $B_i$ gives an upper triangular matrix with $\alpha_i$ on the diagonal and zero immediately above the diagonal. This means that, if $\theta$ is the endomorphism of $F[x]$ given by $\theta(x) = x - f(x)s(x)$, the $m$-fold composition $\phi$ of $\theta$ with itself will satisfy $M_s = \phi(M)$ where $m$ is minimal such that $2^m \geq n_i$ for all $i$. Thus $M_s$ is a polynomial in $M$ over $F$ and the rest follows.

An alternative proof, more conceptual but less constructive, can be obtained by applying "Wedderburn theory" to the ring-with-minimum-condition $\langle M \rangle$ generated by $M$ to write $M$ as the sum of an element from the nilpotent radical and an element from the unique semisimple complement of this radical in $\langle M \rangle$. [see Albert (1939) for details.]. Borel (1991), Chevalley (1951), Humphreys (1972, 1975), Shirvani and Wehrfritz (1986) and Springer (1981) give more details on Jordan decomposition and its application to Lie algebras and algebraic groups.

## 3. Group Algebras

Let $G$ be a finite group of order $n$ and let $F$ be a perfect field. Then any element $\alpha$ of the group algebra $F[G]$ induces a linear mapping from the $F$-vector space $F[G]$ to itself by left multiplication and hence can be considered as an $n$ by $n$ matrix over $F$. We can apply the theory of the previous section to deduce that $\alpha$ can be written in the form $\alpha = \alpha_s + \alpha_n$ where $\alpha_s$ and $\alpha_n$ both lie in $F[G]$ (and are polynomials in $\alpha$ over $F$), $\alpha_s$ is semisimple (satisfies a polynomial with no repeated roots over $F$), $\alpha_n$ is nilpotent, and the two commute. Furthermore this representation is unique. Thus we say that *additive Jordan decomposition* holds in $F[G]$. Similarly, if $\alpha$ in $F[G]$ is invertible, we can write $\alpha = \alpha_s \alpha_u$ where $\alpha_s$ and $\alpha_u$ lie in $F[G]$, $\alpha_s$ is semisimple, $\alpha_u$ is unipotent (i.e., 1 plus a nilpotent), and the two commute - and further each is a polynomial in $\alpha$ over $F$. Thus we also say that *multiplicative Jordan decomposition* holds in $F[G]$.

These results about Jordan decomposition in $F[G]$ can be obtained by working completely internally within $F[G]$, as in the latter part of the previous section, if one desires. There is another way to view the situation which is quite instructive, at least in the case that $F$ is algebraically closed and the characteristic of $F$ does not divide the order of $G$. Then $F[G]$ will be (isomorphic to) a direct sum of matrix rings over $F$ by Wedderburn theory and we can decompose each of the matric components of an element $\alpha$ in $F[G]$ into its semisimple and nilpotent parts in the relevant matrix ring and then piece these parts together to get $\alpha_s$ and $\alpha_n$.

As an application of the last comment above, which will be useful later on, consider a matrix $M$ which is central in the ring $M_n(F)$. Then $M$ is not only semisimple but in fact scalar. Furthermore, if $N$ is any $n$ by $n$ matrix over $F$, $M+N_s$ will be semisimple and we will have $(M + N)_s = M + N_s$ and $(M + N)_n = N_n$. Now consider a central element $\alpha$ of $F[G]$. If $\beta$ is any element of $F[G]$ we can work in each matric component of $\bar{F}[G]$ to conclude that $\alpha$ is semisimple and the additive Jordan decomposition of $(\alpha + \beta)$ is $(\alpha + \beta_s) + \beta_n$.

We will return to group rings in Section 5.

## 4. Matrices over Rings

Throughout this section let $F$ be a field of characteristic zero and let $R$ be a proper subring of $F$ with quotient field $F$. As a warmup for the following section we will consider the following question: if $M$ lies in $M_n(R)$, does it follow that $M_s$ and $M_n$ must also lie in $M_n(R)$? Also, if $M$ in $M_n(R)$ is a unit, i.e., invertible in $M_n(R)$, must $M_s$ and $M_u$ lie in $M_n(R)$? In other words, does (additive or multiplicative) Jordan decomposition hold for matrices with coefficients in $R$? We will see that the size of the matrices plays a critical role in these questions.

First, note that both questions are trivial for $n = 1$. A 1 by 1 matrix is always semisimple.

Now consider the case $n = 2$. Suppose $M$ lies in $\mathbb{M}_2(F)$. If the eigenvalues of $M$ are distinct then $M$ is semisimple so both Jordan decompositions hold trivially. If the eigenvalues of $M$ are equal then $M_s$ is scalar, of the form $\left(\begin{smallmatrix} \alpha & 0 \\ 0 & \alpha \end{smallmatrix}\right)$, and the characteristic polynomial $x^2 - 2\alpha + \alpha^2$ must have coefficients in $R$. If $R$ is integrally closed in $F$ (or if $1/2$ is in $R$ or if $R$ is closed under square roots) then this implies $\alpha$ is in $R$ so $M_s$ is in $\mathbb{M}_2(R)$ and so is $M_n$. Thus additive Jordan decomposition will hold, and similarly so will multiplicative Jordan decomposition since if $M$ is invertible $\alpha$ will be a unit in $R$.

Next suppose that $n = 3$. Suppose that $M$ lies in $\mathbb{M}_3(F)$. If the eigenvalues of $M$ are all distinct, or if they are all equal and $R$ is integrally closed in $F$, then reasoning as above we conclude that both Jordan decompositions will hold. This leaves the case when two of the eigenvalues coincide and the third is different. Consider the matrix

$$M = \begin{pmatrix} 0 & 1 & 0 \\ 0 & 0 & 1 \\ 0 & 0 & a \end{pmatrix}$$

where $a$ is a non-zero element of $R$. This matrix $M$ is similar to the matrix

$$M' = \begin{pmatrix} 0 & 1 & 0 \\ 0 & 0 & 0 \\ 0 & 0 & a \end{pmatrix}$$

But an easy calculation shows that $M'_s = (1/a)M'^2$, and (since $M$ and $M'$ are similar!) this implies that $M_s = (1/a)M^2$. However the upper right corner of $(1/a)M^2$ is the element $1/a$ so taking $a$ to be non-invertible in $R$ shows that additive Jordan decomposition fails in $\mathbb{M}_3(R)$.

More generally any matrix in $\mathbb{M}_3(R)$ with two but not three equal eigenvalues will (unless it is semisimple) be similar to the matrix

$$M = \begin{pmatrix} a & 1 & 0 \\ 0 & a & 0 \\ 0 & 0 & b \end{pmatrix}$$

for some pair of unequal elements $a, b$ in $F$. Here $a$ and $b$ will lie in $R$ if it is integrally closed in $F$, and will be units in $R$ if $M$ is invertible. It is easy to calculate that here we have $M_s = aI + (1/(b-a))(M - aI)^2$. This shows that if $(b - a)$ is invertible in $R$ then $M_s$ will lie in $\mathbb{M}_3(R)$, and the same will hold for any matrix similar to $M$. Hence, multiplicative Jordan decomposition will hold if the difference of any two unequal units in $R$ is again a unit. On the other hand, if $a$ and $b$ are unequal units with $(b - a)$ not a unit, then multiplicative Jordan decomposition will fail for the invertible matrix

$$\begin{pmatrix} a & 1 & 0 \\ 0 & a & 1 \\ 0 & 0 & b \end{pmatrix}$$

since this matrix is similar to $M$ and the upper right corner of its semisimple component is easily calculated to be $1/(b-a)$.

Finally suppose that $n = 4$. Consider the invertible matrices

$$
\begin{pmatrix} 1 & 1 & 0 & 0 \\ 0 & 1 & 1 & 0 \\ 0 & 0 & 1 & 1 \\ 0 & 0 & 0 & 1 \end{pmatrix}, \begin{pmatrix} 1 & 0 & 0 & 0 \\ 0 & 1 & 0 & 0 \\ 0 & 0 & 1 & 0 \\ 0 & 0 & a & 1 \end{pmatrix}
$$

for any non-zero $a$ in $R$. Their product is the invertible matrix

$$
M = \begin{pmatrix} 1 & 1 & 0 & 0 \\ 0 & 1 & 1 & 0 \\ 0 & 0 & 1+a & 1 \\ 0 & 0 & a & 1 \end{pmatrix}
$$

which is similar to the matrix

$$
M' = \begin{pmatrix} 1 & 1 & 0 & 0 \\ 0 & 1 & 0 & 0 \\ 0 & 0 & 1+a & 1 \\ 0 & 0 & a & 1 \end{pmatrix}.
$$

For $M'$ we easily calculate that $M'_n = f(M'-I)$ where $f(x)$ is the polynomial $(-1/a)x(x^2-ax-a)$. Hence, we can calculate $M_n$ in the same way, as $f(M-I)$, and this gives that the upper right hand entry in $M_n$ is $-1/a$. Taking $a$ to be non-invertible in $R$ shows that multiplicative Jordan decomposition must fail.

To summarize, we have sketched a proof of the following result.

**4.1. Proposition** *Let $R$ be an integral domain of characteristic zero which is integrally closed in its quotient field $F$, with $R \neq F$. Then*

(i) *additive Jordan decomposition holds in $M_1(R)$ and in $M_2(R)$ and fails in $M_n(R)$ for $n \geq 3$.*

(ii) *multiplicative Jordan decomposition holds in $M_1(R)$ and in $M_2(R)$; holds in $M_3(R)$ if and only if the units of $R$ (together with 0) form a subfield; and fails in $M_n(R)$ for $n \geq 4$.*

Note for instance that $R = \mathbb{Q}[x]$ is an example where multiplicative Jordan decomposition holds in $M_3(R)$.

We conclude this section with an example in the more general situation where $R$ is taken to be a subring of a non-commutative division ring $D$. What can we say about Jordan decomposition in $M_2(R)$? Consider the matrix $M = \begin{pmatrix} i & 1 \\ 0 & j \end{pmatrix}$ with entries in the integral quaternions. It is easy to check that $M_s = \begin{pmatrix} i & (1+k)/2 \\ 0 & j \end{pmatrix}$. Hence Jordan decomposition fails for the integral quaternions. We do not yet have a general result concerning Jordan decomposition for 2 by 2 matrices over sub-rings of non-commutative division rings.

## 5. Group Rings

In this section let $G$ be a finite group, $F$ a field of characteristic zero, and $R$ a subring of $F$, integrally closed in $F$, with $R \neq F$. We wish to determine if additive (or multiplicative) Jordan decomposition holds in the group ring $R[G]$. In other words, if $\alpha$ is an element of the group algebra $F[G]$ which happens to lie in $R[G]$, then must it be the case that $\alpha_s$ and $\alpha_n$ also lie in $R[G]$? If $\alpha$ in $R[G]$ is invertible in $R[G]$ then must $\alpha_s$ and $\alpha_u$ lie in $R[G]$? These are the questions that originally motivated the results in the previous section.

We note here that, if additive Jordan decomposition holds in $R[G]$, multiplicative Jordan decomposition will also hold. This follows from the previously mentioned identity $(\alpha_s)^{-1} = (\alpha^{-1})_s$ for invertible elements $\alpha$ in $R[G]$. The unit group of $R[G]$ is said to be *splittable* when multiplicative Jordan decomposition holds in $R[G]$.

The results in this section illustrate three main themes. First of all, the larger $R$ is (for fixed $F$), the more likely it is that Jordan decomposition holds in $R[G]$. Next, the closer $F$ is to being algebraically closed, the less likely it is that Jordan decomposition holds in $R[G]$. And, finally, Jordan decomposition is more likely to hold in $R[G]$ if $G$ is close to being abelian. More precisely, the degrees of the irreducible representations of $G$ play a major role in determining the validity of Jordan decomposition in $R[G]$. For use below in this connection, we apply Wedderburn theory to the group algebra $F[G]$ to write it as a direct sum of matrix rings over division rings:

$$F[G] \simeq \oplus_{i=1}^{h} M_{n_i}(D_i) \tag{1}$$

where the center of each $D_i$ is a finite extension of $F$.

Suppose first that $G$ is abelian. Then $F[G]$ is commutative and hence all $n_i$ in (1) are equal to 1. This means that $F[G]$ contains no nilpotent elements, so every element is semisimple, and hence Jordan decomposition holds trivially. More generally, if we know that each $n_i = 1$, we get this conclusion. This happens, for instance, when $R = \mathbb{Z}$ and $G$ is the quaternion group $Q_8$. We will return to this example later in this section.

As a weak converse to the previous remark we can use the results of the previous section to prove the following proposition:

### 5.1. Proposition

(i) *If additive Jordan decomposition holds in $R[G]$, then $n_i \leq 2$ for all $i$ in equation (1).*

(ii) *If multiplicative Jordan decomposition holds in $R[G]$ then $n_i \leq 3$ for all $i$ in equation (1).*

**Proof (sketch).** If, for instance, $n_1 = 3$, we can realize matrices

$$\begin{pmatrix} 0 & b & 0 \\ 0 & 0 & b \\ 0 & 0 & 0 \end{pmatrix}, \begin{pmatrix} 0 & 0 & 0 \\ 0 & 0 & 0 \\ 0 & 0 & c \end{pmatrix}$$

as the first components of elements of $R[G]$ for some non-zero $b$, $c$ in $R$ and hence also realize the matrix

$$\begin{pmatrix} 0 & bc & 0 \\ 0 & 0 & bc \\ 0 & 0 & ab^2c^2 \end{pmatrix}$$

for any non-invertible $a$ in $R$. As in the previous section (with $b = c = 1$) the upper right corner of the semisimple component of this matrix is $1/a$ and this can be shown to lead to a contradiction. See Arora *et al.*, (1998) for details. Similarly, if $n_1 = 4$, we can realize matrices

$$\begin{pmatrix} 0 & b & 0 & 0 \\ 0 & 0 & b & 0 \\ 0 & 0 & 0 & b \\ 0 & 0 & 0 & 0 \end{pmatrix}, \begin{pmatrix} 0 & 0 & 0 & 0 \\ 0 & 0 & 0 & 0 \\ 0 & 0 & 0 & 0 \\ 0 & 0 & c & 0 \end{pmatrix}$$

and hence also realize (as a product) the invertible matrix

$$\begin{pmatrix} 1 & bc & 0 & 0 \\ 0 & 1 & bc & 0 \\ 0 & 0 & 1 + ab^3c^3 & bc \\ 0 & 0 & ab^2c^2 & 1 \end{pmatrix}$$

for any non-invertible $a$ in $R$. The upper right corner of its nilpotent component is $-1/a$ and this leads to a contradiction. Again see Arora *et al.*, (1998) for details.

One conclusion of the proceeding proposition has been shown to imply strong restrictions on the structure of $G$. These can be summarized as follows:

### 5.2. Proposition (Gow and Huppert 1987; Isaacs, 1976)

(i) *If $F$ is algebraically closed and $n_i \leq 2$ for all $i$ in equation (1) then $G$ is either abelian, or has a normal abelian subgroup of index 2, or the center of $G$ has index 8 in $G$.*

(ii) *If $n_i \leq 2$ for all $i$ in equation (1) then $G$ has a normal abelian subgroup of odd order which is a 2-complement.*

As a corollary of proposition 5.1 we can deduce the following proposition, using in addition the known result that, for $F = Q$, the $n_i$ in equation (1) will all equal 1 if and only if $G$ is abelian or of the form $Q_8 \times E \times A$ where $E$ is an elementary abelian 2-group and $A$ is an abelian group of odd order such that the multiplicative order of 2 modulo $|A|$ is odd.

### 5.3. Proposition

(i) *If additive Jordan decomposition holds in $R[G]$ and 2 does not divide the order of $G$ then $G$ is abelian.*

(ii) *If multiplicative Jordan decomposition holds in $R[G]$ and neither 2 nor 3 divides the order of $G$ then $G$ is abelian.*

From here on we will consider Jordan decomposition in $R[G]$ for special rings $R$. First we look at a somewhat artificial situation, but one which provides a prototype for later results. This is an easy consequence of Proposition 4.1.

**5.4. Proposition** *Suppose $F$ is a field such that all $D_i$ in equation (1) are fields, and let $S$ be the polynomial ring $F[x]$. Then additive Jordan decomposition holds in $S[G]$ if and only if all $n_i \leq 2$ and multiplicative Jordan decomposition holds in $S[G]$ if and only if all $n_i \leq 3$.*

Now let $F$ be the algebraic closure of $\mathbb{Q}$ and let $R$ be the ring of algebraic integers in $F$. We have

**5.5. Proposition** *Suppose $R$ is the ring of all algebraic integers. Then additive Jordan decomposition holds in $R[G]$ if and only if $G$ is abelian.*

**Proof (sketch).** Suppose there were a non-abelian $G$ such that additive Jordan decomposition held in $R[G]$. Let $G$ have minimal order with this property. Using 5.2(i) we can reduce to the case that $G$ is either dihedral of order $2p$ ($p$ an odd prime), or dihedral of order 8, or of the form $G = \langle x, t | x^{2^{n-1}} = 1, t^2 = x^2, [x, t] = x^{2^{n-2}} \rangle$. When $G = D_{2p} = \langle x, t | x^p = t^2 = 1, x^t = x^{-1} \rangle$, the element $\alpha = xt + (1+\epsilon)x^2 y + \cdots + (1+\epsilon + \cdots + \epsilon^{p-2})x^{p-1}t$ where $\epsilon$ is a primitive $p^{th}$ root of unity will not have its Jordan components in $R[G]$. When $G = D_8 = \langle x, t | x^4 = t^2 = 1, [x, t] = x^2 \rangle$, $\alpha = x + t$ will not have its Jordan components in $R[G]$. (Here, for instance, we have $\alpha_s = (1/2)(x + x^3) + (1/2)(t + x^2 t)$.) Finally, in the last case (which includes the quaternion group $Q_8$), the element $\alpha = x + it$ with $i^2 = -1$ will not have its Jordan components in $R[G]$. See Hales *et al.*, (1990) for details.

We do not know if the analogous result holds for multiplicative Jordan decomposition.

For the rest of this section we will consider the case $R = \mathbb{Z}$, the ring of integers. We have a complete characterization of the groups $G$ for which additive Jordan decomposition holds in this case. First, however, consider the particular example where $G$ is the dihedral group $D_6$ of the triangle, $G = \langle x, t | x^3 = t^2 = 1, x^t = x^{-1} \rangle$. If $\alpha$ lies in $\mathbb{Z}[G]$ we can write $\alpha = \beta + \gamma$ where $\beta$ is central (constant coefficients on conjugacy classes) and $\gamma$ is of the form $a(x - x^2) + bt + cxt - (b + c)x^2 t$, i.e., its coefficients sum to zero on conjugacy classes. Here we know that $2a$, $3b$ and $3c$ are integers. Now $\mathbb{Q}[G] \simeq \mathbb{Q} \oplus \mathbb{Q} \oplus M_2(\mathbb{Q})$ and the image of $\gamma$ in this representation will be of the form $(0, 0, M)$ where $M$ has trace 0. If $M$ is semisimple then $\gamma$ will be semisimple and hence $\alpha$ will also be semisimple. If

$M$ is not semisimple then its determinant will be 0 and in fact it will be nilpotent (square 0) so $\gamma$ will also be nilpotent. But a calculation shows that the determinant of $M$ being zero implies that $a^2 = b^2 + bc + c^2$. The left side of this lies in $\mathbb{Z}/4$ and the right side lies in $\mathbb{Z}/9$ so both sides must be integers. A congruence check then shows that $b$ and $c$ must also be integers, so $\gamma$ lies in $\mathbb{Z}[G]$ and consequently $\alpha$ has additive Jordan decomposition $\beta + \gamma$ in $\mathbb{Z}[G]$.

**5.6. Proposition** *Additive Jordan decomposition holds in $\mathbb{Z}[G]$ if and only if $G$ is either (a) abelian, or (b) of the form $Q_8 \times E \times A$ where $E$ is an elementary abelian 2-group and $A$ is an abelian group of odd order such that the multiplicative order of 2 modulo $|A|$ is odd, or (c) a dihedral group $D_{2p}$ where $p$ is an odd prime.*

**Proof (sketch).** The preceding argument easily generalizes to show that additive Jordan decomposition holds in $\mathbb{Z}[D_{2p}]$ for any odd prime $p$, completing the proof in one direction. For the other direction, suppose that $G$ is a counterexample of minimal order. We first show that $G$ cannot be a 2-group. Then, using 5.1(i) and 5.2(ii), and passing to subquotients, we can show that $G$ must be one of the following types, where $p$ and $q$ denote (not necessarily distinct) odd primes and $C_p$ denotes a cyclic group of order $p$:

$$D_{2pq}, D_{2p} \times C_q, D_{2p} \times C_2, (C_p \times C_p) \rtimes C_2, C_p \rtimes C_4.$$

Here $\rtimes$ denotes split extension with the generator of $C_2$ inverting everything and a generator of $C_4$ acting either as an automorphism of order 2 or 4.

We rule out all but the last of these cases by exhibiting in each case an element $\alpha$ in $\mathbb{Z}[G]$ whose additive Jordan components do not have integer coefficients, facilitating our computations by the representation of elements of $G$ as $2 \times 2$ matrices over $\mathbb{Q}[H]$ where $H$ has index 2 in $G$. For instance, in the first case (when $p = q$), we take $\alpha = x - x^{-1} + (x^{p+1} - x^{-(p+1)})t$ where $x$ and $t$ are elements of order $p^2$ and 2 respectively. In the very last case it turns out that $\mathbb{Q}[G]$ has a matrix component of order 4, violating 5.1(i). See Hales and Passi (1991) for details.

Results for multiplicative Jordan decomposition are less complete at the present time. One difficulty is that additive Jordan decomposition is inherited by subgroups and quotient groups, whereas multiplicative Jordan decomposition is inherited by subgroups but not necessarily quotient groups (the problem is that units need not lift). On the other hand, the property of being invertible is preserved under homomorphisms, and is strongly restrictive on group ring elements (see Isaacs, 1976), and this can provide a powerful tool. Consider for instance the group ring $\mathbb{Z}[D_8]$, where $D_8$ is the smallest group for which additive Jordan decomposition fails over $\mathbb{Z}$. An invertible element $\alpha$ must map to $+/-1$ under each of the four homomorphisms from $\mathbb{Z}[D_8]$ to $\mathbb{Z}$, and this is sufficiently restrictive to guarantee that multiplicative Jordan decomposition holds (Arora *et al.*, 1993). It turns out that $D_8$ is the only dihedral group for which multiplicative (but not additive) Jordan decomposition holds over $\mathbb{Z}$.

**5.7. Proposition** *Multiplicative Jordan decomposition holds in the integral group ring $\mathbb{Z}[D_{2n}]$ if and only if $n$ is either 2, 4, or an odd prime.*

**Proof (sketch).** Let $G$ be a counterexample of minimal order. Then $G$ must be one of the groups $D_{16}$, $D_{4p}$, $D_{2p^2}$ or $D_{2pq}$ where $p$ and $q$ are distinct odd primes. For each of these we can exhibit an element $\alpha$ in $\mathbb{Z}[G]$ which is invertible (presented as the product of a group element and a unipotent or as the product of two unipotents) but whose Jordan components do not have integral coefficients. For instance, for $D_{2p^2}$, we can take $\alpha = x^p(1 + (x - x^{-1})(1 + t))$ where $D_{2p^2} = \langle x, t | x^{p^2} = t^2 = 1, x^t = x^{-1} \rangle$. For $D_{4p}$ we can take $\alpha = [1 + (x^2 - x^{-2})(1 + t)][1 - (x^2 - x^{-2})(1 - x^p)(1 - t)]$. In each case computations are again facilitated by the representation of $G$ as $2 \times 2$ matrices over $\mathbb{Q}[H]$ where $H = \langle x \rangle$.

With groups of order 8 settled we turn to groups of order 12. Since $D_{12}$ was settled above, the only two left to investigate are the alternating group $A_4$ and the group $Q_{12} = \langle x, t | x^3 = t^4 = 1, x^t = x^{-1} \rangle$. It turns out that multiplicative Jordan decomposition does not hold in $A_4$, as is seen by considering the invertible element $\alpha = (12)(34) + (123) - (243)$. On the other hand we have a general result covering $Q_{12}$.

**5.8. Proposition** *Multiplicative Jordan decomposition holds in the integral group ring $\mathbb{Z}[Q_{4p}]$ where $p$ is any odd prime and $Q_{4p} = \langle x, t | x^p = t^4 = 1, x^t = x^{-1} \rangle$, the group of generalized quaternions of order $4p$.*

**Proof (sketch).** Using the complex representations of $Q_{4p}$ one can obtain a canonical form for the nilpotent part of any non-semisimple element $\alpha$ in the group ring. Then, considering the group modulo $\langle x \rangle$ and modulo $\langle t^2 \rangle$, and invoking a theorem of Kleinert (cf. Humphreys, 1975), it can be shown that the coefficients of $\alpha_n$ must be integral if $\alpha$ is invertible (for details Arora *et al.*, 1998).

A similar argument shows that the same result holds for the group of order 16 with presentation $\langle x, t | x^4 = t^4 = 1, x^t = x^{-1} \rangle$.

The proof of Proposition 5.6 involved reduction to several special cases and then demonstrating the failure of additive Jordan decomposition in each case by exhibiting an appropriate element $\alpha$ in the group ring. In Proposition 5.7 above we have shown that multiplicative Jordan decomposition also fails in some of these cases, and Proposition 5.8 shows that it holds for $C_p \rtimes C_4$ with a generator of $C_4$ acting as an automorphism of order 2 on $C_p$. The rest of the cases are covered by the following proposition.

**5.9. Proposition** *Multiplicative Jordan decomposition fails for the integral group ring of each of the following groups (where $p$ and $q$ denote not necessarily distinct odd primes):*

(i) $D_{2p} \times C_q$;

*(ii)* $(C_p \times C_p) \rtimes C_2 = \langle x, y, t | x^p = y^p = t^2 = 1, xy = yx, x^t = x^{-1}, y^t = y^{-1} \rangle$;

*(iii)* $C_p \rtimes C_4 = \langle x, t | x^p = t^4 = 1, x^t = x^e \rangle$ *where the multiplicative order of* $e$ *mod* $p$ *is 4.*

**Proof (sketch).** For (i) and (ii) we again exhibit invertible elements in the group ring whose Jordan components do not have integral coefficients. We can take $\alpha = y(1 + 2(x - x^{-1})(1 + t))$ in (ii), for instance. For (iii) we just use again the fact that $Q[G]$ has a matrix component of order 4, violating 5.1(ii) (cf. *Arora et al.*, 1998).

Each of the examples in Propositions 5.7 and 5.9 for failure of multiplicative Jordan decomposition, namely $D_{16}$, $D_{4p}$, $D_{2p^2}$, $D_{2pq}$, $D_{2p} \times C_q$, $(C_p \times C_p) \rtimes C_2$ and $C_p \rtimes C_4$ (with $t$ acting with order 4), actually fits into an infinite family of groups where multiplicative Jordan decomposition fails. The family in each case is obtained by letting $t$ have order a higher 2-power, say $2^k$, but still acting as an automorphism on $\langle x \rangle$ with the same order (2 or 4) as before.

In conclusion, the main open problem at this time seems to be to complete the characterization of those finite groups $G$ such that $Z[G]$ has multiplicative Jordan decomposition. In the case where all matric components of $Q[G]$ have degree at most 2, the first groups to consider are the 2-groups and the groups $C_p \rtimes C_{2^k}$ with $k \geq 3$ where the action is of order 2. We have some results in the latter case as long as 8 does not divide $p - 1$, but the case for instance of $C_{17} \rtimes C_8$ seems to be more difficult to handle.

Another problem to investigate is the following. Suppose additive Jordan decomposition fails for $G$, say over $Z$. Consider the quotient group $JZ[G]/Z[G]$ where $JZ[G]$ denotes the additive subgroup of $Q[G]$ generated by the Jordan components of all elements of $Z[G]$. This measures the failure of additive Jordan decomposition. What can be said about it in general? There is an obvious analogue of this question in the multiplicative case, a quotient of multiplicative groups which measures the failure of multiplicative Jordan decomposition. This should also be investigated.

# References

Albert, A. A., *Structure of Algebras*, New York, 1939.

Arora, Satya R., Hales, A. W. and Passi, I. B. S., Jordan decomposition and hypercentral units in integral group rings, *Comm. Algebra*, **21**, 25–35, 1993.

Arora, Satya R., Hales, A. W. and Passi, I. B. S., The multiplicative Jordan decomposition in group rings, *J. Alg.*, 209, 533–542, 1998.

Borel, A., *Linear Algebraic Groups*, Springer–Verlag, 1991.

Chevalley, C., *Théorie des Groupes de Lie II: Groupes Algébriques*, Hermann, Paris, 1951.

Gow, R., and Huppert, B., Degree problems of representation theory over arbitrary fields of characteristic 0 - On theorems of N. Ito and J. G. Thompson, *J. Reine Angew. Math.* **381**, 136–147, 1987.

Hales, A. W., Luthar, I. S. and Passi, I. B. S., Partial augmentations and Jordan decomposition in group rings, *Comm. Algebra*, **18**, 2327–2341, 1990.

Hales, A. W. and Passi, I. B. S., Integral group rings with Jordan decomposition, *Arch. Math.*, **57**, 21–27, 1991.

Humphreys, James E., *Introduction to Lie Algebras and Representation Theory*, Springer-Verlag, 1972.

Humphreys, James E., *Linear Algebraic Groups*, Springer-Verlag, 1975.

Isaacs, I. M., *Character Theory of Finite Groups*, Academic Press, 1976.

Kleinert, E., Units in $Z[Q_p]$, *J. Number Theory*, **26**, 227–236, 1987.

Sehgal, S. K., *Units in Integral Group Rings*, Longman, Essex, 1993.

Shirvani, and Wehrfritz, B. A. F., *Skew Linear Groups*, Cambridge University Press, 1986.

Springer, T. A., *Linear Algebraic Groups*, Birkhauser, 1981.

Ctr. for Comm's. Rsch., 4320 Westerra Court, San Diego, CA 92121, U.S.A.

Panjab University, Chandigarh 160 014, India

# On the Normalizer Problem

*Wolfgang Kimmerle*

## 1. Introduction

The object of this note is the structure of the normalizer of a group basis of the group ring $RG$ of a finite group $G$, where $R = \mathbb{Z}$ or more generally in the situation when $R$ is $G$-adapted. This means that $R$ is an integral domain of characteristic zero in which no prime divisor of $|G|$ is invertible.

Denote by $V(RG)$ the group of units of $RG$ with augmentation 1. Let $H$ be a group basis of $RG$. This means simply that $H$ is a subgroup of $V(RG)$ of the same order as $G$. Then the **normalizer problem** [Sehgal, 1993, Problem 43] may be phrased as follows.

$$N_{V(RG)}(H) = H \cdot Z(V(RG)),$$

where $Z(V(RG))$ denotes the centre of $V(RG)$. A unit $x \in N_{V(RG)}(G)$ induces by conjugation an automorphism of $G$. Denote by $\mathrm{Aut}_R(G)$ the automorphisms of $G$ induced by conjugation with units of $RG$. Clearly $\mathrm{Inn}\,G$ is a normal subgroup of $\mathrm{Aut}_R(G)$. The quotient is denoted by $\mathrm{Out}_R(G)$. Now the normalizer problem holds if, and only if,

$$\mathrm{Out}_R(H) = 1$$

for each group basis $H$ of $RG$.

Recently, the normalizer problem has had a big influence on the development of counterexamples to the isomorphism problem of integral group rings. We shall discuss this in section 2. In section 1 we give a brief survey on known positive results. In section 3, we consider group automorphisms which are close to be inner. In particular locally inner automorphisms, i.e., group automorphisms $\sigma$ with the property that for each Sylow $q$-subgroup $Q$ of $G$ there exists an inner automorphism $\gamma_Q$ such that $\gamma_Q \circ \sigma$ is the identity on $Q$, play an important role. Such automorphisms are used to prove the following result.

**Theorem 1** *Let $G$ be a finite group. Suppose that*

(i) *$G$ has no composition factor isomorphic to a cyclic group of order 2.*

(ii) *Locally inner automorphisms of non-abelian composition factors of $G$ are inner.*

*Then*

$$\text{Out}_Z(G) = 1.$$

*If G is soluble then it suffices to assume that G has no chief factor of order 2.*

I thank M. Hertweck for critical comments on an earlier version of this note.


## 2. Some Known Results

The results mentioned in this section are mainly devoted for soluble groups. The starting point is Coleman's lemma. A proof may be found in [Roggenkamp, 1991, Lemma 23]. It gives a complete description in the case when $G$ is a $p$-group.

**Lemma 1** *Let G be a finite group, let P be a p-subgroup of G and let R be a ring such that p is not invertible in R. Then*

$$N_{V(RG)}(P) = N_G(P) \cdot C_{V(RG)}(P),$$

*where $C_{V(RG)}(P)$ denotes the centralizer of P in $V(RG)$.*
**Corollary 1** *Let G be a finite nilpotent group. Then*

$$N_{V(RG)}(G) = G \cdot Z(V(RG)).$$

In the case when $R = \mathbb{Z}$ Jackowski and Marciniak (1987) proved the following.

**Theorem 2** *Let G be a finite group with normal Sylow 2-subgroup. Then*

$$N_{V(\mathbb{Z}G)}(G) = G \cdot Z(V(\mathbb{Z}G)).$$

Clearly Theorem 2 is a result for soluble groups. But it should be remarked that its proof does not require that groups of odd order are soluble. The proof uses the following fundamental result of Krempa (see Jackowski and Marciniak, 1987).

**Theorem 3** *Let G be a finite group. Then $\text{Out}_Z(G)$ is a 2-group.*

Kimmerle (1996) observed that also in the case when $G$ is a soluble finite group with $O_p, (G) = 1$ for $p$ odd outer group automorphisms of $G$ become not inner on the integral group ring. M. Hertweck pointed out to the present author that this holds with no restriction on the prime $p$ for $p$-constrained groups. A group automorphism $\sigma$ is called $p$-central if it is restricted to a Sylow $p$-subgroup the identity. By a result of $F$. Gross, see Theorem 7 below, and by Lemma 1 we get the following.

**Theorem 4 (Hertweck, 1998)** *Let G be a p-constrained group and let R be a ring such that p is not invertible in R. Assume that $O_p, (G) = 1$. Then*

$$N_{V(RG)}(G) = G \cdot Z(V(RG)).$$

Theorem 4 is a result which tends into the direction to soluble groups. In fact each soluble group $G$ may be written as a subdirect product of groups $G/O_p$, $(G)$. Using the description the normalizer problem for an arbitrary soluble group may be studied analogously as in Kimmerle and Roggenkamp (1993) the isomorphism problem. Simple groups are within the class of finite groups opposite to $p$-constrained groups. The next theorem is a consequence of a result of Feit and Seitz. The proof uses the classification of the finite simple groups.

**Theorem 5** *Let $G$ be a finite simple group. Then*

$$N_{V(RG)}(G) = G \cdot Z(V(RG)).$$

All theorems mentioned above are results with respect to the canonical group basis $G$ of $RG$. If $H$ is an arbitrary group basis of $RG$, where $R$ is $G$-adapted, then $H$ and $G$ have isomorphic normal subgroup lattices. Thus the assumptions of the above theorems hold for an arbitrary group basis and a positive solution of the normalizer problem follows. It should be remarked that for the most classes of groups considered in this section different group bases are even isomorphic. For nilpotent groups this has been proved in Roggenkamp and Scott (1987) and Weiss (1987), for $p$-constrained groups with $O_{p'}(G) = 1$ (see Scott, 1987) and for simple groups this follows from Kimmerle *et al.*, (1990).

## 3. Relationship to the Isomorphism Problem

The isomorphism problem of integral group rings is the question whether a ring isomorphism between $\mathbb{Z}G$ and $\mathbb{Z}H$ implies that $G$ and $H$ are isomorphic as groups. This problem was open for about 60 years, even in the case when $G$ is infinite. For a survey we refer to Kimmerle (1996) and Sandling (1985).

The connection between the isomorphism problem and the normalizer problem has been established by the following fundamental observation of M. Mazur. Let $\sigma$ be an automorphism of the finite group $G$. Denote by $G_\sigma$ the semidirect product of $G$ with the infinite cyclic group $C = \langle c \rangle$, where $C$ acts on $G$ via $\sigma$, i.e., $g^c = \sigma(g)$. Then

**Lemma 2 (Mazur, 1995)**

(i) *Let $\sigma \in \mathrm{Aut}G$. Thus $G_{id} \cong G_\sigma$ if and only if $\sigma \in \mathrm{Inn}G$.*

(ii) *Let $R$ be $G$-adapted. Then $RG_\sigma \cong RG_{id}$ as $R$-algebras if and only if $\sigma \in \mathrm{Aut}_R(G)$.*

It was shown by K. W. Roggenkamp and A. Zimmermann (1995a) that there are groups such that outer group automorphisms become inner on $\mathbb{Z}_{\pi(G)}G$. This leads to a counterexample for the isomorphism problem of integral group rings of infinite groups in the sense that if $S$ denotes the ring of algebraic integers in a suitable algebraic number field $K$ then there are non-isomorphic infinite groups

$G_\sigma$ and $G_{id}$ such that $SG_{id}$ and $SG_\sigma$ are isomorphic as $S$-algebras (Roggenkamp and Zimmermann, 1995$b$). K. W. Roggenkamp and A. Zimmermann did not know whether in their example $K$ can be $\mathbb{Q}$. Indeed it is shown in Hertweck (1998) that the automorphism constructed by K. W. Roggenkamp and A. Zimmermann is not inner on $\mathbb{Z}G$.

But M. Hertweck (1998 [Theorem B]) showed the following.

**Theorem 6** *There exists a finite group $X$ with the following properties.*

(i) *$X$ is a semidirect product of the form $G \cdot C_8$. The order of $X$ is*

$$2^{21} \cdot 97^{28}.$$

*Moreover $X$ has derived length 4 and $X$ has a normal Sylow 97-subgroup.*

(ii) *$\text{Out}_\mathbb{Z}(G) \neq 1$.*

(iii) *The isomorphism problem has a negative answer for $\mathbb{Z}X$.*

The subgroup $Y$ of $\mathbb{Z}X$ in Hertweck's theorem is constructed as follows. $Y$ is generated by $G$ and an element $t \cdot c$, where $c$ is the generator of $C_8$ and $t$ is the element of $\mathbb{Z}G$ which normalizes $G$ but induces on $G$ a non-inner isomorphism. Note because $t^c = t^{-1}$ it follows that $Y$ has the same order as $X$. Thus the isomorphism of $\mathbb{Z}X$ and $\mathbb{Z}Y$ is automatically given.

Clearly the group $G$ of Theorem 6 gives a counterexample to the normalizer problem. M. Hertweck even showed that in order to construct only a counterexample to the normalizer problem this may be already done with a group $H$ of order $2^{25} \cdot 97^2$. Moreover, $H$ is metabelian [Hertweck, 1998, Theorem A]. Clearly together with Mazur's result this produces also counterexamples for the isomorphism problem over the integers for infinite groups.

The group $X$ of Theorem 6 has a similar structure as the infinite groups considered by Mazur. It is also a semidirect product of a group $G$ and a cyclic group $C$ and for $G$ the normalizer problem has a negative solution. Assume that $C = \langle c \rangle$ and that $C$ has order 2. If $c$ acts on $G$ as the normalizing unit $t$ of $\mathbb{Z}G$, the element $t \cdot c$ is a central unit of $\mathbb{Z}X$. Ritter and Sehgal classified the groups $G$ for which $\mathbb{Z}G$ has only trivial central units [12]. As a consequence of these results it is easy to see that the normalizer problem for alternating groups has a positive answer because the centre of $V(\mathbb{Z}S_n)$ is trivial for $n \geq 2$.

Finally we look at infinite groups. The following is due to Parmenter and Li (1997).

**Proposition 1** *Let $G$ be generated by an abelian group $H$ and an element $g$ with $g^4 = 1$ and $ghg^{-1} = h^{-1}$ for each $h \in H$. Then*

$$N_{V(\mathbb{Z}G)}(G) = G \cdot Z(V(\mathbb{Z}G)).$$

**Proposition 2** *Let $G$ be periodic. Then the second centre $Z_2(V(\mathbb{Z}G))$ is contained in $N_{V(\mathbb{Z}G)}(G)$.*

Thus, if the normalizer problem holds for a periodic group, note that Proposition 1 determines some such groups, it follows that $Z_2(V(\mathbb{Z}G)) = T \cdot Z(V(\mathbb{Z}G))$, where $T = G \cap Z_2(V(\mathbb{Z}G))$. It should be remarked that in the case of finite groups this equality has been proved by Arora and Passi (1993).

## 4. Locally Inner Automorphisms

The aim of this section is the proof of Theorem 1.

Let $G$ be a finite group and let $S$ be a $G$-adapted ring. Let $x \in N_{V(SG)}(G)$. Denote by $\sigma$ the automorphism of $G$ induced from the conjugation with $x$. Then $\sigma$ has the following properties:-

(i) $\sigma(C) = C$ for each conjugacy class $C$ of $G$.

(ii) For each prime $p$ and each $P \in \mathrm{Syl}_p(G)$ there exists an inner automorphism $\gamma$ of $G$ such that $\gamma \circ \sigma$ is the identity on $P$.

Note $(i)$ is immediate from the fact that $\sigma$ is a central automorphism of $SG$. $(ii)$ follows from Coleman's Lemma.

Thus it makes sense to introduce the following notations. We call a group automorphism of $G$ **locally inner** if it satisfies property $(ii)$ and **almost inner** if it satisfies additionally property $(i)$.

These automorphisms form a group. The group of locally inner automorphisms will be denoted by ALInn$G$. Clearly Inn$G$ is a normal subgroup of ALInn$G$ and we put OLInn$G$ = ALInn$G$/Inn$G$. The group of almost inner automorphisms we denote by AInn$G$ and OInn$G$ is the quotient AInn$G$/Inn$G$. The preceeding remarks show that Aut$_S G \subset$ AInn$G$ provided $S$ is $G$-adapted. The proof of Theorem 3 now actually shows the following.

If each almost inner automorphism of a finite group $G$ is inner then for each $x \in N_{V(SG)}(G)$ there is a group element $g$ such that $x \cdot g$ centralizes $G$.

Now Theorem 1 follows from the following two Propositions and Theorem 3.

**Proposition 3** *Let $G$ be a finite group.*

(i) *Assume that no composition factor of $G$ has order 2.*

(ii) *Assume that 2 does not divide OLInn$S$ for each non-abelian composition factor $S$ of $G$.*

*Then 2 does not divide $|\mathrm{OLInn}G|$.*

**Proposition 4** *Let $G$ be a finite group and let $p$ be an odd prime.*

(i) *Assume that no composition factor of $G$ has order $p$. Then $p$ does not divide $|\mathrm{OLInn}G|$.*

(ii) *If $G$ is soluble then $p$ does not divide $|\mathrm{OLInn}G|$ provided no chief factor of $G$ has order $p$. Moreover this holds also in the case when $p = 2$.*

We sketch the proof of Propositions 3 and 4. Let G with locally inner automorphism $\sigma$ be a counterexample of minimal order. Assume that $\sigma$ has prime power order $p^n$. Note that locally inner automorphism fix each normal subgroup and thus induce automorphisms on quotients. The induced automorphisms are again locally inner.

**Step 1.** We may assume that $G$ has a unique minimal normal subgroup.

This follows from the following fact.

**Lemma 3** *Let M and N be normal subgroups of the finite group G with $M \cap N = 1$. Assume that $G/(M \cdot N)$ has no central element of order p. Let $\sigma \in$ AutG Assume that $\sigma$ induces on $G/M, G/N, G/(M \cdot N)$ automorphisms $\sigma_M, \sigma_N, \sigma_{MN}$ and assume that $\sigma_M$ and $\sigma_N$ are given by conjugation with p-elements $y_M, x_N$. Then $\sigma$ is given by conjugation with a p-element.*

**Step 2.** If $G$ is soluble and if $G$ has a unique minimal normal subgroup then the Fitting subgroup of $G$ has prime power order and we may apply part (i) of the following:-

**Theorem 7**

  (i) *[Gross, 1982, Corollary 2.4] Let G be a p-constrained group and assume that $O_{p'}(G) = 1$. Let $\sigma$ be an automorphism which centralizes a Sylow p-subgroup P of G. Then $\sigma$ is given by conjugation with an element of the centre of P.*

  (ii) *[Gross, 1982, Theorem A] Let G be a finite group with $O_{p'}(G) = 1$ and $O_p(G) = 1$. Assume that p is odd and that $\sigma$ is an automorphism of G which centralizes a Sylow p-subgroup P. Assume further that $\sigma$ has p-power order. Then $\sigma$ is given by conjugation with an element of the centre of P.*

It follows that $\sigma$ has to be inner and part $(ii)$ of Proposition 4 is proved.

**Step 3.** Let $H$ be a direct product of isomorphic non-abelian simple groups $S_i$. Let $X$ be a group containin $H$. Assume that each simple factor $S_i$ of $H$ is normal in $X$ and that $X/H$ has odd order. Let $\sigma$ be an automorphism of $X$ of 2-power order which is $q$-central for each prime $q$ dividing $|H|$. Then $\sigma$ restricted to each factor $S_i$ is locally inner.

For the proof observe that automorphisms $\tau$ of order $2^m$ which fix a Sylow q-subgroup $Q$ of $S_i$ and which are $q$-central are locally given by a conjugation with an element of 2 power order. By assumption such elements are elements of $H$.

**Step 4a.** Assume that $G$ has a unique minimal normal subgroup which is perfect and not simple. We consider first the case when $p$ is odd.

**Lemma 4** *Let p be an odd prime. Assume that G has no composition factor of order p. Suppose that G has a unique minimal normal subgroup M which is perfect. Then a locally inner automorphism $\sigma$ of G whose order is a power of p is inner.*

**Proof.** $M$ is a direct product of non-abelian simple groups $S_i$, $1 \leq i \leq k$ which are all isomorphic to $S_1$. Because $M$ is a minimal normal subgroup of $G$ it follows that $G$ acts transitively on the factors $S_i$ of $M$.

Suppose now that $\sigma$ is locally inner of order a power of $p$ and suppose that $\sigma$ is not inner.

We may modify $\sigma$ by an inner automorphism $\gamma$ such that $\sigma_1 := \gamma \circ \sigma$ fixes elementwise a Sylow $q$-subgroup $Q$ of $G$ for some prime $q$ dividing $|M|$. Then $\sigma_1$ is the identity on the Sylow $q$-subgroup $Q \cap M$ of $M$ and also on $Q \cap S_i$ for each $i$. It follows that $\sigma_1$ fixes each factor $S_i$. Note that the assumption that $\sigma$ is not inner implies that the order of $\sigma_n$ is divisible by $p$. Taking a suitable power of $\sigma_1$ we get a locally inner automorphism $\sigma_2$ which is not inner which fixes each factor $S_i$ and has order a power of $p$.

If $p$ divides $|M|$ then we may choose $q = p$ and by Theorem 7 $(ii)$ the restriction of $\sigma_2$ to $S_i$ is given by conjugation with a central element of $Q \cap S_i$ for each index $i$. Thus we may modify $\sigma_2$ by a conjugation such that the modified automorphism $\sigma_3$ is the identity on $M$. Because $C_G(M) = 1$ it follows that $\sigma_3$ is the identity. This contradicts our assumption that $\sigma$ is not inner.

Assume that $p$ does not divide $|M|$. Then it follows by a Sylow argument that $\sigma_2$ centralizes $M$. As above we get that $\sigma_2$ is inner. **q. e. d.**

**Step 4b.** Assume now that $p = 2$.

**Lemma 5** *Assume that the generalized Fitting subgroup $F^*(G)$ is a direct product of copies of isomorphic non-abelian simple groups $S_i$. Suppose that not all factors $S_i$ are normal in $G$. Let $\sigma$ be an automorphism of $G$ which is $q$-central for each prime $q$ dividing $|F^*(G)|$ and which has order a power of 2. Assume that $G$ has no composition factor of order 2. Then $\sigma$ is inner provided the Lemma holds for proper subgroups of $G$.*

**Proof.** Assume that $\sigma$ is not inner. As in the proof of Lemma 4 we may modify $\sigma$ to an automorphism $\sigma_2$ which has order a power of 2 which fixes each factor $S_i$ of $F^*(G)$ and is $q$-central for the same primes $q$ as $\sigma$.

Let $D = \bigcap_{i=1}^k N_G(S_i)$ and let $R_D$ be a Sylow $r$-subgroup of $D$. Suppose that $r$ divides $|F^*(G)|$. Clearly $R_D = R \cap S_i$ for some Sylow $r$-subgroup of $G$. Let $R_i = R_D \cap S_i$. Note that $\sigma_n(R_i) \subset S_i$. Since $\sigma_2$ is $r$-central, we find $g \in G$ such that $\gamma(g) \circ \sigma_n$ is the identity on $R$. Here, $\gamma_g$ denotes the conjugation with $g$. But $\gamma_g$ fixes $S_i$ because $\gamma_g(\sigma_n(R_i)) = R_i$. Thus $g \in N_G(S_i)$. Note that $g$ is independent of $i$. Consequently, $g \in D$. This means that $\sigma_2$ restricted to $D$ is $r$-central. By assumption $D$ is a proper subgroup of $G$. $D$ is a normal subgroup of $G$. Each composition factor of $D$ is isomorphic to a composition factor of $G$. Moreover, $F^*(D) = F^*(G)$. By assumption we get that $\sigma_2$ restricted to $D$ is inner.

Thus we may modify $\sigma_2$ by a conjugation to an automorphism $\sigma_3$ which is restricted to $D$ and therefore also to $F^*(G)$ the identity. But $C_G(F^*(G)) \subset F^*(G)$ and by assumption $F^*(G)$ does not have a centre. Hence, $C_G(F^*(G)) = 1$ and it follows that $\sigma_3$ is the identity. This yields the contradiction that the automorphism $\sigma$ which we started with is inner.                                              **q. e. d.**

Step 3 handles the case when the subgroup $D$ of the proof of Lemma 5 coincides with $G$. Moreover the hypothesis on the non-abelian composition factors of $G$ in Proposition 3 gives the beginning for an induction depending on the group order.

**Step 5.** We may now assume that $G$ is not soluble and that $G$ has a unique minimal normal subgroup $A$ which is abelian. By Theorem 4 it follows that $C_G(M)$ is not contained in $M$. Thus the generalized Fitting subgroup of $G$ is not a $p$-group and $M$ is contained in the centre of the layer $E(G)$. By induction we may modify $\sigma$ by an inner automorphism such that the modified automorphism $\sigma_m$ induces on $G/M$ the identity. As before we may assume that $p$ divides the order of $\sigma_m$. Thus we are in the position to apply the following.

**Lemma 6** *Let $G$ be a finite group and let $\sigma \in \mathrm{Aut}G$. Let $N$ be a normal perfect subgroup of $G$. Let $Z$ be a $p$-group which is contained in $Z(N)$. Assume that $\sigma$ induces on $G/Z$ the identity. Then*

*(i) $\sigma$ restricted to $N$ is the identity.*

*(ii) The order of $\sigma$ divides the exponent of $Z$.*

**Proof.** For $g \in G$ put $\sigma(g) = \delta(g) \cdot g$. Then $\delta$ is a 1-cocycle from $G$ to $Z$ because $\sigma$ induces on $G/Z$ the identity. Since $Z \subset Z(N)$ the action of $N$ on $Z$ is trivial. But by assumption $N$ is perfect. Thus $\delta$ restricted to $N$ is 1. Now part $(i)$ follows.

For part $(ii)$ note that $\sigma$ fixes $Z$. Then

$$\sigma^k(g) = \sigma^{k-1}(\delta(g) \cdot g) = \delta^k(g) \cdot g.$$

Because $\delta(g) \in Z$ the order of $\sigma$ divides the exponent of $Z$.        **q. e. d.**

Now Lemma 6 shows that $\sigma_m$ has $p$-prime power order and $p$ divides the order of $A$. But by assumption $p$ does not divide $|A|$. This contradiction completes the proof of Proposition 3.

**Remarks 1.**

(a) It seems likely that the hypothesis on the non-abelian composition factors $S$ in Proposition 3 is always valid. If $S$ is an alternating group $A_n$ then an outer automorphism does not centralize a Sylow $r$ subgroup, if $r$ is an odd prime dividing $n$ or $n-1$. In the case when $n = 6$ a direct inspection shows that no subgroup of $\mathrm{Aut}(A_6)$ contains elements of order 6 and elements of order 10. Note that $\mathrm{PGL}(2, 9)$ has an element of order 10.

(b) For simple groups Feit and Seitz proved that each outer automorphism of a simple non-abelian group moves atleast one conjugacy class (Feit and Seitz, 1988). In particular, almost inner automorphisms are inner. For many groups similar arguments as above may be applied inductively also with respect to almost inner automorphisms. This leads to hypotheses of the type that almost inner automorphisms of almost simple groups, i.e., groups $X$ with

$$S = \mathrm{Inn}S \le X \le \mathrm{Aut}S,$$

where $S$ is non-abelian simple, are inner. This might always hold.

(c) Let $S$ be a non-abelian simple group and suppose that $\mathrm{Out}S$ is cyclic. By the result of Feit and Seitz a generating outer automorphism $\sigma$ acts faithfully on the conjugacy classes of $S$. Thus the subgroups of $\langle\sigma\rangle$ have different orbits on the conjugacy classes of $S$. Hence, almost inner automorphisms are inner for each group $X$ with $\mathrm{Inn}S \subset X \subset \mathrm{Aut}S$.

## 5. G-Adapted Coefficient Rings

**Proposition 5** *Let $G$ be a finite group and let $S$ be a ring which is $G$-adapted. Assume that $O_{p'}(G) = O_p(G) = 1$ for each $p \in \pi(G)$. Then $\mathrm{Out}_S G$ is a 2-group.*

**Proof.** Let $\sigma \in \mathrm{Aut}_S G$ and suppose that $\sigma \in \mathrm{Out}_S G$ has order divisible by the prime $p$. Because $S$ is $G$-adapted we may apply Coleman's Lemma for each prime dividing $|G|$.

Thus for each $q \in \pi(G)$ there is an inner automorphism $\gamma$ of $G$ such that $\tau = \gamma \circ \sigma$ is $q$-central, i.e., there is a Sylow $q$-subgroup $Q$ such that $\tau$ restricted to $Q$ is the identity. Because $\tau$ is not inner and because $p$ divides the order of $\sigma$ a suitable power $\tau^r$ is not inner and has order $p^a$ with $a \ge 1$. Clearly $\tau^r$ is $q$-central. If $q$ is odd it follows by [Gross, 1982] that $(p, q) = 1$.

Looking at $LG$ where $L$ is a field containing the quotient field of $S$ and which is a splitting field of $G$ we see that $\sigma$ preserves the conjugacy classes of $G$. Thus by [Jackowski and Marciniak, 1987, Proposition 2.4] each prime dividing the order of $\sigma$ divides as well the order of $G$. In particular, $p$ cannot be an odd prime not dividing the order of $G$. Now the Proposition follows immediately.     **q. e. d**

### Remarks 2.
By the Theorem of Feit and Thompson a finite group satisfying the hypotheses of Proposition 5 has even order.

Proposition 3 and 4 yield immediately the following.

**Proposition 6** *Let $G$ be a finite group. Assume that all composition factors of $G$ are non-abelian. Let $S$ be a $G$-adapted ring. Assume that locally inner automorphisms of 2 power order of the composition factors of $G$ are inner. Then*

$$\mathrm{Out}_S(G) = 1.$$

# References

Arora, S. R. and Passi, I. B. S. Central height of the unit group of an integral group ring, *Comm. Alg.*, **21**, 3673–3683, 1993.

Conway, J. H. Curtis, R. T. Norton, S. P. Parker, R. A. and Wilson, R. A. *Atlas of Finite Groups*, Oxford University Press, London/New York, 1985.

Feit, W. and Seitz, G. M. On finite rational groups and related topics, *Illinois J. Math.*, **33**, No.1, 103–131, 1988.

Gross, F. Automorphisms which centralize a Sylow *p*-subgroup, *J. Alg.*, **77**, 202–233, 1982.

Hertweck, M. Eine Lösung des Isomorphieproblems für ganzzahlige Gruppenringe von endlichen Gruppen. *Dissertation*, Universität Stuttgart, 1998.

Jackowski S. and Marciniak, Z. Group automorphisms inducing the identity map on cohomology, *J. p. appl. Alg.*, **44**, 241–250, 1987.

Kimmerle, W. On automorphisms of $\mathbb{Z}G$ and the Zassenhaus conjectures, *CMS Conf. Proc.*, **18**, 383–397, 1996.

Kimmerle, W. Lyons, R. Sandling, R. and Teague, D. Composition factors from the group ring and Artin's theorem on orders of simple groups, *Proc. London Math. Soc.*, **60**(3), 89–122, 1990.

Kimmerle, W. and Roggenkamp, K. W. Projective limits of group rings, *J. pure appl. Alg.*, **88**, 119–142, 1993.

Mazur, M. On the isomorphism problem for infinite group rings, *Expositiones Mathematicae*, Spektrum Akademischer Verlag (Heidelberg) **13**, 433–445, 1995.

Parmenter, M. Talk given at Groups 97 in Bath, England.

Ritter, J. and Sehgal, S. K. Integral group rings with trivial central units, *Proc. AMS* **108**, Nr. 2, 327–329, 1990.

Roggenkamp, K. W. The isomorphism problem for integral group rings of finite groups, *Proc. int. Cong. Math.*, Kyoto 1990, Springer 369–380, 1991.

Roggenkamp, K. W. and Scott, L. L. Isomorphisms of *p*-adic group rings, *Ann. Math.*, **126**, 593–647, 1987.

Roggenkamp, K. W. and Zimmermann, A. Outer group automorphisms may become inner in the integral group ring, *J. pure appl. Alg.*, **103**, 91–99, 1995a.

Roggenkamp, K. W. and Zimmermann, A. A counterexample for the isomorphism problem of polycyclic groups. *J. pure appl. Alg.*, **103**, 101–103, 1995b.

Sandling, R. The isomorphism problem for group rings, a survey, *Lect. Not. Math.*, Nr. 1148, Springer 256–289, 1985.

Sehgal S. K. *Units in Integral Group Rings*, Longman Scientific and Technical, Essex, 1993.

Scott, L. L. Recent Progress on the Isomorphism Problem, *Proc. Symp. pure Math.*, **47**, 259–274, 1987.

Weiss, A. *p*-adic rigidity of *p*-torsion, *Ann. Math.*, **127**, 317–332, 1987.

Mathematisches Institut B, Universität Stuttgart, Pfaffenwaldring 57, D–70550 Stuttgart, German·

# Galois Cohomology of Classical Groups

## R. Parimala

In this article, we survey recent results of Eva Bayer-Fluckiger and the author on the Galois cohomology of classical groups over fields of virtual cohon. Jlogical dimension 2. Number fields are examples of such fields. We begin by describing a well-known classification theorem for quadratic forms over number fields in terms of the so-called classical invariants (§ 2). We explain in § 3 how this classification leads to Hasse principle for principal homogeneous spaces for Spin $q$ over number fields. In § 4 and § 7, we state the conjecture of Serre concerning the triviality of principal homogeneous spaces under semi-simple, simply connected, linear algebraic groups over perfect fields of cohomological dimension 2 and its real analogue due to Colliot-Thélène and Scheiderer in the form of a Hasse principle, if the field has virtual cohomological dimension $\leq 2$. As in the case of Spin $q$ over number fields, a main step in the proof of these conjectures is a classification theorem of hermitian forms over involutorial division algebras defined over fields of virtual cohomological dimension $\leq 2$, which is described in § 6 and § 7.

We refer to [Se1] for basic facts on Galois cohomology and [Sch] for hermitian forms over involutorial division algebras.

## 1. Non-abelian Galois Cohomology

Let $k$ be a field, $k_s$, a separable closure of $k$ and $\Gamma = \mathrm{Gal}(k_s/k)$. Let $G$ be a linear algebraic group defined over $k$. The action of $\Gamma$ on $G(k_s)$ is continuous for the profinite topology on $\Gamma$ and the discrete topology on $G(k_s)$. Let $Z^1(\Gamma, G(k_s)) = \{f \in \Gamma \to G(k_s), f \text{ continuous}, f(st) = f(s).sf(t), f \text{ or } s, t \in \Gamma\}$ denote the set of continuous 1- cocycles from $\Gamma$ to $G(k_s)$. We define an equivalence on $Z^1(\Gamma, G(k_s))$ by setting $f \sim g \Leftrightarrow$ there exists $a \in G(k_s)$ such that $f(s) = a^{-1}.g(s).sa$ for every $s \in \Gamma$. Let $H^1(k, G) = H^1(\Gamma, G(k_s)) = Z^1(\Gamma, G(k_s))/\sim$. The set $H^1(k, G)$ is a pointed set with the distinguished element $f_0$ defined by $f_0(s) = 1 \forall s \in \Gamma$.

If $(V, x)$ is a pair consisting of a finite dimensional vector space $V$ over $k$ with a tensor $x$, $G = \mathrm{Aut}(V, x)$ is a linear algebraic group defined over $k$. The set $H^1(k, G)$ classifies the set of isomorphism classes of pairs $(V', x')$, where $V'$ is a vector space over $k$ with a tensor $x'$ such that $(V', x') \otimes_k k_s$ is isomorphic to

$(V, x) \otimes_k k_s$. In particular, $H^1(k, PGL_n)$ classifies isomorphism classes of central simple algebras of degree $n$ over $k$. Let char $k \neq 2$ and $O_n$ the orthogonal group $\{X \in GL_n, XX^t = \text{Identity}\}$. Then $H^1(k, O_n)$ classifies the set of isomorphism classes of non-degenerate quadratic forms of dimension $n$ over $k$.

In general, $H^1(k, G)$ classifies isomorphism classes of principal homogeneous spaces for $G$, which are trivial over $k_s$. A principal homogeneous space for $G$ is a $k$-scheme $P$ together with a $G$-action $P \times G \rightarrow P$ such that the morphism $P \times G \rightarrow P \times P, (x, g) \rightarrow (x, xg)$ is an isomorphism. Given a $k_s$-rational point $x_0$ of a principal homogeneous space $P$ for $G$, the map $f : \Gamma \rightarrow G(k_s)$, defined by the relation $sx_0 = x_0 f(s)$ is a 1-cocycle in $Z^1(\Gamma, G(k_s))$ and $[P] \rightarrow [f]$ gives a bijection between the set of isomorphism classes of principal homogeneous spaces for $G$, trivial over $k_s$ and $H^1(k, G)$. If $G$ is smooth, then $H^1(k, G)$ classifies isomorphism classes of all principal homogeneous spaces for $G$ ([Se1], Chap. III, § 1).

## 2. Classical Invariants for Quadratic Forms

Let $k$ be a field with char $k \neq 2$. Let $q : V \rightarrow k$ be a non-degenerate quadratic form on a finite dimensional vector space $V$ over $k$. The dimension of $q$, denoted $\dim q$ is the dimension of the vector space $V$ over $k$. Let $b_q : V \times V \rightarrow k, b_q(v, w) = q(v + w) - q(v) - q(w)$ be the associated bilinear form. Let $\{e_i\}$ be a basis for $V$ and $a_{ij} = b_q(e_i, e_j), 1 \leq i, j \leq n$. Then $\det(a_{ij}) \in k^*/k^{*2}$ is independent of the choice of the basis. We define the discriminant of $q$ by

$$disc(q) = (-1)^{n\frac{(n-1)}{2}} \det (a_{ij}) \in k^*/k^{*2}.$$

The *Clifford algebra* $C(q)$ of $(V, q)$ is the $k$-algebra $T(V)/I$, where $T(V)$ is the tensor algebra of $V$ and $I$ the two-sided ideal generated by $\{v \otimes v - q(v), v \in V\}$. The gradation $T_0(V) = \bigoplus_{i \text{ even}} \otimes^i V, T_1(V) = \bigoplus_{i \text{ odd}} \otimes^i V$ induces a $\mathbb{Z}/2$-gradation $C(q) = C_0(q) \oplus C_1(q)$ on $C(q)$. If $\dim q$ is even, $C(q)$ is a central simple algebra over $k$ and if $\dim q$ is odd, $C_0(q)$ is a central simple algebra over $k$. The class of $C(q)$ or $C_0(q)$ according as $\dim q$ is even or odd, in the Brauer group $Br(k)$ of $k$ is called the *Clifford invariant* of $q$, denoted by $c(q)$.

We define the total signature attached to a quadratic form. An ordering $v$ of $k$ is a decomposition $k = P_v \cup (-P_v)$ satisfying

$$P_v \cap (-P_v) = 0, x, y \in P_v \text{ implies } x + y, xy \in P_v, k^{*2} \subset P_v.$$

We call $P_v$ the set of *positive elements* for $v$. Let $\Omega$ be the set of all orderings of $k$. Then $\Omega$ can be identified with the conjugacy classes of involutions in $\text{Gal}(k_s/k)$ and has a topology induced from the profinite topology on $\text{Gal}(k_s/k)$. Let $q$ be a non-degenerate quadratic form over $k$. Let $< a_1, \cdots, a_n >$ be the diagonal matrix representing $q$ with respect to a basis of $V$. Let $v \in \Omega$. We define $sgn_v(q)$ to be the number of positive $a_i$ minus the number of negative $a_i$. We define $sgn : \Omega \to V$ by $sgn(v) = sgn_v(q)$. This function is continuous on $\Omega$ for the discrete topology on **Z** and is called the *total signature* of $q$.

The invariants: dimension, discriminant, Clifford invariant and total signature are called the *classical invariants* of a quadratic form. If $k = \mathbf{R}$, dimension and signature determine the isomorphism class of any non-degenerate quadratic form over $k$. If $k$ is a $p$-adic field, i.e., a finite extension of $Q_p$, dimension, discriminant and the Clifford invariant determine the isomorphism class of a non-degenerate quadratic form over $k$. Using Hasse-Minkowski theorem, one concludes that a non-degenerate quadratic form over a number field is determined upto isomorphism by its dimension, discriminant Clifford invariant and signature. Hence the natural question: Over what kind of fields $k$ do the classical invariants classify quadratic forms?. Elman and Lam give an answer to this question in ([EL], Th. 3). Let $W(k)$ denote the Witt group of $k$ and $I(k)$ the ideal of even dimensional forms in $W(k)$. Let $I^n(k) = (I(k))^n$.

**Theorem. (Elman-Lam ([EL], Th. 3))** *The classical invariants determine the isomorphism class of a quadratic form over $k$ if and only if $I^3(k)$ is torsion free.*

## 3. Hasse Principle for Spin Group over Number Fields

By a quadratic form or a hermitian form, in the rest of the paper, we mean a non-degenerate quadratic or hermitian form. Let $q$ be a quadratic form over a field $k$ with char $k \neq 2$. Then $H^1(k, O(q))$ is in bijection with the set of isomorphism classes of all quadratic forms over $k$ with dimension equal to the dimension of $q$. The map $SO(q) \hookrightarrow O(q)$ induces an injection of sets $H^1(k, SO(q)) \hookrightarrow H^1(k, O(q))$, the image being the set of isomorphism classes of quadratic forms with the same dimension and discriminant as $q$. We have an exact sequence of algebraic groups

$$1 \to \mu_2 \to \text{Spin } (q) \to SO(q) \to 1,$$

$\mu_2$ denoting the group of order 2. This gives rise to an exact sequence of pointed sets

$$1 \to k^*/sn(SO(q)(k)) \xrightarrow{\iota} H^1(k, \text{Spin}(q)) \to H^1(k, SO(q)).$$

Here $sn : SO(q)(k) \to k^*/k^{*2}$ denotes the spinor norm map. The image of the map $H^1(k, \text{Spin}(q)) \to H^1(k, SO(q))$ is the set of isomorphism classes of quadratic forms with the same dimension, discriminant and Clifford invariant as $q$. Let $k$ be a number field. We shall now describe how classification of quadratic forms in terms of classical invariants leads to Hasse principle for principal homogeneous spaces for Spin $q$, if dim $q \geq 3$: the map

$$H^1(k, \text{Spin}(q)) \to \prod_{v \in \Omega_\infty} H^1(k_v, \text{Spin}(q))$$

is injective, $\Omega_\infty$ denoting the set of real places of $k$ and $k_v$ denoting the completion of $k$ at $v$.

Let $\xi \in H^1(k, \mathrm{Spin}(q))$ be such that $\xi_v = 0$ for each $v \in \Omega_\infty$. Let $q'$ be a quadratic form over $k$ corresponding to the image of $\xi$ in $H^1(k, SO(q))$. Then $q'$ has the same classical invariants as $q$ and by the classification theorem stated in § 2, $q'$ is isomorphic to $q$ and the image of $\xi$ in $H'(k, SO(q))$ is trivial. Let $\alpha \in k^*$ be such that $\iota(\bar{\alpha}) = \xi$. In view of the commutative diagram

$$
\begin{array}{ccc}
k^*/sn(SO(q)(k)) & \xrightarrow{\iota} & H^1(k, \mathrm{Spin}\,q) \\
\downarrow & & \downarrow \\
\displaystyle\prod_{v \in \Omega_\infty} k_v^*/sn(SO(q_v)(k_v)) & \xrightarrow{(\iota v)} & \displaystyle\prod_{v \in \Omega_\infty} H^1(k_v, \mathrm{Spin}(q)),
\end{array}
$$

since $\xi_v = 0$ for all $v \in \Omega_\infty$, $\alpha_v \in sn(SO(q_v)(k_v))$ for all $v \in \Omega_\infty$. An element $x \in k_v^*$ is a spinor norm from $SO(q_v)$ if and only if either $q_v$ is isotropic or $x$ is positive in $k_v$. Given $\alpha > 0$ at all $v \in \Omega_\infty$ where $q_v$ is anisotropic, we show that $\alpha \in sn(SO(q)(k))$.

If $q$ is isotropic, $sn(SO(q)(k)) = k^*/k^{*2}$. We assume that $q$ is anisotropic over $k$. By the condition on $\alpha$, $\langle 1, -\alpha \rangle \otimes q$ is isotropic over $k_v$ for all $v \in \Omega_\infty$. Since $\dim q \geq 3$, $\dim(\langle 1, -\alpha \rangle \otimes q) \geq 6$ and hence $\langle 1, -\alpha \rangle \otimes q$ is isotropic at all finite completions of $k$. Thus, by Hasse-Minkowski theorem, $\langle 1, -\alpha \rangle \otimes q$ is isotropic. Hence there exist non zero values $a, b$ of $q$ such that $a - \alpha b = 0$ i.e., $\alpha = ab^{-1}$.

Let $x, y \in V$ be such that $q(x) = a$, $q(y) = b^{-1}$. Let $\tau_x, \tau_y$ denote the reflections with respect to $x$ and $y$ respectively. Then $sn(\tau_x \tau_y) = \alpha$. This completes the proof that the map $H^1(k, \mathrm{Spin}\,(q)) - \to \prod_{v \in \Omega_\infty} H^1(k_v, \mathrm{Spin}\,(q))$ has trivial *kernel*. One uses a standard twisting argument ([Se1], Proposition 35, p. 46) to conclude that this map is indeed injective.

We have the following more general theorem, due to Kneser ([kn]) for classical groups, Harder ([Ha1], [Ha2]) for exceptional groups other than $E_8$ and Chernousov for $E_8$ ([Ch]).

**Theorem.** *Let $k$ be a number field. Let $G$ be a semi-simple simply connected linear algebraic group defined over $k$. Then the map*

$$
H^1(k, G) \to \prod_{v \in \Omega_\infty} H^1(k_v, G)
$$

*is injective. In particular if $k$ is a totally imaginary number field, $H^1(k, G) = 0$. If $k$ is a $p$-adic field, Kneser also proves that $H^1(k, G) = 0$.*

## 4. Fields of Cohomological Dimension 2 and a Conjecture of Serre

A field $k$ is said to have *cohomological dimension* (abbreviated $cd(k)$) $\leq n$ if $H^i(\Gamma, A) = 0$ for $i \geq n + 1$ for all finite $\Gamma$-modules $A$. Totally imaginary number fields, *p*-adic fields and function fields of surfaces over **C** are examples of fields of cohomological dimension 2. Serre conjectured the following in 1962 ([Se1]).

**Conjecture I.** Let $k$ be a perfect field with $cd(k) \leq 1$ and $G$ a connected linear algebraic group defined over $k$. Then $H^1(k, G) = 0$.

**Conjecture II.** Let $k$ be a perfect field with $cd(k) \leq 2$ and $G$ a semi-simple simply connected linear algebraic group defined over $k$. Then $H^1(k, G) = 0$.

Conjecture I was settled in the affirmative by Steinberg in the mid 60's ([St]). The solution to conjecture II for fields of arithmetic type was stated in § 4. The first major breakthrough towards the solution of conjecture II for arbitrary fields is the following theorem of Merkurjev-Suslin ([MSu1]).

**Theorem.** *Let $k$ be a perfect field. The following are equivalent:*

1) *$cd(k) \leq 2$.*

2) *For every finite extension $L$ of $k$ and for every finite dimensional central division algebra $D$ over $L$, the reduced norm map Nrd: $D^* \to L^*$ is surjective.*

Given a central simple algebra $A$ over $k$, if $\varphi : k_s \otimes_k A \xrightarrow{\sim} M_n(k_s)$ is a splitting over the separable closure of $k$, for $a \in A$, $\det \varphi(1 \otimes a) = \mathrm{Nrd}(a)$ belongs to $k$ and is independent of the splitting $\varphi$. This function Nrd is called the reduced norm on $A$.

**Corollary.** *Conjecture II is true for $G = SL(n, D)$, $D$ a finite dimensional central division algebra over $k$, $SL(n, D)$ denoting the subgroup of reduced norm one elements in $GL(n, D)$.*

**Proof.** We have an exact sequence of algebraic groups

$$1 \to SL(n, D) \to GL(n, D) \xrightarrow{Nrd} G_m \to 1$$

which gives the following exact sequence of pointed sets:

$$1 \to k^*/_{Nrd(D^*)} \to H^1(k, SL(n, D)) \to H^1(k, GL(n, D)).$$

The set $H^1(k, GL(n, D))$ classifies isomorphism classes of vector spaces of dimension $n$ over $D$ and is trivial. Hence there is a bijection of sets

$$k^*/Nrd(D^*) \simeq H^1(k, SL(n, D)).$$

The latter set is trivial if and only if $k^* = Nrd(D^*)$.

We have the following theorem ([BaP1]).

**Theorem.** *Let $k$ be a perfect field with $cd(k) \leq 2$. Let $G$ be a semi-simple simply connected linear algebraic group defined over $k$, of classical type or of type $G_2$ or $F_4$. Then $H^1(k, G) = 0$.*

The proof of this theorem relies very much on the results of Merkurjev- Suslin stated in this section. The pattern of proof for classical groups is similar to the one indicated for Spin $q$ in § 4. We prove a classification theorem for hermitian forms over involutorial division algebras over fields $k$ with $cd(k) \leq 2$. We then use certain results of Merkurjev on norm principle for algebraic groups ([M2]) to complete the proof of the theorem for classical groups. If $G$ is of type $G_2$, the set $H^1(k, G)$ classifies isomorphism classes of Cayley algebras over $k$ ([Se2]). Since Cayley algebras are determined upto isomorphism by their norm which is a 3-fold Pfister form, the theorem follows from the fact that $I^3(k) = 0$ for fields with $cd(k) \leq 2$. If $G$ is of type $F_4$, the set $H^1(k, G)$ classifies the set of isomorphism classes of exceptional central simple Jordan algebras over $k$. Reduced norm surjectivity for division algebras over $cd(k) \leq 2$ fields yields, using Tits' rational construction of these algebras ([J], Ch. 9) that these algebras are 'reduced'. By a theorem of Springer ([Se2], §9), these algebras are determined upto isomorphism by their trace forms and the proof of the theorem once again appeals to the classification theorem for quadratic forms.

In the next section we give a list of groups of "classical type".

## 5. Classical Groups

Let $K$ be a field with char $K \neq 2$. Let $D$ be a finite dimensional central division algebra over $K$ with an involution $\sigma$. Let $k = K^\sigma = \{x \in K, \sigma x = x\}$. We say that $\sigma$ is of first kind if $K = k$ and $\sigma$ is of second kind otherwise. Suppose $\sigma$ is of first kind. Let $\varphi : k_s \otimes_k D \xrightarrow{\sim} M_n(k_s)$ be a splitting for $D$. The transport of $1 \otimes \sigma$ under $\varphi$ gives an involution $X \rightarrow AX^t A^{-1}$ on $M_n(k_s)$ with $A^t = \epsilon A, \epsilon = \pm 1$. We say that $\sigma$ is *orthogonal* if $\epsilon = 1$ and *symplectic* if $\epsilon = -1$.

This definition is independent of the splitting chosen and can be internally characterised as follows: Let $D^+ = \{x \in D, \sigma x = x\}$. Then $\sigma$ is orthogonal if $\dim_k D^+ = n(n + 1)/2$ and symplectic if $\dim_k D^+ = n(n - 1)/2$.

Let $(V, h)$ be a (non-degenerate) hermitian form over $(D, \sigma)$ i.e., $V$ is a right $D$-vector space and $h : V \times V \rightarrow D$ a biadditive map satisfying

$$h(va, wb) = \sigma(a)h(v, w)b, \quad h(w, v) = \sigma(h(v, w))$$

for $v, w \in V, a, b \in D$. Further, given $v \in V, v \neq 0$, there exists $w \in V$ such that $h(v, w) \neq 0$. For a choice of a basis of $V$, $h$ is represented by a matrix $T = T(h)$ in $GL_n(D)(n = \dim_D V)$ with $\sigma(T)^t = T$. Let

$$U(h) = \{X \in GL_n(D), XT(\sigma X)^t = T\}$$

$$SU(h) = \{X \in U(h), Nrd(X) = 1\}.$$

If $\sigma$ is of orthogonal type, one has the group Spin $(h)$ which coincides with the Spin group of a quadratic form if $D = k$.

The following is the list of simply connected, absolutely simple, classical groups defined over $k$ (cf. [We]).

1) $SL(n, D)$, $D$ a finite dimensional central division algebra over $k$ (with $n \cdot$ degree $(D) \geq 2$).

2) $SU(h)$, $(V, h)$ a hermitian form over $(D, \sigma)$, $\sigma$ a $K/k$-involution of unitary type (with $n \cdot$ degree $(D) \geq 2$).

3) $U(h)$, $(V, h)$ a hermitian form over $(D, \sigma)$, $\sigma$ of symplectic type.

4) Spin $(h)$, $(V, h)$ a hermitian form over $(D, \sigma)$, $\sigma$ of orthogonal type (with $n \cdot$ degree $(D) \geq 3$).

## 6. Classification of Hermitian-forms

Let $(D, \sigma)$ be a division algebra with an involution of either kind. Let $(V, h)$ be a hermitian form over $(D, \sigma)$. We define the dimension of $h$, denoted by $dim\ h$ to be the dimension of $V$ over $D$. For a central simple algebra $A$ over a field $E$, we define $E^{*+}$ to be the set of elements of $E^*$ which are reduced norms from $A$. Let $T(h) = T$ be the matrix representing $h$ with respect to a basis of $V$ over $D$.

Let $N = nm$ where $m = dim_D V$ and $n^2 = dim_k D$. We define

$$\begin{aligned}
\text{Disc}(h) &= (-1)^{N(N-1)/2} Nrd(T) \in k^*/k^{*+2} \\
\text{disc}(h) &= (-1)^{N(N-1)/2} Nrd(T) \in k^*/k^{*2}
\end{aligned}$$

if $\sigma$ is of first kind and

$$\begin{aligned}
\text{Disc}(h) &= (-1)^{N(N-1)/2} Nrd(T) \in k^*/N_{K/k}(K^{*+}) \\
\text{disc}(h) &= (-1)^{N(N-1)/2} Nrd(T) \in k^*/N_{K/k}(K^*)
\end{aligned}$$

if $\sigma$ is of second kind. If $cd(k) \leq 2$, in view of ([MSu1]), $Nrd : D \to K$ is surjective and Disc $=$ disc. One also has an invariant, called the *Clifford invariant* $c(h)$ of a hermitian form if $\sigma$ is of orthogonal type, which takes values in $Br_2(k)/[D]$, $Br_2(k)$ denoting the 2-torsion subgroup of the Brauer group of $k$. This invariant coincides with the Clifford invariant of quadratic forms if $D = k$. We have the following classification theorems ([BaP1]) for hermitian forms over involutorial division algebras over fields with $cd(k) \leq 2$.

**Theorem.** *Let $k$ be a field with char $k \neq 2$. Let $I^3(k) = 0$. Let $(D, \sigma)$ be a central division algebra over $k$ with an involution of first kind.*

A) *If $\sigma$ is of symplectic type, a hermitian form over $(D, \sigma)$ is determined upto isomorphism by its dimension.*

B) *If σ is of orthogonal type, a hermitian form over $(D, \sigma)$ is determined, upto isomorphism by its dimension, discriminant and Clifford invariant.*

**Theorem.** *Let $K/k$ be a quadratic extension and $D$ a central division algebra over $K$ with a $K/k$ involution $\sigma$ of unitary type. Suppose $cd(k) \leq 2$. Then hermitian forms over $(D, \sigma)$ are classified by dimension and discriminant.*

A key inductive step in the proof of the theorem is provided by an exact sequence of Witt groups of hermitian forms, constructed by Parimala - Sridharan - Suresh in ([BaP1]. Appendix II). Besides the theorems of Merkurjev - Suslin ([MSu1]) we also use the following results in the proof of the classification theorem.

**Theorem.** **(Eva Bayer-Fluckiger, Lenstra [BaL])** *Let $(V, h)$ be a hermitian form over $(D, \sigma)$. If $(V, h) \otimes_k L$ is hyperbolic where $L$ is a finite odd degree extension of $k$, then $(V, h)$ is hyperbolic.*

**Theorem.** **(Yanchevskii [Y])** *Let $(D, \sigma)$ be of unitary type and $cd(k) \leq 2$. Given $\lambda \in k^* \cap Nrd(D^*)$, there exists $x_1, \ldots, x_r \in D^*$ with $\sigma(x_i) = x_i$ and $\lambda = Nrd(x_1 x_2, \ldots, x_r)$.*

## 7. Real Versions of Serre's Conjectures

Colliot-Thélène and Scheiderer ([CT], [Sch2]) formulated the following real analogues of the conjectures of Serre, which we shall call the *Hasse principle* conjectures.

We say that a field $k$ has *virtual cohomological dimension* (abbreviated $vcd(k)$) $\leq n$ if $cd(k(\sqrt{-1})) \leq n$. Number fields and function fields of surfaces over **R** are examples of fields with $vcd \leq 2$.

**Hasse Principle Conjecture I** (HPCI) Let $k$ be a perfect field with $vcd(k) \leq 1$. Let $G$ be a connected linear algebraic group defined over $k$. Then the map

$$H^1(k, G) \rightarrow \prod_{v \in \Omega} H^1(k_v, G)$$

is injective, $\Omega$ denoting the space of orderings of $k$.

**Hasse Principle Conjecture II** (HPCII) Let $k$ be a perfect field with $vcd(k) \leq 2$. Let $G$ be a semi-simple simply connected linear algebraic group defined over $k$. Then the map

$$H^1(k, G) \rightarrow \prod_{v \in \Omega} H^1(k_v, G)$$

is injective.

We refer to ([CT], [Sch2]) for the history of HPCI.

Let $vcd(k) \leq 2$ and $D$ a central division algebra over $k$. HPCII states that an element $\lambda \in k^*$ is a reduced norm form $D$ if and only if it is positive at all orderings $v$ of $k$ where $D$ is ramified, i.e., $D \otimes_k k_v$ is not split. This, in the case of number fields, is due to Hasse - Maass - Schilling ([R], p. 289).

We have the following ([BaP2]).

**Theorem.** *Let $k$ be a perfect field with $vcd(k) \leq 2$. Let $G$ be a semi-simple, simply connected linear algebraic group defined over $k$ of classical type or of type $G_2$ or $F_4$. Then the map*

$$H^1(k, G) \to \prod_{v \in \Omega} H^1(k_v, G)$$

*is injective.*

Once again, the main steps in proving HPC II for classical groups is to prove the theorem for $SL(1, D)$ and prove a classification theorem for hermitian forms in terms of invariants.

Let $D$ be a division algebra with an involution $\sigma$ of either kind. Let $(V, h)$ be a hermitian form over $(D, \sigma)$. One can define a signature $sgn(h) : \Omega \to \mathbf{Z}$ which coincides with the following, in special cases.

1) If $D = k, \sigma = $ identity, $h$ a quadratic form over $k$, $sgn(h)$ is the total signature map described in § 2.

2) If $D = K, [K : k] = 2$ and $h$ a $K/k$ hermitian form. Then $Tr(h) : V \to k$, defined by $Tr(h)(v) = h(v, v)$ is a quadratic form over $k$ and $sgn(h)$ coincides with the signature of the quadratic form $Tr(h)$.

3) Let $D$ be a quaternion division algebra over $k$ and $\sigma$ the standard symplectic involution on $D$. Then, $Tr(h) : V \to k$ defined by $Tr(h)(v) = h(v, v)$, is a quadratic form over $k$ and $sgn(h)$ is the signature of the quadratic form $Tr(h)$.

If $(D, \sigma)$ is of orthogonal type, one has an invariant $R(h)$ called the Rost invariant associated to $h$ which takes values in $H^3(k, \mu_4^{\otimes 2})/H^1(k, \mathbf{Z}/2) \cup [D]$ which coincides with the Arason invariant $e_3(h) \in H^3(k, \mathbf{Z}/2)$ if $D = k([A])$.

We have the following classification theorems.

**Theorem.** *Let $k$ be a field with char $k \neq 2$ and $I^3(k)$ torsion free. Let $(D, \sigma)$ be a division algebra with an involution of first kind.*

A) *If $\sigma$ is symplectic, hermitian forms over $(D, \sigma)$ are determined, upto isomorphism, by dimension and signature.*

B) *If $\sigma$ is of orthogonal type, hermitian forms over $(D, \sigma)$ are determined upto isomorphism, by dimension, Discriminant, Clifford and Rost invariants and signature.*

**Theorem.** *Let $k$ be a field with char $k \neq 2$ and $vcd(k) \leq 2$. Let $(D, \sigma)$ be of unitary type over $K/k$. Then, hermitian forms over $(D, \sigma)$ are determined upto isomorphism by dimension, Discriminant and signatures.*

When one looks at the special case of the above theorem for number fields, one wonders why the Rost invariant never surfaces in the classification of hermitian forms in the orthogonal case. Further, in all the classification over number fields only discriminant is needed instead of Discriminant. This can be explained by the fact that the number fields are examples of the so-called SAP fields, i.e., fields with strong approximation property: given an open and closed subset $X$ of the space $\Omega$ of orderings of $k$, there exists $\alpha \in k^*$ such that $X = \{x \in \Omega, \alpha > 0 \text{ at } x\}$. If $k$ satisfies the SAP property, the map

$$H^3(k, \mu_4^{\otimes 2})/H^1(k, \mathbf{Z}/2) \cup [D] \rightarrow \prod_{v \in \Omega} H^3(k_v, \mu_4^{\otimes 2})/H^1(k_v, \mathbf{Z}/2) \cup [D_v]$$

is injective (cf. [BaP2], proof of 7.5) and the Rost invariant can be dispensed with for the classification problem.

## References

[A]　　　J. Arason: Cohomologische Invarianten quadratischer Formen, *J. Algebra* **36**, 448–491 (1975).

[B]　　　H.-J. Bartels: Invarianten hermitescher Formen über Schiefkörpern, *Math. Ann.* **215**, 269–288 (1975).

[BaL]　　E. Bayer-Fluckiger, H.W. Lenstra, Jr.: Forms in odd degree extensions and self-dual normal bases, *Amer. J. Math.* **112**, 359–373 (1990).

[BaP1]　E. Bayer-Fluckiger, R. Parimala: Galois Cohomology of the Classical Groups over Fields of Cohomological Dimension $\leq$ 2, *Invent. Math.* **122**, 195–229 (1995).

[BaP2]　E. Bayer-Fluckiger, R. Parimala: Classical Groups and the Hasse Principle *Annals of Mathematics* **147**, 651–693 (1998).

[Ch]　　V.I. Chernousov: On the Hasse principle for groups of type $E_8$, *Dokl. Akad. Nauk SSSR* **306**, 25, 1059–1063 (1989).

[CT]　　J.-L. Colliot-Thélène: Groupes linéaires sur les corps de fonctions de courbes réelles, *J. Reine Angew Math.* **474**, 139–167 (1996).

[D]　　　A. Ducros: Principe de Hasse pour les espaces principaux homogènes sous les groupes classiques sur un corps de dimension cohomologique virtuelle au plus 1, *Manuscripta Math.* **89**, 335–354 (1996).

[EL]　　R. Elman, T.Y. Lam: Classification Theorems for Quadratic Forms over Fields, *Comment. Math. Helv.* **49**, 373–381 (1974).

[Ha1]　　G. Harder: Über die Galoiskohomologie halbeinfacher Matrizengruppen, I, *Math. Z.* **90**, 404–428 (1965).

[Ha2]　　G. Harder: Über die Galoiskohomologie halbeinfacher Matrizengruppen, II, *Math. Z.* **92**, 396–415 (1966).

[J]　　　N. Jacobson: Structure and representations of Jordan algebras, A.M.S. Colloquium Publ. XXXIX, Providence (1968).

[K]      M. Kneser: *On Galois Cohomology of Classical Groups*, Tata Lecture Notes **47** (1969).

[M1]     A. Merkurjev: On the norm residue symbol of degree 2, *Dokladi Akad. Nauk. SSSR* **261** (1981), 542–547, English translation: *Soviet Math. Dokladi* **24**, 546–551 (1981).

[M2]     A. Merkurjev: Norm principle for algebraic groups, *St. Petersburg J. Math.* **7**, 243–264 (1996).

[MSu1]   A. Merkurjev, A. Suslin: On the norm residue homomorphism of degree three, *Izvestiya Akad. Nauk. SSSR* **54** (1990), English translation: *Math. USSR Izvestiya* **36**, 349–367 (1991).

[MSu2]   A. Merkurjev, A. Suslin: K-cohomology of Severi-Brauer varieties and the norm-residue homomorphism, *Izvestiya Akad. Nauk. SSSR* **46** (1982), English translation: *Math. USSR Izvestiya* **21**, 307–340 (1983).

[R]      I. Reiner: Maximal Orders: Academic press, London-New York-San Francisco (1975).

[Sc]     W. Scharlau: Quadratic and Hermitian Forms, Grundlehren Math. Wiss. **270**, Springer-Verlag, Berlin (1985).

[Sch1]   C. Scheiderer: Classification of hermitian forms and semisimple groups over fields of virtual cohomological dimension one, *Manuscripta Math.* **89**, 373–394 (1996).

[Sch2]   C. Scheiderer: Hasse principles and approximation theorems for homogeneous spaces over fields of virtual cohomological dimension one, *Invent. Math.* **125**, 307–365 (1996).

[Se1]    J.-P. Serre: Cohomologie Galoisienne. Lecture Notes in Mathematics **5**, Springer-Verlag (1964 and 1994).

[Se2]    J.-P. Serre: Cohomologie Galoisienne: progrès et problèmes, Séminaire Bourbaki, exposé $n°783$, 1993/94.

[Se3]    J.-P. Serre: Cohomologie Galoisienne des groupes algébriques linéaires, *Colloque sur la théorie des groupes algébriques*, Bruxelles (1962), 53–68 (= Coll. papers, *Springer Verlag* Vol. II, 53).

[St]     R. Steinberg: Regular elements of semi-simple algebraic groups, *Publ. Math. IHES* **25**, 49–80 (1965).

[Su]     A. Suslin: Algebraic K-theory and norm residue homomorphism, *Journal of Soviet mathematics*, **30**, 2556–2611 (1985).

[We]     A. Weil: Algebras with involutions and the classical groups, *J. Ind. Math. Soc.* **24**, 589–623 (1961). (Coll. Papers II, [1960 b], 413–447.)

[Y1]     V. Yanchevskii: Simple algebras with involution and unitary groups, *Math. Sbornik* **93** (1974), English translation: *Math. USSR Sbornik*, **22**, 372–384 (1974).

[Y2]     V. Yanchevskii: The commutator subgroups of simple algebras with surjective reduced norms, *Dokl. Akad. Nauk SSSR* **221** (1975), English translation: *Soviet Math. Dokl.* **16**, 492–495 (1975).

School of Mathematics, Tata Institute of Fundamental Research, Homi Bhabha Road, Bombay 400 005, India. E-mail: parimala@tifrvax.tifr.res.in

# Central Units in Integral Group Rings

*M.M. Parmenter*[1]

## Introduction

This paper is intended to give a survey of recent work on central units in integral group rings. For units in general, the definitive reference is the book by Sehgal (1993) while a survey paper by Jespers contains additional very recent results. Both of these sources contain results on central units (in fact, Jespers devotes a chapter to the topic), but our work complements theirs in two ways. Firstly, we describe some results contained in papers which were not available to the other authors. Secondly, we choose to emphasize some topics which are mentioned either very briefly or not at all in their work. Nevertheless, we acknowledge that there is considerable overlap, especially between our survey and that of Jespers, and would like to thank him for supplying us with a preprint.

Notation will follow that in Sehgal (1993)-specifically, $\mathcal{U}(\mathbb{Z}G)$ denotes the group of units of an integral group ring $\mathbb{Z}G$. Many results have been proved about $\mathcal{U}(\mathbb{Z}G)$ when $G$ is Abelian, and these could all be thought of as results about central units. We do not intend to discuss that body of work here, and refer the reader to Karpilovsky (1983) for further information. One major result in the area, however, is a theorem of Bass and Milnor which states that if $G$ is finite Abelian, then the Bass cyclic units (to be defined later) generate a subgroup of finite index in $\mathcal{U}(\mathbb{Z}G)$. The following generalization, also due to Bass (1966), will be needed later.

**Theorem 1.** *Let $G$ be a finite group and let $j : \mathcal{U}(\mathbb{Z}G) \to K_1(\mathbb{Z}G)$ denote the natural map. The images of the Bass cyclic units of $\mathbb{Z}G$ generate a subgroup of finite index in $K_1(\mathbb{Z}G)$.*

We will also need the following important observation.

**Theorem 2. (Bass, 1976 and Sehgal, 1978)** *If $G$ is a torsion group and $\alpha$ is a central unit of finite order in $\mathcal{U}(\mathbb{Z}G)$, then $\alpha$ is trivial (i.e., $\alpha \in \pm G$).*

The reader will doubtless note that our paper is extremely short as survey papers go! This is simply a reflection of the fact that very little has been done on central

[1] This research was supported in part by NSERC grant A-8775.

units in integral group rings, and one of the purposes of writing this survey is to suggest that more study is warranted. To this end, we will mention a number of open problems as we proceed.

## 1. Existence

When $G$ is finite, the fundamental problem of determining precisely when non-trivial central units exist in $\mathcal{U}(\mathbb{Z}G)$ was settled by Ritter and Sehgal (1990).

**Theorem 3. (Ritter and Sehgal, 1990 and Sehgal, 1993)** *Let $G$ be a finite group. All central units of $\mathbb{Z}G$ are trivial if and only if for every $g \in G$ and every natural number $j$ relatively prime to $|G|$, the element $g^j$ is conjugate to $g$ or $g^{-1}$*

One direction of Theorem 3 can be proved using Theorem 1 and an additional result from Jespers et al., (1996). To see this, assume that the conditions hold - we will show that all central units are trivial.

Recall that a typical Bass cyclic unit is of the form

$$b = (1 + g + g^2 + \cdots + g^{i-1})^m + \frac{1 - i^m}{|g|}\hat{g},$$

where $(i, |g|) = 1$ and $m = \phi(|g|)$. The given conditions then imply that (using the notation of Theorem 1) $j(b)$ is equal to the image of a group element in the Abelian group $K_1(\mathbb{Z}G)$. It follows from Theorem 1 that if $c$ is any unit in $\mathbb{Z}G$, then $c^r \in Ker\ j$ for some integer $r$. However, it was shown by Jespers et al., (1996) that all central units in $Ker\ j$ are trivial. It follows that if $c$ is any central unit in $\mathbb{Z}G$, then $c$ must be of finite order and hence trivial by Theorem 2.

The argument just presented is essentially the same as one given by Li (1997).

It is not known whether Theorem 3 can be extended to groups which are not finite. The following is Research Problem 26 in Sehgal (1993).

**Open Problem 1**
Characterize groups $G$ for which all central units of $\mathbb{Z}G$ are trivial.

## 2. Complete Descriptions

One type of complete description of the centre of $\mathcal{U}(\mathbb{Z}G)$ would be to obtain its precise group theoretic structure. When $G$ is finite, Theorem 2 and the following show that this problem reduces to the Wedderburn decomposition of $\mathbb{Q}G$.

**Theorem 4. (Aleev, 1994)** *If $G$ is finite, the group of central units of $\mathcal{U}(\mathbb{Z}G)$ and the unit group of the integer ring of the centre of $\mathbb{Q}G$ have the same torsion free rank.*

Another possibility would be to try to find a complete set of generators for the centre of $\mathcal{U}(\mathbb{Z}G)$. Since this is an intractable problem when $G$ is Abelian (for examples where this has been done, see Aleev, 1994 and Karpilovsky, 1983), we wouldn't expect to be able to do this except in some very special cases. The following are the only results we know of where a complete set of generators for a nontrivial centre has been obtained.

**Theorem 5. (Aleev (1994) and Li and Parmenter (1997))** *The centre of $\mathcal{U}(\mathbb{Z}A_5)$ is $\pm\langle v \rangle$ where $v = 49 + 26C_1 + 10C_2 - 16C_4$. Here $C_1$ is the sum of elements from $A_5$ conjugate to (12345), $C_2$ is the sum of elements conjugate to (13524) and $C_4$ is the sum of elements conjugate to (12)(34).*

**Theorem 6. (Aleev, 1994)** *The centre of $\mathcal{U}(\mathbb{Z}A_6)$ is $\pm\langle v \rangle$ where $v = 18433\,C_0 - 2304(C_2 + C_3) + 3728C_5 - 1424C_6$. Again, the $C_i$ are suitable conjugacy class sums in $\mathbb{Z}A_6$.*

We would like to remark that Li and Parmenter were unaware of Aleev (1994) when they carried out their work.

It would still be of interest to compute more examples like those in the previous two theorems.

**Open Problem 2**

Find a finite set of generators for the centre of $\mathcal{U}(\mathbb{Z}G)$ in some other cases.

## 3. Finite Index

Much more reasonable than asking for a complete set of generators would be to try to construct a generating set for a subgroup of finite index in the centre of $\mathcal{U}(\mathbb{Z}G)$. In the following theorem, Ritter and Sehgal found such a finite generating set whenever $G$ is finite.

**Theorem 7. (Ritter and Sehgal (1993) and Sehgal (1993))** *Let $G$ be a finite group of exponent $s$, $F = Q(\xi)$ where $\xi$ is a primitive $s$'th root of unity, $\mathcal{G}$ the Galois group of $F$ over $Q$, $O$ the ring of algebraic integers of $F$, $r$ the number of invertible elements in the ring $O/|G|O$ and $E$ a finite subset of units of $O$ generating a subgroup of finite index in $O - \{0\}$.*

*If $X(G)$ denotes the set of all irreducible characters of $G$, then the elements*

$$\prod_{\sigma \in \mathcal{G}} (1 + (\epsilon^\sigma - 1)\epsilon_{\chi^\sigma})^r$$

*with $\chi \in X(G)$ and $\epsilon \in E$, lie in the centre of $\mathcal{U}(\mathbb{Z}G)$ and generate a subgroup of finite index in the centre of $\mathcal{U}(\mathbb{Z}G)$.*

These generators are not explicitly exhibited inside $\mathbb{Z}G$, nor are they particularly easy to compute in practice. For finite nilpotent groups, a simpler construction was described in Jespers *et al.*, (1996).

Let $Z_i$ denote the $i$'th centre of a finite group $G$. If $b$ is a Bass cyclic unit in $\mathbb{Z}G$, define $b_{(1)} = b$, and, for $k \geq 1$, $b_{(k+1)} = \prod_{x \in Z_{k+1}} b_{(k)}^x$.

**Theorem 8. (Jespers *et al.*, (1996))** *Let $G$ be a finite nilpotent group of class $n$. Then $\langle b_{(n)} | b$ Bass cyclic$\rangle$ is of finite index in the centre of $\mathcal{U}(\mathbb{Z}G)$.*

By continuing this conjugation procedure further, such a generating set can be constructed whenever $G$ is finitely generated nilpotent (Jespers *et al.*, 1996). Moreover, the proof of Theorem 8 can be applied to some other classes of groups (e.g. dihedral or quaternion groups).

Using an entirely different method, Giambruno and Jespers have found a finite set of generators for a subgroup of finite index in the centre of $\mathcal{U}(\mathbb{Z}A_n)$ for all $n$. In general, however, the following is still of interest.

**Open Problem 3**

Find a "simple" set of generators for a subgroup of finite index in the centre of $\mathcal{U}(\mathbb{Z}G)$ for some classes of groups where the method of Theorem 8 doesn't apply.

Even more basic is the following difficulty that we perceive with central units. In the case of units in general, several important families (Bass cyclic, bicyclic etc.,) have been described and a number of recent papers have demonstrated that these families can be used to prove many significant results about $\mathcal{U}(\mathbb{Z}G)$. We don't seem to be able to find simple examples of central units so easily.

**Open Problem 4**

Find some new interesting families of central units in integral group rings.

In passing from finite nilpotent to finitely generated nilpotent groups, Jespers *et al.*, (1996) required a particular case of the following result, stated here as it appears in Jespers and Juriaans. Further extensions of this result would be of interest.

**Theorem 9.** *Let $G$ be a group such that $T$, the set of torsion elements of $G$, forms a finite subgroup. Suppose that $G/T$ is ordered. Then any central unit $u$ in $\mathbb{Z}G$ can be decomposed as $u = vg$ where $v \in \mathbb{Z}T$ and $g \in G$.*

## 4. Generalizations

Let $(1) = Z_0(\mathcal{U}) \leq Z_1(\mathcal{U}) \leq \cdots \leq Z_n(\mathcal{U}) \leq \cdots$ be the upper central series of the unit group $\mathcal{U} = \mathcal{U}(\mathbb{Z}G)$. The following theorem was proved for finite groups by Arora, Hales and Passi (1993) and extended to torsion groups by Li.

**Theorem 10.** *Let $G$ be a torsion group. Then $Z_3(\mathcal{U}) = Z_2(\mathcal{U})$.*

This leaves open the question of how $Z_1(\mathcal{U})$ and $Z_2(\mathcal{U})$ are related, but that was settled for finite groups by Arora and Passi (1993).

**Theorem 11.** *Let G be a finite group. Then* $Z_2(\mathcal{U}) = T \cdot Z_1(\mathcal{U})$ *where T is the torsion subgroup of* $Z_2(\mathcal{U})$.

Moreover $T$ was completely determined by Arora, Hales and Passi (1993), and this, together with Theorem 11, led Arora and Passi to the interesting conclusion that finite $Q^*$-groups are precisely those finite groups $G$ with the property that $\mathcal{U}(\mathbb{Z}G)$ is of central height 2.

Unlike Theorem 10, it is not known whether Theorem 11 can be extended to torsion groups.

**Open Problem 5**

Is $Z_2(\mathcal{U}) = T \cdot Z_1(\mathcal{U})$ still true when $G$ is a torsion group?

Another generalization of the idea of central unit is the notion of the $n$-centre of a group. Baer (1952) defined the $n$-centre $Z(G, n)$ as the set of elements $x \in G$ such that $(xy)^n = x^n y^n$ and $(yx)^n = y^n x^n$ for all $y \in G$. Later, Kappe and Newell (1991) proved that only one of these conditions is needed in the definition. They also described many fundamental properties of the $n$-centre, the most basic of which is that it is always a characteristic subgroup. The reader should consult Kappe and Newell (1991) for further details.

The following result, due to Li, completely characteristizes the $n$-centre of $\mathcal{U}(\mathbb{Z}G)$ whenever $G$ is a torsion group.

**Theorem 12.** *Let G be a torsion group. Then*

$$
Z(\mathcal{U}(\mathbb{Z}G), n) = \begin{cases} \mathcal{U}(\mathbb{Z}G) & \text{if} \quad n = 0, 1 \\ Z_2(\mathcal{U}(\mathbb{Z}G)) & \text{if} \quad n > 1 \text{ and } n \equiv 0, 1 \pmod 4 \\ Z_1(\mathcal{U}(\mathbb{Z}G)) & \text{if} \quad n \equiv 2, 3 \pmod 4 \end{cases}
$$

### References

Aleev, R.Z., Higman's central unit theory, units of integral group rings of finite cyclic groups and Fibonacci numbers, *Int. J. Alg. Comput.* **4**, 309–358, 1994.

Arora, Satya, R., Hales, A.W. and Passi, I.B.S., Jordan decomposition and hypercentral units in integral group rings, *Comm. Alg.* **21**, 25–35, 1993.

Arora, Satya, R. and Passi, I.B.S., Central height of the unit group of an integral group ring, *Comm. Alg.* **21**, 3673–3683, 1993.

Baer, R., Endlichkeitskriterien für kommutatorgruppen, *Math. Ann.* **124**, 161–177, 1952.

Bass, H., The Dirichlet unit theorem, induced characters and Whitehead groups of finite groups, *Topology* **4**, 391–410, 1966.

Bass, H., Euler characteristics and characters of discrete groups, *Invent. Math.* **35**, 155–196, 1976.

Giambruno, A. and Jespers, E., Central idempotents and units in rational group algebras of alternating groups, *Int. J. Alg. Computation* (*to appear*).

Jespers, E., Units in integral group rings - a survey, *Proc. of Int. Conf. Methods in Ring Theory*, Marcel Dekker (*to appear*).

Jespers, E. and Juriaans, S.O., Isomorphisms of integral group rings of infinite groups, *Preprint*.

Jespers, E., Parmenter, M.M. and Sehgal, S.K., Central units of integral group rings of nilpotent groups, *Proc. Amer. Math. Soc.* **124**, 1007–1012, 1996.

Kappe, L.C. and Newell, M.L., On the *n*-centre of a group, *Proc. Groups St. Andrews 1989*, Cambridge Univ. Press, 339–352, 1991.

Karpilovsky, G., *Commutative Group Algebras*, Marcel Dekker, New York, 1983.

Li Yuanlin, Units in Integral Group Rings, *Ph.D. Thesis*, Memorial University of Newfoundland, 1997.

Li Yuanlin, The hypercentre and the *n*-centre of the unit group of an integral group ring, *Canad. J. Math. (to appear)*.

Li Yuanlin and Parmenter, M.M., Central units of the integral group ring $\mathbb{Z}A_5$, *Proc. Amer. Math. Soc.* **125**, 61–65, 1997.

Ritter, J. and Sehgal, S.K., Integral group rings with trivial central units, *Proc. Amer. Math. Soc.* **108**, 327–329, 1990.

Ritter, J. and Sehgal, S.K., Units of group rings of solvable and Frobenius groups over large rings of cyclotomic integers, *J. Alg.* **158**, 116–129, 1993.

Sehgal, S.K., *Topics in Group Rings*, Marcel Dekker, New York, 1978.

Sehgal, S.K., *Units in Integral Group Rings*, Longman Scientific and Technical, Essex, 1993.

Department of Mathematics and Statistics, Memorial University of Newfoundland,
St. John's, Newfoundland, Canada A1C 5S7

# Alternative Loop Rings and Related Topics

*César Polcino Milies**

## 1. Introduction

Let $R$ be a commutative (and associative) ring with unity and let $L$ be a loop (roughly speaking, a loop is a group which is not necessarily associative, see Definition 3.1). The *loop algebra* of $L$ over $R$ was introduced in 1944 by R.H. Bruck (1944) as a means to obtain a family of examples of nonassociative algebras and is defined in a way similar to that of a group algebra; i.e., as the free $R$-module with basis $L$, with a multiplication induced distributively from the operation in $L$.

In 1983, E.G. Goodaire (1983) defined *RA loops* as those loops whose loop algebra over a ring with no 2-torsion is alternative but not associative (it follows, as a consequence of his characterization, that if the loop algebra over a ring with no 2-torsion is alternative, then the loop algebra of the given loop, over all such rings, is also alternative) and O. Chein and E.G. Goodaire (1986) gave a full description of those loops, in 1990. In a subsequent paper they also defined *RA2 loops* as those loops whose loop algebra over a ring of characteristic 2 is alternative. In particular, they showed that RA loops are also RA2 loops.

In the last fifteen years, alternative loop rings have been extensively studied. Since these rings are "close" to associative rings, it is no surprise that many properties of group rings also hold for alternative loop rings and actually some of the well-known conjectures for group rings have been verified for loop rings of RA loops.

Most of the results obtained since 1983 were surveyed by Goodaire and Polcino Milies (1995) and later organized into a book [Goodaire *et al.*, (1996)]. In this paper, we shall give a general outline of some aspects of the theory and then concentrate on results obtained after the publication of Goodaire *et al.*, (1996). For the main results two references will be given: the original article where they appear for the first time and a reference to Goodaire *et al.*, (1996).

*The author was partially supported by a research grant from CNP$_q$., Proc. 300243/79-0(RN)

## 2. Loop Rings: Historical Background

In 1843, Sir William Rowan Hamilton discovered the set of *quaternions* which is the first example of a non commutative division ring. They can be defined as the set of all elements of the form $a + bi + cj + dk$ where $a, b, c, d$ are real numbers and $i, j, k$ are regarded simply as symbols. The addition of two such elements is defined "coordinatewise":

$$(a+bi+cj+dk)+(a'+b'i+c'j+d'k) = (a+a')+(b+b')i+(c+c')j+(d+d')k.$$

To define multiplication, using the distributive laws and assuming that real coefficients commute with the basis elements, it is enough to show how to multiply these basis elements among themselves. It is done as shown in the following table:

|       | 1 | $i$  | $j$  | $k$  |
|-------|---|------|------|------|
| 1     | 1 | $i$  | $j$  | $k$  |
| $i$   | $i$ | $-1$ | $k$ | $-j$ |
| $j$   | $j$ | $-k$ | $-1$ | $i$ |
| $k$   | $k$ | $j$ | $-i$ | $-1$ |

The day after his discovery, Hamilton wrote a letter to his friend J.T. Graves communicating his results. A few months latter, Graves discovered an algebra with a basis consisting of eight elements, the *octonions*. Hamilton himself observed that the associative law holds for quaternions, but is not valid for octonions. Graves did not publish his results, so this algebra was again discovered independently, in 1845, by Arthur Cayley; for this reason they are also known as *Cayley numbers*. The road for new generalizations was opened. In his *Lectures on Quaternions*, published in 1853, Hamilton introduced *Biquaternions*, defined in a way similar to the real quaternions, but assuming complex coefficients. He showed that this algebra contains zero divisors and, thus, cannot be a division ring. In this same work, he develops an idea initiated in one of his papers in 1848: the *hypercomplex numbers*.

Once again, a system of hypercomplex numbers is defined imitating the definition of quaternions: it is the set of all symbols of the form

$$\alpha = x_1 e_1 + x_2 e_2 + \cdots + x_n e_n,$$

where $x_1, x_2, \ldots, x_n$ are real (or complex) numbers and $e_1, e_2, \ldots, e_n$ are symbols called the *units* of the system. As in the case of quaternions, the sum of two such elements is defined coordinatewise. Once again, to define the product of two hypercomplex numbers, assuming the distributive laws and that coefficients commute with the units, it suffices to decide how to multiply units among themselves.

Since the product of two units must be again an element of the system, it must be of the form:

$$e_i e_j = \sum_{k=1}^{n} \gamma_k(i, j) e_k.$$

Hence, the structure of this algebra is determined by the choice of the coefficients $\gamma_k(i, j)$; this is why they are known as the *structural constants* of the hypercomplex system.

In the meantime, the theory of permutation groups was being developed in the continent. A series of papers on the subject, published in the period 1844–1846 by A. Cauchy called the attention of Cayley who saw the general notion behind the particular cases and gave the first definition of an abstract group in a paper in the *Philosophical Magazine* in 1854. In this very same paper he also formulated for the first time the idea of a *group ring*. He had used multiplicative notation for the operation of the abstract group. Following the trend of the times, he noted that the elements of the group could be taken as basic units of a hypercomplex system; i.e., given a finite group $G = \{g_1, g_2, \ldots, g_n\}$, he considered the set of all the elements of the form

$$x_1 g_1 + x_2 g_2 + \cdots + x_n g_n,$$

where he took $x_1, x_2, \ldots, x_n$ to be either real or complex numbers. Then the product of two such elements $\alpha = \sum_{i=1}^n x_i g_i$, $\beta = \sum_{i=1}^n y_i g_i$ is given by:

$$\alpha\beta = \sum_{i,j}(x_i y_j)(g_i g_j).$$

This paper had no immediate influence on the mathematicians of the time and group rings remained unknown for some time.

Group rings were studied again by Theodor Molien when he realized that this was a natural setting to apply some of his earlier work. He had developed a criteria for semisimplicity of hypercomplex systems that depended on the structural constants. Since group rings are a special case of this construction where all the structural constants are equal to either 0 or 1, it was the easiest system to work with.

The connections between group representation theory and the structure theory of algebras—obtained through group rings—were widely recognized after a most influential paper by Emmy Noether (1933), some joint work of hers with Richard Brauer (1927), and a paper by Brauer (1929). From then on, group rings were intensively studied and became of special interest also to ring theorists.

As mentioned in the introduction, loop rings were first defined by R.H. Bruck in 1944 as a means to construct exemples of non-associative algebras. In that very same paper he proved a non associative analog to the well-known Maschke's theorem. In 1955, L. Paige proved that in characteristic different from 2, in a commutative loop algebra, the very weak identity, $x^2 x^2 = x^3 x$, implies full associativity (Paige, 1955). In other words, there are no "interesting" nonassociative commutative loop algebras which are not already group algebras but, in 1983, E.G. Goodaire showed that there do exist alternative loop algebras which are not group algebras (Goodaire, 1983).

Alternative rings arose in the work of Ruth Moufang (1933). Given a projective plane, one can label the points and the lines with elements from a set $R$ and then define the addition and multiplication of elements of $R$ in terms of incidence relations in the plane (see Hall, 1959, Chapter 20, for an introduction to projective planes and their coordinatization). One can then relate various geometrical properties of the plane to algebraic properties of the "ring", $(R, +, \cdot)$. In particular, it can be shown that a plane is desarguesian if and only if it can be coordinated by a planar alternative division ring. Much of Moufang's attention was directed at the multiplicative structure of an alternative division ring. Just as the non-zero elements of a field form a group under multiplication, so the non-zero elements of an alternative division ring form a *Moufang loop* under multiplication.

## 3. Definitions and Basic Facts

We begin with some of the main definitions.

**Definition 3.1.** A loop is a set $L$ together with a (closed) binary operation $(a, b) \mapsto ab$ for which there is a two-sided identity element 1 and such that the right and left translation maps

$$R_x : a \mapsto ax \quad \text{and} \quad L_x : a \mapsto xa$$

are bijections for all $x \in L$. This requirement implies that, for any $a, b \in L$, the equations $ax = b$ and $ya = b$ have unique solutions.

In a loop $L$ we can define the commutator of two elements in a way which is similar to their definition in groups. We also define the associator of three elements.

**Definition 3.2.** Given elements $a, b, c$ in a loop $L$, the commutator $(a, b)$ and associator $(a, b, c)$ are the elements (uniquely) defined by the following equations:

$$ab = ba(a, b) \quad \text{(loop) commutator}$$
$$(ab)c = [a(bc)](a, b, c) \quad \text{(loop) associator}$$

The *commutator subloop* is the subloop generated by the set of all commutators and the *associator subloop* is the subloop generated by all associators. The *nucleus* and *centre* of $L$ are the subloops $\mathcal{N}(R)$ and $\mathcal{Z}(R)$, respectively, defined by

$$\mathcal{N}(L) = \{x \in L | (a, b, x) = (a, x, b) = (x, a, b) = 1, \text{ for all } a, b \in L\},$$
$$\mathcal{Z}(L) = \{x \in \mathcal{N}(L) | (a, x) = 1 \text{ for all } a \in L\}.$$

**Definition 3.3.** A loop $L$ is called a Moufang Loop if it satisfies any of the following three identities (which are equivalent).

$$((xy)x)z = x(y(xz)) \quad \text{left Moufang identity}$$
$$((xy)z)y = x(y(zy)) \quad \text{right Moufang identity}$$
$$(xy)(zx) = (x(yz))x \quad \text{middle Moufang identity}$$

Now we turn to rings.

**Definition 3.4.** A (not necessarily associative) ring is a set $R$ with two operations, denoted $+$ and $\cdot$, such that $(R, +)$ is an abelian group, $(a, b) \mapsto a \cdot b$ is a binary operation on $R$, and both distributive laws hold: $a(b + c) = ab + ac$, $(a + b)c = ac + bc$, for all $a, b, c \in R$. If, in addition, $(R, +)$ is a module over a commutative, associative ring $\Phi$ such that $\alpha(ab) = (\alpha a)b = a(\alpha b)$ for all $\alpha \in \Phi$ and all $a, b \in R$, then $(R, +, \cdot)$ is said to be a (nonassociative) algebra.

A ring $R$ is alternative if

$$x(xy) = (xx)y \text{ and } (xy)y = x(yy) \text{ for all } x, y \in R.$$

It can be shown that the Moufang identities hold in an alternative ring $R$. Consequently, it follows that the set $\mathcal{U}(R)$ of all invertible elements is a Moufang loop.

One of the most useful properties of alternative rings is the fact that if three elements associate, then the subring which they generate is associative. Thus alternative rings are *diassociative* in the sense that the subring generated by any two elements is always associative. Similarly, Moufang loops are *diassociative*: the subloop generated by any pair of elements is always associative, and thus, a group.

For the proofs of the above cited and other basic facts about loops and alternative rings we refer the reader to [Goodaire *et al.*, 1996, Chapters I and II].

**Definition 3.5.** Let $L$ be a loop and let $R$ be a commutative associative ring with 1. The loop ring of $L$ over $R$ is the free $R$-module $RL$ with basis $L$ and multiplication given by extending, via the distributive laws, the multiplication in $L$. In other words, the elements of $RL$ are formal sums, $\sum_{g \in L} \alpha_g g$, where the $\alpha_g \in R$ are almost all 0 and unique in the sense that

$$\sum \alpha_g g = \sum \beta_g g \quad \text{implies} \quad \alpha_g = \beta_g \text{ for all } g \in L.$$

Addition and multiplication are given by

$$\sum \alpha_g g + \sum \beta_g g = \sum (\alpha_g + \beta_g) g$$

$$\left( \sum \alpha_g g \right) \left( \sum \beta_g g \right) = \sum \left( \sum_{hk=g} \alpha_h \beta_k \right) g.$$

By an *alternative loop ring*, we mean a loop ring which happens also to be alternative. As a subloop of the loop of units of an alternative loop ring $RL$, the loop $L$ which defines $RL$ must of course be Moufang, as noted earlier. That such (nonassociative) loops actually exist was first shown by E.G. Goodaire (1983). We quote the following.

**Theorem 3.6.** *Let L be a loop. Then L is a loop with an alternative loop ring if and only if it has the following properties:*

  (i) *if three elements associate in some order then they associate in all orders; and*

  (ii) *if $g, h, k \in L$ do not associate, then $gh.k = g.kh = h.gk$.*

It follows from this characterization that if the loop ring of a loop $L$ over a ring $R$ whose characteristic is not 2 is alternative then the loop ring $RL$ is alternative for *any* ring $R$ with *char* $(R) \neq 2$.

**Definition 3.7.** An *RA* (ring alternative) loop is a loop whose loop ring $RL$ over some ring $R$ of characteristic not 2 is alternative, but not associative.

**Theorem 3.8.** *Let L be an RA Loop. Then:-*

  (i) *$g^2 \in \mathcal{N}(L), \forall g \in L$.*

  (ii) *$\mathcal{N}(L) = \mathcal{Z}(L)$.*

  (iii) *For any pair of elements $g, h \in L$ we have that $(g, h) = 1$ if and only if $(g, h, L) = 1$.*

  (iv) *Given $g, h, k \in L$, if $(g, h, k) \neq 1$ then $(g, h, k) = (g, h) = (g, k) = (h, k)$ is a central element of order 2.*

  (v) *The commutator and associator subgroups coincide and are a central subgroup of order 2.*

Using the theorem above, it is not hard to show

**Corollary 3.9.** *The direct product $L \times K$ of loops is an RA loop if and only if precisely one of L and K is an RA loop while the other is an abelian group.*

The following theorem, which is due to O. Chein and E.G. Goodaire (1986, Section 3) (see also [Goodaire, *et al.*, 1996, Theorem IV.3.1]) forms the basis for the theory of RA loops and gives a construction for RA loops.

**Theorem 3.10.** *A loop L is RA if and only if it is not commutative and, for any two elements a and b of L which do not commute, the subloop of L generated by its centre together with a and b is a group G such that*

  (i) *for any $u \notin G$, $L = G \cup Gu$ is the disjoint union of G and the coset Gu;*

  (ii) *G has a unique nonidentity commutator s, which is necessarily central and of order 2;*

  (iii) *the map*
$$g \mapsto g^* = \begin{cases} g & \text{if g is central} \\ sg & \text{otherwise} \end{cases}$$
  *is an involution of G (i.e., an antiautomorphism of order 2);*

*(iv) multiplication in L is defined by*

$$g(hu) = (hg)u,$$
$$(gu)h = gh^*u$$

*and*

$$(gu)(hu) = g_0h^*g,$$

*where $g, h \in G$ and $g_0 = u^2$ is a central element of G such that $g_0^* = g_0$.*

The loop described by this theorem shall be denoted by $M(G, *, g_0)$.

**Corollary 3.11.** *Let L be an RA loop. Then $L/Z(L) \cong C_2 \times C_2 \times C_2$, where $C_2$ denotes the cyclic group of order 2. For any elements $a, b \in L$ which do not commute and any $u \in L$ which does not associate with a and b, we have that the soobloop $G = \langle a, b, Z(L) \rangle$ is a group and $L = M(G, *, u)$, where L is the involution defined in the theorem above; in particular $G/Z(G) \cong C_2 \times C_2$.*

## 4. Classifying RA Loops

After Corollary 3.9 we see that, to classify RA loops we need only to consider the indecomposable ones and it is very easy to show that these are 2-loops (see [Goodaire *et al.*, Corollary V.1.3]). We shall first study groups G such that $G/Z(G) \cong C_2 \times C)_2$ and then see how to construct indecomposable RA loops from them. Groups G such that $G/Z(G) \cong C_p \times C_p$, for an integral prime $p$, were studied by G. Leal and C. Polcino Milies. We quote one result, specializing for $p = 2$.

**Theorem 4.1. [Leal and Polcino Milies (1993), Lemma 1.1]** *A group G is such that $G/Z(G) \cong C_2 \times C_2$ if and only if G can be written in the form $G = D \times A$, where A is abelian and D is an indecomposable 2-group generated by its centre and two elements $x$ and $y$ which satisfy*

*(i) $Z(D) = C_{2^{m_1}} \times C_{2^{m_2}} \times C_{2^{m_3}}$, where $C_{2^{m_i}}$ is cyclic of order $2^{m_i}$ for $i = 1, 2, 3, m_1 \geq 1$ and $m_2, m_3 \geq 0$;*

*(ii) $(x, y) \in C_{2^{m_1}}$; and*

*(iii) $x^2 \in C_{2^{m_1}} \times C_{2^{m_2}}$ and $y_2 \in C_{2^{m_1}} \times C_{2^{m_2}} \times C_{2^{m_3}}$.*

Using this description, E. Jespers, G. Leal and C. Polcino Milies (1995) classified all finite groups of this type.

**Theorem 4.2.** *Let G be a finite group. Then $G/CZ(G) \cong C_2 \times C_2$ if and only if G can be written in the form $G = D \times A$, where A is abelian and $D = \langle Z(D), x, y \rangle$ is of one of the following five types of indecomposable 2-groups:*

| Type | $\mathcal{Z}(D)$ | $D$ |
|---|---|---|
| $D_1$ | $\langle t_1 \rangle$ | $(x, y, t_1 \mid (x,y) = t_1^{2^{m_1-1}}, x^2 = y^2 = t_1^{2^{m_1}})$ |
| $D_2$ | $\langle t_1 \rangle$ | $(x, y, t_1 \mid (x,y) = t_1^{2^{m_1-1}}, x^2 = y^2 = t_1, t_1^{2^{m_1}} = 1)$ |
| $D_3$ | $\langle t_1 \rangle \times \langle t_2 \rangle$ | $(x, y, t_1, t_2 \mid (x,y) = t_1^{2^{m_1-1}}, x^2 = t_1^{2^{m_1}} = t_2^{2^{m_2}} = 1, y^2 = t_2)$ |
| $D_4$ | $\langle t_1 \rangle \times \langle t_2 \rangle$ | $(x, y, t_1, t_2 \mid (x,y) = t_1^{2^{m_1-1}}, x^2 = t_1, y^2 = t_2, t_1^{2^{m_1}} = t_2^{2^{m_2}} = 1)$ |
| $D_5$ | $\langle t_1 \rangle \times \langle t_2 \rangle \times \langle t_3 \rangle$ | $(x, y, t_1, t_2, t_3 \mid (x,y) = t_1^{2^{m_1-1}}, x^2 = t_2, y^2 = t_3, t_1^{2^{m_1}} = t_2^{2^{m_2}} = t_3^{2^{m_3}} = 1)$ |

Then, it is possible to describe all finite indecomposable RA loops.

**Theorem 4.3. (Jespers *et al.*, 1995, and Goodaire *et al.*, 1996, theorem V.3.1)**
*Let $L = M(G, *, g_0)$ be a finite indecomposable RA loop. Then $G$ is either one of the five groups specified in Theorem 4.2 or the direct product $D_5 \times \langle w \rangle$ of $D_5$ and a cyclic group $\langle w \rangle$ and $L$ is one of the following seven types of loops:*

| Type | $G$ | $x^2$ | $y^2$ | $g_0$ |
|---|---|---|---|---|
| $L_1$ | $D_1$ | 1 | 1 | 1 |
| $L_2$ | $D_2$ | $t_1$ | $t_1$ | $t_1$ |
| $L_3$ | $D_3$ | 1 | $t_2$ | 1 |
| $L_4$ | $D_4$ | $t_1$ | $t_2$ | $t_1$ |
| $L_5$ | $D_5$ | $t_2$ | $t_3$ | 1 |
| $L_6$ | $D_5$ | $t_2$ | $t_3$ | $t_1$ |
| $L_7$ | $D_5 \times \langle w \rangle$ | $t_2$ | $t_3$ | $w$ |

## 5. The Isomorphism Problem and Related Questions

A classical question in the study of group rings is the so-called *isomorphism problem*, whose analogue in the case of alternative loop rings can be stated as follows: *given a ring $R$, when does the loop ring $RL$ determine $L$; i.e., if $L_1$ is another loop, when does the isomorphism $RL \cong RL_1$ imply that $L \cong L_1$?*

In the case of group rings, it is very well known that the most significant context in which to study this question is that of integral group rings, the main reason being that this is the strongest hypothesis possible: if two groups $G$ and $H$ are such that $\mathbb{Z}G \cong \mathbb{Z}H$ then it is rather easy to see that also $RG \cong RH$ for every ring with unity $R$. In this setting, the question has a positive answer, due to E.G. Goodaire and C. Polcino Milies (Goodaire *et al.*, 1988, 1996, Theorem IX.1.1).

**Theorem 5.1.** *Let $L_1$ and $L$ be finite RA loops such that $\mathbb{Z}L_1 \cong \mathbb{Z}L$. Then $L_1 \cong L$.*

In the case of groups, this problem has motivated several other questions. To introduce them, we need some notation. First, let us remark that the map $\varepsilon : \mathbb{Z}L \to \mathbb{Z}$

given by $\sum_{g\in G} a_g g \mapsto \sum_{g\in G} a_g$ is a homomorphism, called the *augmentation mapping*. An automorphism $\phi : \mathbb{Z}L \to \mathbb{Z}L$ is called *normalized* if $\varepsilon(\alpha) = \varepsilon(\phi(\alpha))$, for all $\alpha \in \mathbb{Z}L$. It is easy to see that given any isomorphism $\phi : \mathbb{Z}L \to \mathbb{Z}L$ the map $\phi : \mathbb{Z}L \to \mathbb{Z}L$ defined by $\bar{\phi}(\sum_{g\in L} a_g g) = \sum_{g\in L} a_g \phi(g)/\varepsilon(g)$ is a normalized isomorphism. Because of this fact, all isomorphisms consider can be assumed to be normalized, without loss of generality.

It is easy to see that given any automorphism $\sigma : g \mapsto g^\sigma$ of $G$, it can be extended linearly to an automorphism $\bar{\sigma} : \sum_{g\in L} a_g g \mapsto \sum_{g\in L} a_g g^\sigma$ of $\mathbb{Z}L$.

Also, if $\gamma \in QL$ is an invertible element in the rational loop algebra such that $\gamma^{-1}g\gamma \in \mathbb{Z}L$ for all $g \in L$, then the map $\phi_\gamma : \mathbb{Z}L \to \mathbb{Z}L$ given by $\phi_\gamma(g) = \gamma^{-1}g\gamma$ is again an automorphism of $\mathbb{Z}L$. In the case of group rings, it has been conjectured that all automorphism can be obtained from automorphisms of this kind. In the case of alternative loop rings, the result is, in fact, true; however, one must be careful, since inner automorphism are defined in a different way in the context of alternative algebras. We recall that given an alternative algebra $A$ and an element $x \in A$, we define the translation maps $R_x : A \to A$ and $L_x : A \to A$ by:

$$R_a(x) = xa; \quad L_a(x) = ax; \quad \forall x \in A.$$

An *inner automorphism* of $A$ is any automorphism in the group generated by the set $\{R_a, L_a \mid a \text{ is a unit in } A\}$. It can be shown that if $A$ is associative, then this concept of inner automorphism coincides with the usual one. With this definition in mind, we can state:

**Theorem 5.2. (Goodaire *et al.*, 1996, see also [Theorem IX.3.2])** *Let $L$ be a torsion RA loop and $\theta$ a normalized automorphism of $\mathbb{Z}L$. Then there exists an inner automorphism $\phi$ of the rational loop algebra $QL$ and an automorphism $\sigma$ of $L$ such that $\theta = \phi \circ \bar{\sigma}$.*

Given a normalized isomorphism $\phi : \mathbb{Z}L \to \mathbb{Z}L$, it follows readily that the image $\phi(L)$ is a finite subloop of normalized units; i.e., units whose augmentation is equal to 1. The set of all normalized units in $RL$ will be denoted by $V(RL)$. If $\gamma \in QL$ is an invertible element in the rational loop algebra such that $\gamma^{-1}g\gamma \in \mathbb{Z}L$ for a given element $g \in L$, then certainly $\gamma^{-1}g\gamma \in V(\mathbb{Z}L)$.

Once again, let us refer first to the associative case. In the early seventies H.J. Zassenhaus made several conjectures regarding units and finite subgroups of units contained in $V(\mathbb{Z}G)$, when $G$ is a finite group. Essentially, he stated that all finite subgroups of units in $V(\mathbb{Z}G)$ can be obtained from subgroups of $G$ by conjugation with units in the rational group algebra. More precisely, the conjectures are as follows.

(ZC1) Every normalized unit of finite order $u \in V(\mathbb{Z}G)$ is rationally conjugate to an element $g \in G$; i.e., there exists a unit $\gamma \in QG$ such that $\gamma^{-1}u\gamma \in G$.

(ZC2) Let $H$ be a subgroup of normalized units in $\mathbb{Z}G$ such that $|H| = |G|$. Then $H$ is rationally conjugate to $G$.

(*ZC*3) Let $H$ be any finite subgroup of normalized units in $\mathbb{Z}G$. Then $H$ is rationally conjugate to a subgroup of $G$.

These conjectures have been established for various kinds of groups and, though (*ZC*1) remains open in general, it was shown by K.W. Roggenkamp and L. Scott that there exists a metabelian group of order $2^6 \cdot 3 \cdot 5 \cdot 7$ which is a counterexample to *ZC*2 (see L. Klinger, 1991).

Clearly, (*ZC*3) implies the other two. In the case of finite RA loops, the conjectures were decided in the affirmative by E.G. Goodaire and C. Polcino Milies: (*ZC*1) was studied in 1989 and then this result was used to obtain (*ZC*3) in 1996. Of course, once again the non-associative version of these conjectures depends on the notion of inner automorphisms of an alternative algebra. A detailed exposition can be found [Goodaire *et al.*, 1996, Theorem IX.4.13].

## 6. Zorn's Vector Matrix Algebra

We recall that, for loop algebras of RA loops over fields whose characteristic does not divide the order of the given loop, an extension of Mashcke's theorem holds (for a proof, Goodaire *et al.*, 1996, Theorem VI.4.2):

**Theorem 6.1. (Mashcke)** *Let $L$ be a finite RA loop and let $K$ be a field. If the characteristic of $K$ does not divide the order of $L$, then $FL$ has zero Jacobson radical; in particular, $FL$ is semiprime.*

**Corollary 6.2.** *Let $L$ be a finite RA loop and let $K$ be a field. If the characteristic of $K$ does not divide the order of $L$, then $FL$ is the direct sum of ideals which are simple rings.*

It is possible to give a more detailed description of the structure of $KL$ in the semisimple case. In what follows, $s$ denotes the unique non-identity commutator of $L$.

**Lemma 6.3.** *Let $L$ be a finite RA loop and let $K$ be a field such that char $(K) \nmid |L|$. Then:*

$$KL = KL\left(\frac{1+s}{2}\right) \oplus KL\left(\frac{1-s}{2}\right).$$

*where $KL\left(\frac{1+s}{2}\right) \cong K[L/L']$ and $KL\left(\frac{1-s}{2}\right) \cong \Delta(L, L')$.*

It is easy to see that $KL\left(\frac{1+s}{2}\right)$ is a direct sum of component which are both associative and commutative (since $L/L'$ is an abelian group) and that the simple components of $KL(\frac{1-s}{2})$ are neither associative nor commutative. However, they can be well described, starting from their centres.

**Theorem 6.4.** *Let $L = M(G, *, g_0)$ be an RA loop and let $K$ be a field such that* char $(K) \nmid |L|$. *Then*

(i) $\mathcal{Z}(\Delta(G, G')) \cong \mathcal{Z}(\Delta(L, L')) \cong K[\mathcal{Z}(L)] \cdot (1 - \hat{G}') \cong \bigoplus_{i=1}^{m} K_i$, *where each field $K_i$ is an extension of $K$ by a primitive $n^{th}$ root of unity;*

(ii) $\Delta(G, G') \cong \bigoplus_{i=1}^{m} \mathcal{B}_i$, *where each $\mathcal{B}_i$ is an algebra of generalized quaternions over the field $K_i$; and*

(iii) $\Delta(L, L') \cong \bigoplus_{i=1}^{m} \mathcal{A}_i$, *where each $\mathcal{A}_i$ is a Cayley-Dickson algebra over the field $K_i$.*

These descriptions have been used by G. Leal and C. Polcino Milies (1993) and by L.G.X. de Barros (1993a & b) to study the isomorphism problem over fields.

It is well-known that any Cayley-Dickson algebra over a field $K$ of characteristic different from 2, which is not a division ring, is isomorphic to the algebra we define below (Goodaire *et al.*, 1996, Corollary I.4.17).

Consider the set of $2 \times 2$ matrices of the form

$$\begin{bmatrix} a & x \\ y & b \end{bmatrix},$$

where $a, b \in K$ and $x, y \in K^3$, the set of ordered triples of elements of $K$.

We add two such matrices entrytwise and multiply them according to the following rule:

$$\begin{bmatrix} a_1 & x_1 \\ y_1 & b_1 \end{bmatrix} \begin{bmatrix} a_2 & x_2 \\ y_2 & b_2 \end{bmatrix}$$
$$= \begin{bmatrix} a_1 a_2 + x_1 \cdot y_2 & a_1 x_2 + b_2 x_1 - y_1 \times y_2 \\ a_2 y_1 + b_1 x_2 + x_1 \times x_2 & b_1 b_2 + y_1 x_2 \end{bmatrix},$$

where . and $\times$ denote the usual dot and cross products respectively in $K^3$. In this way, we obtain an alternative algebra known as *Zorn's vector matrix algebra over K* which will be denoted by $M(K)$. It is also called the *split Cayley-Dickson algebra over K*. We remark that $M(K)$ is not only a simple ring but it is also irreducible in the sence that it contains no proper one-sided ideals [Merlini Giuliani, 1998, Theorem 2.1.]

We shall denote by $GLL(K)$ the set of all invertible elements of the algebra $M(K)$. It is natural to consider loops of the form $GLL(K)$ as a non-associative analog of linear groups so we shall call $GLL(K)$ the *general linear loop over K*. Since $M(K)$ is alternative, it follows that $GLL(K)$ is a Moufang loop.

We recall that it is possible to define a determinant mapping in $M(K)$ by:

$$\alpha = \begin{bmatrix} a & x \\ y & b \end{bmatrix} \mapsto \det(\alpha) = ab - x.y.$$

An element $\alpha = \begin{bmatrix} a & x \\ y & b \end{bmatrix} \in M(K)$ is invertible if and only if $\det(\alpha) \neq 0$ and, in this case, its inverse is

$$\alpha^{-1} = (\det \alpha)^{-1} \begin{bmatrix} b & -x \\ -y & a \end{bmatrix}.$$

The set $SLL(K) = \{\alpha \in M(K)| \det(\alpha) = 1\}$ is a normal subloop of $GLL(K)$ and will be called the *special linear loop*. It is easy to see that the centre of $SLL(K)$ is $Z(SLL(K)) = \{I, -I\}$. We also define the *projective special linear loop* to be the quotient loop:

$$PLL(K) = SLL(K)/Z(SLL(K)).$$

Notice that, if $K$ is a finite field with $|K| = q$ the it can be easily shown that:

$$|GLL(F_q)| = q^3(q^4 - 1)(q - 1) \qquad |SLL(F_q)| = q^3(q^4 - 1),$$

$$|PSLL(F_q)| = \frac{q^3(q^4-1)}{2}, \text{ if } q \neq 2^n, \quad |PSLL(F_q)| = q^3(q^4 - 1), \text{ if } q = 2^n.$$

We wish to exibit a set of generators for $GLL(K)$. To do so, we introduce more notation.

We consider the following elements:

$$e_{11} = \begin{bmatrix} 1 & 0 \\ 0 & 0 \end{bmatrix}, \quad e_{22} = \begin{bmatrix} 0 & 0 \\ 0 & 1 \end{bmatrix},$$

$$e_{12}^{(i)} = \begin{bmatrix} 0 & i \\ 0 & 0 \end{bmatrix}, \quad e_{12}^{(j)} = \begin{bmatrix} 0 & j \\ 0 & 0 \end{bmatrix}, \quad e_{12}^{(k)} = \begin{bmatrix} 0 & k \\ 0 & 0 \end{bmatrix},$$

$$e_{21}^{(i)} = \begin{bmatrix} 0 & 0 \\ i & 0 \end{bmatrix}, \quad e_{21}^{(j)} = \begin{bmatrix} 0 & 0 \\ j & 0 \end{bmatrix}, \quad e_{21}^{(k)} = \begin{bmatrix} 0 & 0 \\ k & 0 \end{bmatrix}.$$

A streightforward computation shows that they multiply according to the following rules:

$$e_{ii}^2 = e_{ii}, e_{ii}e_{jj} = 0; \qquad e_{12}^{(h)}e_{12}^{(f)} = e_{21}^{h \times f}, e_{21}^{(h)}e_{21}^{(f)} = e_{12}^{-(h \times f)}$$

$$e_{ii}e_{ij}^{(h)} = e_{ij}^{(h)}e_{jj} = e_{ij}^{(h)}; \qquad e_{jj}e_{ij}^{(h)} = e_{ij}^{(h)}e_{ii} = 0;$$

$$(e_{ij}^{(h)})^2 = 0 = e_{ij}^{(h)}e_{ij}^{(f)}, \text{ if } h \neq f; \quad e_{ij}^{(h)}e_{ji}^{(h)} = e_{ii}, \text{ if } e_{ij}^{(h)}e_{ji}^{(f)} = 0, h \neq f;$$

Also, we denote $i = (1, 0, 0), j = (0, 1, 0), k = (0, 0, 1) \in K^3$. Given a scalar $\lambda \in K$ we define the *transvection* $\tau_{ij}^{(h)}(\lambda)$, where $i, j = 1, 2, i \neq j$ and $h = i, j$ or $k$, as

$$\tau_{ij}^{(h)}(\lambda) = I + \lambda e_{ij}^{(h)} \in GLL(K)$$

Notice that our definition of transvection is more restricted than the usual one in the associative case (see Definition 4.1).

For an element $\mu \in K$, $\mu \neq 0$, we define $D(\mu) \in GLL(K)$ as

$$D(\mu) = \begin{bmatrix} 1 & 0 \\ 0 & \mu \end{bmatrix}.$$

The set of matrices of the form above, together with the set of all transvections give a set of generators of the general linear loop. More precisely, we have the following.

**Theorem 6.5. [Merlini Giuliani, 1998, Theorem 3.2]**

$$GLL(K) = \langle \tau_{ij}^{(h)}(\lambda), D(\mu) | h = i, j, k, \lambda, \mu \in K \rangle.$$

As an immediate consequence, we obtain the following.

**Corollary 6.6.** *The special linear loop is generated by the set of all transvections; i.e.:*

$$SLL(K) = \langle \tau_{ij}^{(h)}(\lambda) \mid h = i, j, k; i, j = 1, 2, \lambda \in K \rangle.$$

Using this set of generators, it is possible to prove that all loops of the form $PSLL(K)$ are simple [Merlini Giuliani, 1998, Theorem 2.3.6], a fact that was first obtained by L. Paige in 1956. Later M.W. Liebeck (1987) showed that every finite simple Moufang loop is isomorphic to a loop of this type.

It is of particular interest to decide when the torsion units of an alternative algebra form a subgroup, because of its connections with several properties of the loop of units. If this is the case, for briefness, we shall say that the units have *tpp* (torsion product property). For Zorn's algebra, we have the following.

**Proposition 6.7. (Goodaire and Polcino Milies, 1995 a, Theorem 2.5)** *Let F be a field of characteristic $p > 0$. Then Zorn's vector matrix algebra $M(F)$ over F has tpp if and only if F is algebraic over its prime field P.*

**Proposition 6.8. (Polcino Milies, unpublished, Proposition 2.4)** *Let R be a commutative ring with unity, of characteristic 0. Then, $M(R)$ does not have tpp.*

The behavior of alternative division rings is also known.

**Proposition 6.9. (Goodaire and Polcino Milies, 1995a, Proposition 2.3)** *Let D be an alternative division ring of characteristic $p > 0$. Then D has tpp if and only if any two elements of finite order in D commute. In this case, $E = TU(D) \cup \{0\}$ is a central subfield of D.*

**Proposition 6.10. (Polcino Milies, unpublished, Theorem 2.2)** *Let D be an alternative division ring of characteristic 0. Then D has tpp if and only if $TU(D)$ is central in D.*

Gathering these results it is possible to decide when the torsion units of an alternative artinian algebra form a subgroup.

**Theorem 6.11.** *Let A be a semisimple alternative artinian algebra over a field K of characteristic $p > 0$. Then A has tpp if and only if its simple components are either division rings whose torsion units are central, or isomorphic to full matrix rings or to Zorn's vector matrix algebras over fields which are algebraic over their corresponding prime fields.*

**Theorem 6.12.** *Let A be a semisimple alternative artinian algebra over a field K of characteristic 0. Then A has tpp if and only if all its torsion units are central. In this case A is a direct sum of division rings with this property.*

The case when the radical is nonzero is closely related to the semisimple case. However, when the characteristic of the given field is 0 we can only give a complete answer when the algebra is finite dimensional over its ground field.

**Theorem 6.13.** *Let A be an alternative artinian algebra over a field K of characteristic $p > 0$. Then A has tpp if and only if $A/J(A)$ has tpp.*

**Theorem 6.14.** *Let A be an alternative algebra finite dimensional over a field K of characteristic 0. Then, A has tpp if and only if $TU(A)$ is central. In this case, all nilpotent elements of A belong to $J(A)$.*

The particular case of alternative loop algebras over fields was studied by Goodaire and Polcino Milies (1995a)

**Theorem 6.15.** *Let L be an RA loop with torsion subloop T and let F be a field of characteristic $p > 0$. Then FL has tpp if and only if one of the following conditions holds:-*

(i) *$L = T$ and F is algebraic over its prime field $\mathcal{P}$.*

(ii) *$p = 2$.*

(iii) *$p$ is odd, T is an abelian group and, if T is not central, then the algebraic closure $\bar{\mathcal{P}}$ of $\mathcal{P}$ in F is finite and for all $x \in L$ and all $p'$-elements $a \in T$, we have $xax^{-1} = a^{p^r}$ for some positive integer r, where $[\bar{\mathcal{P}} : \mathcal{P}] \mid r$.*

**Theorem 6.16.** *Let L be an RA loop with torsion subloop T and let F be a field of characteristic 0. Then FL has tpp if and only if*

(i) *T is an abelian group, and*

(ii) *for each $t \in T$ and each $x \in L$, there exists a positive integer i such that $xtx^{-1} = t^i$ and, if t is noncentral element $t \in T$, and $\zeta$ denotes a root of unity of order $o(t)$, then there exists map $\sigma \in Gal(K(\zeta) : K)$ such that $\sigma(\zeta) = \zeta^j$.*

## 7. Some Remarks on Nilpotency

The question of deciding when an alternative loop algebra $RL$ contains non-trivial nilpotent elements is complitely decided in the case when $R = K$, a field.

**Theorem 7.1. (Goodaire *et al.*, 1996, Theorem XIII.2.1)** *Let $K$ be a field of characteristic $p \geq 0$ and let $L$ be an RA loop with torsion subloop $T$. Then $KL$ has no nonzero nilpotent elements is and only if the following conditions are satisfied.*

(i) *$L$ does not contain elements of order $p$ (in particular, $p \neq 2$).*

(ii) *Every subloop of $T$ is normal in $L$.*

(iii) *If $p = 0$ and $T$ contains a noncentral element of order $n$ then, for any primitive root of unity $\xi_n$ of the same order over $K$, the map $\sigma$ defined by $\sigma(\xi_n) = \xi_n^{(n/2)+1}$ is in $Gal\,(K(\xi_n) : K)$. Furthermore,*

    (a) *$T$ is an abelian group, or*

    (b) *$T$ is a Hamiltomian group and, if it contains an element of odd order $k$, then the field $K(\xi_n)$ does not contain a solution of the equation $x^2 + y^2 = -1$, or*

    (c) *$T$ is a Hamiltonian RA 2-loop and $K$ does not contain a solution of the equation $x^2 + y^2 + z^2 + w^2 = -1$*

(iv) *If $p > 2$, then $T_{p'}$ (the set of all elements in $T$ whose order is not divisible by $p$) is an abelian group and, if it is not central, then the algebraic closure $\overline{F}_p$ of the prime field $F_p$ of $K$ in $K$ is finite and, for all $t \in T_{p'}$ and $x \in L$, there exists a positive multiple $r$ of $[\overline{F}_p : F_p]$ such that $xtx^{-1} = t^{p^r}$.*

The theorem above shows that the answer to the question depends strongly on the structure of the coefficient ring. However, jn the case when the torsion subloop $T$ is central in $L$ a general result was given by A. Zatelli (1993, Theorem 3.3.1). This result can also be regarded as an extension of a similar property which holds in the case of group rings of finite groups (see Sehgal, 1978, Proposition VI. 1.12).

**Theorem 7.2.** *Let $L$ be an RA loop such that $T$ is central in $L$. Then $RL$ contains no nonzero nilpotent elements if and only if $R$ contains no nonzero nilpotent elements and for every $g \in T$ we have that $o(g)$ is not a zero divisor in $R$.*

It is possible to prove such a general result because of the particular structure of RA loops. The following technical lemma is a key step in the proof.

**Lemma 7.3.** *Let $L = L(G, *, g_0)$ be an RA loop such that $R[\mathcal{Z}(G)]$ contains no nonzero nilpotent elements and let $X$ be a transversal of $\langle s \rangle$ in $\mathcal{Z}(G)$. If $x \in RL$ is of the form*

$$x = (1 - s)[(x_0 + x_1 a + x_2 b + x_3 ab) + (x_4 + x_5 a + x_6 b + x_7 ab)u],$$

*with $x_i \in RX$ $i = 0, 1, \ldots, 7$ and $x^2 = 0$, then $x_0 = 0$.*

Well-known results concerning the nilpotency of augmentation ideals also extend in a natural way to alternative loop algebras.

**Theorem 7.4. (Zatelli, 1993, Theorem 4.4.1)** *Let $H$ be a normal subloop of an RA loop $L$. Then, the augmentation ideal $\Delta_R(L : H)$ is nilpotent if and only if $H$ is a finite 2-subloop of $L$ and $R$ is a ring of characteristic $2^m$, for some positive integer $m \geq 1$.*

**Corollary 7.5.** *Let $L$ be an RA loop. Then, the augmentation ideal $\Delta(L)$ is nilpotent if and only if $L$ is a finite 2-loop and $R$ is a ring of characteristic $2^m$, for some positive integer $m \geq 1$.*

Finally, we recall that in the study of group rings, the existence of idempotent ideals in $\mathbb{Z}G$ depends on the solvability of the given group. It was shown by T. Akasaki (1972, 1973) that a finite group $G$ is solvable if and only if $\mathbb{Z}G$ contains no idempotent ideals which are contained in the augmentation ideal. This result was improved by K.W. Roggenkamp (1974) and extended by P.F. Smith (1976, Theorem 2.2), who proved that if $G$ is a polycyclic group, then $\mathbb{Z}G$ contains no non trivial idempotent ideals. A similar result holds for RA loops.

**Theorem 7.6.** *Let $L$ be an RA loop such that $L/L'$ is finitely generated and let $I$ be an idempotent ideal of $\mathbb{Z}L$. Then either $I = 0$ or $I = \mathbb{Z}L$.*

## References

Akasaki, T. *Idempotent Ideals in Integral Group Rings*, J. Alg., **23**, 343–346, 1972.

Akasaki, T. *Idempotent Ideals in Integral Group Rings II*, Arch. Math., **24**, 126–128, 1973.

Artin, E. *Geometric Algebra*, Interscience, New York, 1957.

de Barros, L.G.X. *Isomorphisms of rational loop algebras*, Comm. Alg., **21**(11), 3977–3993, 1993a.

de Barros, L.G.X. *On Semisimple Alternative Loop Algebras*, Comm. Alg., **21**(11), 3995–4011, 1993b.

Brauer, R. *Über Systeme Hypercomplexer Zahlen*, Math. Z., **30**, 79–107, 1929.

Brauer, R. and Noether, E. *Über minimale Zerfällungskörper irreduzibler Darstellungen*, Sitz. Preuss. Akad. Wiss. Berlin, 221–228, 1927.

Bruck, R.H. *Some Results in the Theory of Linear Non-Associative Algebras*, Trans. Amer. Math. Soc., **56**, 141–199, 1944.

Chein, O. and Goodaire, E.G. *Loops whose Loop Rings are Alternative*, Comm. Alg., **14**, 293–310, 1986.

Chein, O. and Goodaire, E.G. *Loops whose Loop Rings in characteristic 2 are Alternative,* Comm. Alg., **18**(3), 659–688, 1990.

Goodaire, E.G. *Alternative Loop Rings, Publ. Math. Debrecen,* **30**, 31–38, 1983.

Goodaire, E.G. Jespers, E. and Polcino Milies, C. *Alternative Loop Rings,* North Holland Math. Studies 184, Elsevier, Amsterdam, 1996.

Goodaire, E.G. and Polcino Milies, C. *Isomorphisms of Integral Alternative Loop Rings,* Rend. Circ. Mat. Palermo, **XXXVII**, 126–135, 1988.

Goodaire, E.G. and Polcino Milies, C. *Torsion Units in Alternative Loop Rings, Proc. Amer. Math. Soc.,* **107**, 7–15, 1989.

Goodaire, E.G. and Polcino Milies, C. *Finite subloops of units in an alternative loop ring,* Proc. Amer. Math. Soc., **124**(4), 995–1002, 1996.

Goodaire, E.G. and Polcino Milies, C. *On the Loop of Units of an Alternative Loop Ring,* Nove J. Alg. Geom., **3**(3), 199–208, 1995a.

Goodaire, E.G. and Polcino Milies, C. *Ring Alternative Loops and their Loop Rings, Resenhas Inst. Mat. Est. Univ, São Paulo,* **2**(1), 47–82, 1995b.

Hall, M. Jr., *The Theory of Groups,* MacMillan, New York, 1959.

Jespers, E. Leal, G. and Polcino Milies, C. *Classifying indecomposable RA loops, J. Alg.,* **176**, 5057–5076, 1995.

Klinger, L. *Construction of a Counterexample to a Conjecture of Zassenhaus, Comm. Alg.,* **19**, 2303–2330, 1991.

Leal, G. and Polcino Milies, C. *Isomorphic Group (and Loop) Algebras, J. Alg.,* **155**, 195–210, 1993.

Merlini Giuliani, M.L. *Loops de Moufang Lineares, PhD. Thesis,* Instituto de Matemática e Estatística, Universidade de São Paulo, 1998.

Liebeck, M.W. The classification of finite simple Moufang loops, *Math. Proc. Camb. phil. Soc.,* **102**, 33–47, 1987.

Moufang, R. *Alternativekörper und der Satz vom vollständigen Viersteit (D9), Abh. Math. Sem. Univ. Hamburg* **9**, 207–222, 1933.

Noether, E. *Nichtkommutative Algebra, Math. Z.,* **37**, 513–541, 1933.

Paige, L.J. *A theorem on commutative power associative loop algebras, Proc. Amer. Math. Soc.,* **6**, 279–280, 1955.

Paige, L.J. *A class of simple Moufang loops, Proc. Amer. Math. Soc.,* **7**, 471–482, 1956.

Polcino Milies, C. *The torsion product property in alternative algebras II, Comm. Alg., to appear.*

Roggenkamp, K.W. *Integral Group Rings of Solvable Finite Groups have no Idempotent Ideals, Arch. Math.* **25**, 125–128, 1974.

Sehgal, S.K. *Topics in Group Rings,* Marcel Dekker, New York, 1978.

Sehgal, S.K. *Units in Integral Group Rings,* Longman Scientific & Technical, Essex, 1993.

Smith, P.F. *A Note on Idempotent Ideals in Group Rings, Arch. Math.,* **XXVII**, 22–27, 1976.

Weiss, A. *Units in Integral Group Rings, J. Reine. Angew. Math.,* **415**, 175–187, 1991.

Zatelli, A. *Elementos Nilpotentes em Anéis de Loop Alternativos, PhD. Thesis,* Instituto de Matemática e Estatística, Universidade de São Paulo, 1993.

Zhevlakov, K.A. Slin'ko, A.M. Shestakov, I.P. and Shirshov, A.I. *Rings That Are Nearly Associative,* Academic Press, New York, 1982.

Instituto de Matemática e Estatística, Universidade de São Paulo, Caixa Postal 66281, Cep 05315-970, São Paulo, Brasil. E-mail: polcino@ime.usp.br

# L-values at Zero and the Galois Structure of Global Units

*Jürgen Ritter*

## 1. Introduction

This article intends to present a comprehensive survey of the striking interplay between the Galois structure of the group of units in a number field and the values at zero of Artin $L$-functions. The algebraic ingredients come from integral representation theory, the ones from number theory include the Main Conjecture of Iwasawa theory. In fact, the discussion of recently defined invariants which go along with the unit group seems to propose possible generalizations of the Main Conjecture and fits very well into the framework of rather general conjectures regarding $L$-values by providing first affirmative answers. To begin with, we collect the principal ideas.

## 2. Set-Up, Problems and Results

The following notation is used throughout.

$K/k$ is a finite Galois extension of number fields with group $G$,

$S_\infty$ the set of all infinite (real and complex) primes of $K$ and $S = S_\infty \cup S_f$ a union of $S_\infty$ and a fixed set $S_f = \{\mathfrak{p}_1, \ldots, \mathfrak{p}_r\}$ of non-archimedean primes $\mathfrak{p}_i$ of $K$, which is closed under the action of $G$. We call $S$ large, if $S_f$ contains the set of ramified primes in $K/k$ and, moreover, if $\mathfrak{p}_1, \ldots, \mathfrak{p}_r$ generate the ideal class group $cl$ of $K$.

$E_S = \{u \in K : w_\mathfrak{p}(u) = 0 \text{ for all non-archimedean } \mathfrak{p} \notin S_f\}$ is the group of $S$-units in $K$. In particular, $E = E_{S_\infty} = \mathfrak{o}^\times$ is the full unit group of $K$, i.e., the group of all invertible elements in the ring $\mathfrak{o}$ of integers of $K$. Above, $w_\mathfrak{p}$ denotes the normalized valuation with respect to $\mathfrak{p}$. It induces the $\mathfrak{p}$-value $|x|_\mathfrak{p} = (N\mathfrak{p})^{-w_\mathfrak{p}(x)}$ for $x \in K$, where $N\mathfrak{p} = |\mathfrak{o}/\mathfrak{p}|$ is the number of elements in the residue field $\mathfrak{o}/\mathfrak{p}$.

$\mathbb{Z}S$ is the $\mathbb{Z}$-free module on the basis $\mathfrak{p} \in S$, on which $G$ acts by acting on $S$, and $\Delta S = \{\sum_{\mathfrak{p} \in S} z_\mathfrak{p} \mathfrak{p} \in \mathbb{Z}S : \sum_{\mathfrak{p} \in S} z_\mathfrak{p} = 0\}$.

## 2a. Dirichlet's Unit Theorem; Isogenies

Since $S$ is $G$-stable, $E_S$ is a $\mathbb{Z}G$-module. It has torsion submodule $\mu$, the group of roots of unity in $K$. Moreover, the Dirichlet unit theorem shows $\chi_{E_S} = \chi_{\Delta S}$ for the characters $\chi_{E_S}$, $\chi_{\Delta S}$ of the $\mathbb{C}G$-modules spanned by $\Delta S$ and $E_S$, respectively, namely on account of the $\mathbb{R}G$-isomorphism (cf. [Ta2, p. 25])

$$\lambda : \mathbb{R} \otimes_{\mathbb{Z}} E_S \to \mathbb{R} \otimes_{\mathbb{Z}} \Delta S, u \mapsto - \sum_{\mathfrak{p} \in S} \log |u|_{\mathfrak{p}} \mathfrak{p}.$$

That indeed $\sum_{\mathfrak{p} \in S} \log |u|_{\mathfrak{p}} = 0$ follows from the product formula (cf. [La, p. 99]), by which $\prod_{\text{all } \mathfrak{p}} |u|_{\mathfrak{p}} = 1$, and from $u \in E_S$, i.e., $|u|_{\mathfrak{p}} = 1$ for $\mathfrak{p} \notin S$. The minus sign above is of no importance here; it will become important at a later stage, though.

Because of the Deuring-Noether theorem (cf. [CR0, p. 200]), $\mathbb{R} \otimes_{\mathbb{Z}} E_S \simeq \mathbb{R} \otimes_{\mathbb{Z}} \Delta S$ implies $\mathbb{Q} \otimes_{\mathbb{Z}} E_S \simeq \mathbb{Q} \otimes_{\mathbb{Z}} \Delta S$. By restricting an isomorphism $\mathbb{Q} \otimes_{\mathbb{Z}} \Delta S \xrightarrow{\sim} \mathbb{Q} \otimes_{\mathbb{Z}} E_S$ to $\Delta S$ and by picking a fundamental set of $S$-units $u_1, \ldots, u_{n_s}$ in $E_S$ (a $\mathbb{Z}$-basis of $E_S$ modulo $\mu$) we arrive at a $\mathbb{Z}G$-map $\Delta S \to \frac{1}{N} \langle u_1, \ldots, u_{n_s} \rangle$ with some denominator $N \in \mathbb{N}$. Multiplying by $N$ then yields a map $\varphi : \Delta S \rightarrowtail E_S$ which is injective since $\Delta S$ has no torsion. By no means, however, is it a straightforward matter to construct such $\mathbb{Z}G$-monomorphisms $\varphi : \Delta S \rightarrowtail E_S$. In fact, only in very special cases one has been successful in this respect, and in almost each case $S$ has just been $S_\infty$. Note that every $\varphi$ – which will be referred to as an isogeny – displays a $\mathbb{Z}G$-submodule of finite index in $E_S$ which is isomorphic to the (known) module $\Delta S$.

For large sets $S$ the two modules $E_S$ and $\Delta S$ have even more in common. This is due to the link given by so-called Tate sequences arising from class field theory, which will be turned to after a digression.

## 2b. The Numbers $A_\varphi$; Stark's Conjecture

The analytic class number formula (cf. [La, p. 161]) reveals a new aspect of $E$. It combines analytic and algebraic data, namely the residue at $s = 1$ of the zeta function $\zeta(s)$ of $K$ with the class group $cl$ of $K$, the ramification in $K/\mathbb{Q}$ and, finally, the unit group $E$. Recall that $\zeta(s) = \prod_{\text{finite } \mathfrak{p}} (1 - \frac{1}{N\mathfrak{p}^s})^{-1}$ is holomorphic for $\Re(s) > 1$ and can be extended to a meromorphic function on all of $\mathbb{C}$, which then has a single, simple pole at $s = 1$ with residue

$$\text{res}_{s=1} \zeta(s) = \frac{2^{r_1} (2\pi)^{r_2} h R}{|\mu| \cdot \sqrt{|d|}}.$$

Here, $r_1$ and $r_2$ are the numbers of real and complex primes of $K$, respectively, $h = |cl|$ is the class number, $d$ the absolute discriminant of $K$ and $R$ the regulator, so the modulus of a principal minor of the matrix $(\log |u_j|_{\mathfrak{p}})_{1 \le j \le n_\infty, \mathfrak{p} \in S_\infty}$, where $u_1, \ldots, u_{n_\infty}$ now is a set of fundamental ($S_\infty$-) units in $E$.

A direct consequence of the analytic class number formula is the equation (cf. [Ta2, p. 21])

$$c = -hR/|\mu|$$

for the leading coefficient $c$ of the Taylor expansion of $\zeta(s)$ at $s = 0$.

Turning back to our set $S$, we introduce the $S$-zeta function $\zeta_S(s) = \prod_{\mathfrak{p} \notin S_f} (1 - \frac{1}{N\mathfrak{p}^s})^{-1}$ and its leading coefficient $c_S$ at $s = 0$. Then $c_S = -h_S R_S/|\mu|$, where now $h_S$ is the $S$-class number, so the order of the $S$-class group $cl_S$ of $K$ which is the ideal class group of the ring $o_S = \{x \in K : w_{\mathfrak{p}}(x) \geq 0 \text{ for } \mathfrak{p} \notin S_f\}$ of $S$-integers in $K$, and where the $S$-regulator $R_S$ is a quantity attached to $S$ in the same way as $R$ to $S_\infty$. Note that $R_S/c_S$ is rational; in particular, $c_S$ takes off everything transcendental in $R_S$.

How can one bring the group $G$ into play? On the one hand, this is done by replacing $\zeta_S(s)$ by the Artin $L$-functions $L_S(s, \chi)$ built to the complex characters $\chi$ of $G$,

$$L_S(s, \chi) = \prod \det\left(1 - \frac{\phi_{\mathfrak{p}}}{(N\mathfrak{p}_k)^s}|V_\chi^{I_{\mathfrak{p}}}\right)^{-1}.$$

The product runs over all non-archimedean primes $\mathfrak{p}_k$ of $k$ which have no divisor in $S_f$; the prime $\mathfrak{p}$ then is a divisor of $\mathfrak{p}_k$ in $K$ and $\phi_{\mathfrak{p}}$ its Frobenius automorphism in $G_{\mathfrak{p}}/I_{\mathfrak{p}}$, where $G_{\mathfrak{p}}$ and $I_{\mathfrak{p}}$ denote the decomposition and inertia subgroup of $\mathfrak{p}$, respectively. Finally, $V_\chi^{I_{\mathfrak{p}}}$ is the submodule of $I_{\mathfrak{p}}$ fixed points in a $\mathbb{C}G$-module $V_\chi$ affording the character $\chi$.[1]

On the other hand, the group $G$ is brought in by replacing the $S$-regulator $R_S$ by the Stark-Tate regulator $R_\varphi(\chi)$ (cf. [Ta2, p. 26]), which is attached to a given isogeny $\varphi : \Delta S \longrightarrow E_S$ and to $\chi$:

$$R_\varphi(\chi) \overset{\text{def}}{=} \det(\lambda\varphi \mid \text{Hom}_{\mathbb{C}G}(V_{\check{\chi}}, \mathbb{C} \otimes_{\mathbb{Z}} \Delta S)).$$

Here, $V_{\check{\chi}}$ is a $\mathbb{C}G$-module having character $\check{\chi}$, the contragredient to $\chi$, i.e., $\check{\chi}(g) = \chi(g^{-1})$ for $g \in G$. If one defines $c_S(\chi)$ to be the leading coefficient at $s = 0$ of $L_S(s, \chi)$, then as a $\chi$-analogue of the analytic class number formula we might regard any quantification of

$$A_\varphi(\chi) \overset{\text{def}}{=} R_\varphi(\chi)/c_S(\chi).$$

**Stark's Conjecture (ST).** $A_\varphi(\chi^\sigma) = A_\varphi(\chi)^\sigma$ for all $\sigma \in \text{Aut}(\mathbb{C})$. (Cf. [Ta2, p. 27].)

In particular, $A_\varphi(\chi)$ should be algebraic and belong to the field $\mathbb{Q}(\chi)$ spanned by the values of $\chi$. This is true whenever $K$ *is abelian over $\mathbb{Q}$ or over an imaginary quadratic field* $\mathbb{Q}(\sqrt{-d})$ (see [Ta2, IV]). Moreover:-

i) $A_\varphi(1) = \pm\frac{|\text{coker } \varphi|}{h_S}$ (cf. [Ta2, p. 45]).

---

[1] Recall that $L_S(s, 1)$ is the zeta function of $k$ (with all Euler factors $\mathfrak{p}_i \cap k, 1 \leq i \leq r$ removed) and $L_S(s, \rho)$ the corresponding one of $K$, if $\rho$ denotes the character of the regular representation of $G$.

ii) (ST) is true if so for just one pair $(S, \varphi)$ (cf. [Ta2, p. 27]).

iii) (ST) is true if $\chi$ is rational-valued.

The last statement is a consequence of Tate's work on the Stark conjecture (cf. [Ta2, II]). It introduces a completely new idea which we want to turn to next.

## 2c. Tate's q-ideal

Let $F$ be a finite Galois extension of $\mathbb{Q}$ with group $\Gamma$ and big enough such that every complex character $\chi$ of $G$ can be realized over $F$. From now on $V_\chi$ denotes an $FG$-module affording $\chi$. The field $F$ will be kept fixed in what follows.

Pick an $\mathfrak{o}_F G$-lattice $M_\chi$ on $V_\chi$, where $\mathfrak{o}_F$ is the ring of integers in $F$, and attach to an isogeny $\varphi : \Delta S \rightarrowtail E_S$ the $\mathfrak{o}_F$-map

$$\varphi_M : \mathrm{Hom}_{\mathfrak{o}_F}(M_\chi, \mathfrak{o}_F \otimes_{\mathbb{Z}} \Delta S)_G \xrightarrow{\hat{G}} \mathrm{Hom}_{\mathfrak{o}_F}(M_\chi, \mathfrak{o}_F \otimes_{\mathbb{Z}} \Delta S)^G$$
$$\xrightarrow{\varphi} \mathrm{Hom}_{\mathfrak{o}_F}(M_\chi, \mathfrak{o}_F \otimes_{\mathbb{Z}} E_S)^G$$

which is induced by $\varphi$ and by the $G$-norm $\hat{G}$ between $G$-coinvariant and $G$-invariant elements of $\mathrm{Hom}_{\mathfrak{o}_F}(M_\chi, \mathfrak{o}_F \otimes_{\mathbb{Z}} \Delta S)$. Kernel and cokernel of $\varphi_M$ are finite $\mathfrak{o}_F$-modules and therefore possess composition series. As usual, for a finite $\mathfrak{o}_F$-module $X$ with composition factors $C_{\mathfrak{q}}$ of type $\mathfrak{o}_F/\mathfrak{q}$ the order $|X|_{\mathfrak{o}_F}$ of $X$ is the product of the $\mathfrak{q}$, taken over all factors $C_{\mathfrak{q}}$.

**Definition.** $\quad q_\varphi(\chi) = |\mathrm{coker}\ \varphi_M|_{\mathfrak{o}_F}/|\ker \varphi_M|_{\mathfrak{o}_F}$

This is Tate's q-ideal in $\mathfrak{o}_F$. Surprisingly enough, $q_\varphi(\chi)$ only depends on the character $\chi$, not on the special choice of the lattice $M_\chi$, provided only that the set $S$ is large (remember that $S$ is hidden in $\varphi$).

Tate's proof of this fact makes use of a 4-term exact sequence (cf. [Ta1] and section 2d.)

$$E_S \rightarrowtail A \to B \twoheadrightarrow \Delta S$$

with finitely generated $\mathbb{Z}G$-modules $A, B$ such that $A$ has finite projective dimension and $B$ is projective. Each such sequence and each embedding $\iota : M_1 \rightarrowtail M_2$ of $\mathfrak{o}_F G$-lattices leads to a commutative diagram, as shown below, which serves as the key ingredient for checking the independence of $q_\varphi(\chi)$ from the choice of $M_\chi$:

$$
\begin{array}{ccccccc}
\mathrm{Hom}(M_1, \mathfrak{o}_F \otimes E_S)^G & \rightarrowtail & \mathrm{Hom}(M_1, \mathfrak{o}_F \otimes A)^G & \xrightarrow{v_1} & \mathrm{Hom}(M_1, \mathfrak{o}_F \otimes B)_G & \twoheadrightarrow & \mathrm{Hom}(M_1, \mathfrak{o}_F \otimes \Delta S)_G \\
\downarrow \iota & & \downarrow \iota & & \downarrow \iota & & \downarrow \iota \\
\mathrm{Hom}(M_2, \mathfrak{o}_F \otimes E_S)^G & \rightarrowtail & \mathrm{Hom}(M_2, \mathfrak{o}_F \otimes A)^G & \xrightarrow{v_2} & \mathrm{Hom}(M_2, \mathfrak{o}_F \otimes B)_G & \twoheadrightarrow & \mathrm{Hom}(M_2, \mathfrak{o}_F \otimes \Delta S)_G
\end{array}
$$

(cf. [Ta2, pp. 58–61]). Here, the maps $v_1, v_2$ are induced by the map $A \to B$ in the Tate sequence and by the isomorphism $\mathrm{Hom}(N, C)_G \xrightarrow{\hat{G}}{\simeq} \mathrm{Hom}(N, C)^G$ for any $\mathfrak{o}_F G$-module $C$ with finite projective dimension and any $\mathfrak{o}_F G$-lattice $N$.

Once one knows that $q_\varphi(\chi)$ only depends on $\chi$ (and $\varphi$) one can discuss its behaviour with respect to induction and inflation of characters. As a consequence one gets information on $q_\varphi(\chi)$ for a permutation character $\chi$ from the trivial character (cf. [Ta2, II§7]), whence on $q_\varphi(\chi)$ for rational-valued characters as well:

**Theorem (Tate).** $q_\varphi(\chi) = A_\varphi(\chi) \circ_F$, if $\mathbb{Q}(\chi) = \mathbb{Q}$.

This immediately implies $iii)$ in (2b).

## 2d. Tate Sequences and Chinburg's $\Omega$-invariant

The derivation of an exact sequence $E_S \rightarrowtail A \to B \twoheadrightarrow \Delta S$ requires a large set $S$ and results from matching the local fundamental classes of $K/k$ with the global one (cf. [CF, p. 131, p. 196]). Two facts are needed, firstly, that for large $S$ the natural map $J_S \to C_K$ from the $S$-idèle group $J_S = \prod_{\mathfrak{p} \notin S} U_\mathfrak{p} \times \prod_{\mathfrak{p} \in S} K_\mathfrak{p}^\times$ of $K$ to the idèle class group $C_K$ of $K$ is surjective and, secondly, that the unit group $U_\mathfrak{p}$ in the completion $K_\mathfrak{p}$ of $K$ at $\mathfrak{p}$ is cohomologically trivial (i.e., of finite projective dimension) if $\mathfrak{p}$ is a non-ramified finite prime (cf. [CF, p. 134]), in particular, if $\mathfrak{p} \notin S$. – We sketch the construction. For simplicity, we assume that the union $\bigcup_{\mathfrak{p} \in S} G_\mathfrak{p}$ of the decomposition groups $G_\mathfrak{p}$ of the $\mathfrak{p} \in S$ is all of $G$.

The local fundamental class at $\mathfrak{p} \in S$ is the generator of $H^2(G_\mathfrak{p}, K_\mathfrak{p}^\times)$ which has Hasse invariant $1/|G_\mathfrak{p}|$. Via the canonical isomorphisms (cf. [Br, p. 61])

$$H^2(G_\mathfrak{p}, K_\mathfrak{p}^\times) \cong H^1(G_\mathfrak{p}, \mathrm{Hom}(\Delta G_\mathfrak{p}, K_\mathfrak{p}^\times)) \cong \mathrm{Ext}^1_{G_\mathfrak{p}}(\Delta G_\mathfrak{p}, K_\mathfrak{p}^\times),$$

with the first one induced by $\Delta G_\mathfrak{p} \rightarrowtail \mathbb{Z} G_\mathfrak{p} \twoheadrightarrow \mathbb{Z}$, it gets translated into a distinguished short exact sequence

$$K_\mathfrak{p}^\times \rightarrowtail V_\mathfrak{p} \twoheadrightarrow \Delta G_\mathfrak{p}$$

in which $V_\mathfrak{p}$ is a cohomologically trivial $G_\mathfrak{p}$-module (cf. [GW1, §11]). The global analogue is

$$C_K \rightarrowtail \mathfrak{V} \twoheadrightarrow \Delta G,$$

with $\mathfrak{V}$ a cohomologically trivial $G$-module (cf. [RW2, §1]).

For $\mathfrak{p} \notin S$ we place $U_\mathfrak{p} \rightarrowtail U_\mathfrak{p} \twoheadrightarrow 0$ beside the first displayed sequence above. By inducing up all these local short exact sequences and then taking the direct product over all primes $\mathfrak{p}$ one arrives at

$$
\begin{array}{ccccc}
J_S & \rightarrowtail & V & \twoheadrightarrow & \underset{\mathfrak{p} \in S \bmod G}{\oplus} \mathrm{ind}^G_{G_\mathfrak{p}} \Delta G_\mathfrak{p} \\
\downarrow & & \downarrow & & \downarrow \\
C_K & \rightarrowtail & \mathfrak{V} & \twoheadrightarrow & \Delta G
\end{array}
$$

with surjective vertical maps.[2] The first half of a Tate sequence is now obtained from filling in a surjective map $\theta : V \twoheadrightarrow \mathfrak{V}$ in the above diagram, which makes it

---

[2]The right one is so because of our above assumption on the union of the decomposition groups.

commute, and then taking kernels; the existence of $\theta$ is due to the matching of the global and local fundamental classes. The second half is obtained from piecing together, in the same manner, the local sequences $\Delta G_{\mathfrak{p}} \rightarrowtail \mathbb{Z}G_{\mathfrak{p}} \twoheadrightarrow \mathbb{Z}$ (for $\mathfrak{p} \in S$) and the global sequence $\Delta G \rightarrowtail \mathbb{Z}G \twoheadrightarrow \mathbb{Z}$ (cf. [RW2]).

The just given construction provides a unique extension class $\tau_S \in \text{Ext}^2_G(\Delta S, E_S)$ which will be referred to as Tate's canonical class. Enlarging $S$ to $S'$ yields canonical maps $\Delta S \rightarrowtail \Delta S'$, $E_S \rightarrowtail E_{S'}$, so a triangle (cf. [We, p. 27])

$$\text{Ext}^2_G(\Delta S, E_S) \qquad\qquad\qquad\qquad \text{Ext}^2_G(\Delta S', E_{S'})$$
$$\searrow \qquad\qquad\qquad \swarrow$$
$$\text{Ext}^2_G(\Delta S, E_{S'})$$

in which the images of $\tau_S$ and $\tau_{S'}$ coincide.

As will be seen later, there are similar 4-term exact sequences for arbitrary sets $S$.

Chinburg [Ch1] has used a sequence $E_S \rightarrowtail A \rightarrow B \twoheadrightarrow \Delta S$ belonging to $\tau_S$ for the definition of a new invariant

$$\Omega_S = [A] - [B] \in K_0(\mathbb{Z}G).$$

As usual, $K_0(\mathbb{Z}G)$ denotes the Grothendieck group of finitely generated projective $\mathbb{Z}G$-modules in which $[A]$ is regarded as $[P_1] - [P_2]$, if $P_2 \rightarrowtail P_1 \twoheadrightarrow A$ is a projective resolution of $A$ (cf. [Se, p. 152]). Because of Schanuel's lemma (cf. [Br, p. 192]), $[A]$ is indeed well-defined.

**Theorem (Chinburg [Ch1]).** $\Omega_S$ *only depends on* $\tau_S$ *($S$ is large)*.

We henceforth write $\Omega$ rather than $\Omega_S$. Note that $\Omega$ already lies in the class group $Cl(\mathbb{Z}G)$ which is the kernel of the rank map $K_0(\mathbb{Z}G) \rightarrow \mathbb{Z}$, namely since $E_S$ and $\Delta S$, and so $A$ and $B$, have the same rank.

## 2e. Fröhlich's Hom language

Even though it may appear as if, by introducing $\Omega$, we had turned away from $A_\varphi$ as well as from the Galois structure of $E_S$, we have, in fact, reached the stage where we can combine both at the same time. This is made possible by Fröhlich's Hom language (cf. [Fr1, p. 18 ff.]) which allows to play off $F$-idèlic valued functions on characters $\chi$ of $G$ against algebraic objects – in our case, $A_\varphi$ against $\Omega$ – and which will be recalled next.

Let $J_F = \{(\dots, x_q, \dots) : x_q \in F_q, w_q(x_q) = 0 \text{ almost everywhere}\}$ denote the idèle group of $F$ and $R(G)$ the ring of complex characters of $G$. Both objects carry natural $\Gamma$-actions; the latter because $F$ is a splitting field for $G$. Fröhlich's Hom language provides a canonical epimorphism

$$\text{Hom}^+_\Gamma(R(G), J_F) \twoheadrightarrow Cl(\mathbb{Z}G)$$

with the +sign indicating that the function values at symplectic characters have to be positive at the real primes of $F$. Remember that an irreducible character $\chi$ is called symplectic if it is real-valued and has Schur index 2 over $\mathbb{R}$.

Now, rather than introducing the epimorphism itself, we prefer to display a preimage in $\mathrm{Hom}_\Gamma^+(R(G), J_F)$ of a given generator $[P_1] - [P_2]$ of $Cl(\mathbb{Z}G)$, where $P_1, P_2$ are projective $\mathbb{Z}G$-modules having the same rank, say $t$. The procedure is based on Swan's theorem by which projective $\mathbb{Z}G$-modules are locally free (cf. [CR1, p. 676]). In particular, $\mathbb{Q} \otimes_\mathbb{Z} P_1 \simeq \mathbb{Q} \otimes_\mathbb{Z} P_2 \simeq (\mathbb{Q}G)^t$. These isomorphisms shall be regarded as identifications so that $P_1$ and $P_2$ become full lattices on $(\mathbb{Q}G)^t$. As a result, they coincide locally almost everywhere because both have finite index in their sum. At each prime $p$ with $\mathbb{Z}_p \otimes_\mathbb{Z} P_1 \neq \mathbb{Z}_p \otimes_\mathbb{Z} P_2$ we still have an isomorphism $i_p : \mathbb{Z}_p \otimes_\mathbb{Z} P_1 \xrightarrow{\sim} \mathbb{Z}_p \otimes_\mathbb{Z} P_2$ which, on account of $\mathbb{Q}_p \otimes_\mathbb{Z} P_1 = \mathbb{Q}_p \otimes_\mathbb{Z} P_2$ is regarded as an element in $GL(t, \mathbb{Q}_pG)$. Setting $i_p = 1$ whenever $\mathbb{Z}_p \otimes_\mathbb{Z} P_1 = \mathbb{Z}_p \otimes_\mathbb{Z} P_2$, we obtain the idèle $(\dots, i_p, \dots)$ in $J(GL(t, \mathbb{Q}G))$ satisfying $(\dots, i_p, \dots)P_1 = P_2$. Then $\mathbf{i} = (\dots, \det T_\chi(i_p), \dots)$ is an idèle in $J_F$, if $T_\chi$ is a matrix representation of $G$ over $F$ with character $\chi$. The function $[\chi \mapsto \mathbf{i}]$ is the preimage of $[P_1] - [P_2]$. If $\chi$ is symplectic and q a real prime of $F$, then $T_\chi(F_qG)$ is a matrix ring over the real quaternions, and applying det (i.e., the reduced norm) yields positive values only.

The kernel of $\mathrm{Hom}_\Gamma^+(R(G), J_F) \twoheadrightarrow Cl(\mathbb{Z}G)$ takes care of the many choices we have made in the above process. It equals $\mathrm{Hom}_\Gamma^+(R(G), F^\times) \cdot \mathrm{Det}\, U(\mathbb{Z}G)$ with the latter group consisting of the maps $\chi \mapsto \det T_\chi(\mathbf{i})$ where $\mathbf{i}$ is a unit idèle of $\mathbb{Z}G$, so an idèle whose $p$-component is a unit in $\mathbb{Z}_pG$, for all $p$.

If $\Lambda$ happens to be an order in $\mathbb{Q}G$ containing $\mathbb{Z}G$, one gets an analoguous Hom description for $Cl(\Lambda)$.

$$\frac{\mathrm{Hom}_\Gamma^+(R(G), J_F)}{\mathrm{Hom}_\Gamma^+(R(G), F^\times)\mathrm{Det}\, U(\Lambda)} \simeq Cl(\Lambda).$$

Moreover, if $\Lambda_{\max}$ is a maximal order, then $\mathrm{Det}\, U(\Lambda_{\max}) = \mathrm{Hom}_\Gamma^+(R(G), U_F)$ with $U_F$ denoting the unit idèles of $F$ (the finite idèles of $F$ having a unit at each component), and so the elements in $Cl(\Lambda_{\max})$ can be represented by $F$-ideal-valued functions, that is, there is a surjection (cf. [Fr3, p. 126 ff.])

$$\mathrm{Hom}_\Gamma^*(R(G), I_F) \twoheadrightarrow Cl(\Lambda_{\max}).$$

Here, $I_F = J_F/U_F$ is the group of ideals in $F$ and the $*$ indicates to restrict to functions $f$ satisfying, for each irreducible character $\chi$, that $f(x)$ is an ideal in $F$ which is extended from an ideal in $\mathbb{Q}(\chi) \subset F$.

## 2f. Chinburg's Conjectures

We have finally reached the stage where we can turn back to $\Omega$ and $A_\varphi$. Two conjectures of Chinburg will guide us.

**Conjecture (SST).**[3] *If (ST) holds true, then* $A_\varphi(\check\chi)\mathfrak{o}_F = \mathfrak{q}_\varphi(\chi)$.

Evidence for this comes from Tate's theorem above; observed however the shift from $\chi$ to $\check\chi$ (if $\chi$ is rational-valued, $\chi = \check\chi$).

---

[3](cf. [Ta2, II, 7]); (SST) is short for Strong Stark Conjecture.

**Conjecture (RN).**[4] $\Omega \in Cl(\mathbb{Z}G)$ *is represented by* $\chi \mapsto A_\varphi(\check{\chi})W_{K/k}(\check{\chi})$ *in* $\mathrm{Hom}_\Gamma^+(R(G), J_F)$.

Here, $W_{K/k}$ is the Artin root number class, so by definition that element in $\mathrm{Hom}_\Gamma(R(G), J_F)$ that assigns to a non-symplectic $\chi$ the idèle 1 and to a symplectic $\chi$ the idèle having component 1 at the finite primes and the Artin root number $W(\chi)$ at $\infty$ which is a once and for all fixed infinite prime in $F$ (cf. [Ch3, p. 18]). Note that the infinite primes of $F$ are all in one $\Gamma$-orbit. The Artin root number $W(\chi)$ is taken from the functional equation (cf. [Ta2, pp. 18, 19])

$$\Lambda(1 - s, \chi) = W(\chi)\Lambda(s, \check{\chi})$$

with, as usual, $\Lambda(s, \chi)$ denoting the function $L(s, \chi)$ completed at infinity by $\Gamma$-factors.

Except for knowing that $W_{K/k}(\chi)$ has trivial components at the finite primes we only need to know one further property, namely $W_{K/k}(\chi)A_\varphi(\chi) > 0$ at $\infty$ for all symplectic $\chi$ (cf. [Ch3, p. 19]). So $A_\varphi W_{K/k}$ is indeed in $\mathrm{Hom}_\Gamma^+(R(G), J_F)$.

### 2g. Results

We conclude this section by listing results which have been achieved so far.

(1) *Let* $\Lambda_{\max}$ *be a maximal order in* $\mathbb{Q}G$ *containing* $\mathbb{Z}G$. *Under the natural epimorphism* $Cl(\mathbb{Z}G) \twoheadrightarrow Cl(\Lambda_{\max})$ *the invariant* $\Omega$ *is sent to an element represented by* $[\chi \mapsto q_\varphi(\chi)] \in \mathrm{Hom}_\Gamma^*(R(G), I_F)$.

This is a strong result, indeed; it has been proved by Chinburg [Ch1] and relates $\Omega$ and $q_\varphi$ as far as possible. Note that $\Omega$ does not depend on $\varphi$; the above result now suggests that there might exist some quantity $\Omega_\varphi$ arising from $\Omega$ and $\varphi$. And in fact, such $\Omega_\varphi$ can be defined in the Grothendieck group $K_0T(\mathbb{Z}G)^5$ of all finite $\mathbb{Z}G$-modules having finite projective dimension, see [GRW]. The natural map $K_0T(\mathbb{Z}G) \to Cl(\mathbb{Z}G)$ takes $\Omega_\varphi$ to $\Omega$, and it is only in the $K_0T$ set-up where (1) can still be sharpened; we come back to this in section 5.

A second remark is appropriate. $\Omega$ depends on Tate's canonical class $\tau_S$ but $q_\varphi$ does not. Hence, by passing to a maximal order one forgets about the extension class of the Tate sequence. If $\Omega$, in $Cl(\mathbb{Z}G)$, is replaced by the set $\{[A] - [B]\}$ of differences of all cohomologically trivial $\mathbb{Z}G$-modules $A$, $B$ fitting into an exact sequence $E_S \rightarrowtail A \to B \twoheadrightarrow \Delta S$, then one obtains a coset of a certain subgroup in $Cl(\mathbb{Z}G)$ (a so-called $B$-group) which lies in the kernel $D(\mathbb{Z}G)$ of $Cl(\mathbb{Z}G) \to Cl(\Lambda_{\max})$. This group has been studied in [GW2].

(2) *Assume that* $K$ *is contained in a cyclotomic field* $\mathbb{Q}(\zeta_m)$ *in which 2 does not ramify. Then* $(SST)$ *is true.*

This is shown in [RW3]. We shall sketch the proof in section 4.

---

[4] [Ch2]; (RN) is short for Root Number Conjecture.
[5] cf. [Fr2, p. 14 ff.].

By putting (1) and (2) together we arrive at the representing homomorphism $[\chi \mapsto A_\varphi(\check{\chi}) \circ_F]$ for $\Omega$ in $Cl(\Lambda_{max})$. On the other hand, Chinburg's root number conjecture (RN) asserts that $[\chi \mapsto A_\varphi(\check{\chi}) W_{K/k}(\check{\chi})]$ represents $\Omega$ in $Cl(\mathbb{Z}G)$. Now, since $W_{K/k}(\check{\chi})$ has component 1 at all finite primes, the idèle $A_\varphi(\check{\chi}) W_{K/k}(\check{\chi})$ has ideal content $A_\varphi(\check{\chi}) \circ_F$, and thus

$$(RN) \text{ is true modulo } D(\mathbb{Z}G), \text{ i.e., in } Cl(\Lambda_{max}),$$

$$\text{provided that } K/k \text{ is as in (2)}.$$

There is a special case where we even reach (RN) itself; in fact, for nothing. This is when $D(\mathbb{Z}G) = 1$, so for example when $G$ is a cyclic group of prime order (cf. [CR2, p. 253]).

Observe that (2) is restricted to one of the two known situations in which the Stark conjecture (ST) is valid. The other one, when $K$ is an abelian extension of an imaginary quadratic field $\mathbb{Q}(\sqrt{-d})$, $0 < d \in \mathbb{Z}$, has been studied in [Bw] where the following is shown:

(SST) *holds true if $\mathbb{Z}$ gets replaced by a semi-localization in which all prime divisors of $|G|$ and of the class number $h_d$ of $\mathbb{Q}(\sqrt{-d})$ are invertible.*

The restriction is due to difficulties with the Main Conjecture of Iwasawa theory when working over a base field bigger than $\mathbb{Q}$.[6]

(3) *(RN) is correct in infinitely many cases where $k = \mathbb{Q}$ and $G$ is the quaternion group of order 8. It is also correct for abelian field extensions $K/\mathbb{Q}$ of prime power conductor.*

This comes from [Ch3] and [Gc], respectively. Some related work is in [Bu1] and [BH]. We do not go into this.

With (1), (2) and (3) we have reached the state of affairs with respect to (SST) and (RN). However, a new approach towards (RN) has emerged from introducing the generalized $\Omega$-invariants $\Omega_\varphi \in K_0 T(\mathbb{Z}G)$ which have already been mentioned and will be turned to in section 5.

The last result in our list concerns the Galois structure of $E_S$, if $S$ is large.

(4) *Assume $S$ to be large. Then there is a character $\varepsilon$ on $H^2(G, \text{Hom}(\Delta S, \mu))$, which is constructed by means of class field theoretical data, such that the stable isomorphism class of $E_S$ is determined by $\mu$, $S$, $\varepsilon$ and $\Omega$.*

The result is taken from [GW2]. Remember that two $\mathbb{Z}G$-modules $M_1$, $M_2$ are said to be stably isomorphic, if $M_1 \oplus L \simeq M_2 \oplus L$ for some finitely generated free $\mathbb{Z}G$-module $L$. In the next section, following [RW1] (which actually initiated [GW2]), we discuss (4) in a special situation where the character $\varepsilon$ is trivial. We

---

[6]Equally well, the assumption on the behaviour of the prime 2 in (2) arises from the incompleteness of the proof of the Main Conjecture of Iwasawa theory. – With respect to the Main Conjecture of Iwasawa theory see [MW], [Wil, 2].

also look at the Galois structure of $E_S$ for arbitrary $S$, but only for very limited fields $K$ [Du].

We conclude our exposition, in section 6, by adding a few remarks on the relation of the presented material, in particular of (RN) and its refined version introduced in section 5, with more general conjectures concerning cohomology and $L$-values made by Bloch and Kato.

## 3. $E_S$ for special $K$

The main restriction on $K$ in this section is that its group $\mu$ of roots of unity shall be cohomologically trivial. This is the case whenever $[K : k]$ is odd and $K$ is real; in particular, when $K$ is an abelian extension of $\mathbb{Q}$ of odd degree. An immediate consequence is the splitting of the sequence $\mu \rightarrowtail E_S \twoheadrightarrow U_S \stackrel{\text{def}}{=} E_S/\mu$, which results from the torsion freeness of $U_S$. Indeed, $\operatorname{Hom}(U_S, \mu) \rightarrowtail \operatorname{Hom}(U_S, E_S) \twoheadrightarrow \operatorname{Hom}(U_S, U_S)$ is exact and $H^1(G, \operatorname{Hom}(U_S, \mu))$ vanishes by [Se, IX, 5]. Hence, in order to describe $E_S$ it suffices to describe $U_S$.

**Remark.** As $\mu$ is cohomologically trivial, so is $\operatorname{Hom}(\Delta S, \mu)$. Therefore the character $\varepsilon$ in (4) above vanishes.

### 3a. Sufficiently Large Sets S

We not only require $S$ to be large but also to satisfy

one prime in $S_f$, say $\mathfrak{p}_1$, "splits" in $K$, i.e., $\mathfrak{p}_1^g \neq \mathfrak{p}_1$ for $1 \neq g \in G$,

$\sum_{i=2}^r \operatorname{ind}_{G_{\mathfrak{p}_i}}^G (1)$ contains every irreducible character of $G$ (where $\operatorname{ind}_{G_{\mathfrak{p}_i}}^G$ denotes induction from the decomposition group $G_{\mathfrak{p}_i}$ of $\mathfrak{p}_i$ up to $G$).

Let then $E_S \rightarrowtail A \to B \twoheadrightarrow \Delta S$ denote an exact sequence in which $A$ has finite projective dimension and $B$ is projective. Note that we do not specify the extension class. By adding a suitable projective module to $A$ as well as to $B$ we arrive at $B$ being free, say $B \simeq (\mathbb{Z}G)^t$.

Obviously $E_S$ and $A$ have the same torison, namely $\mu$. Dividing it out yields

$$U_S \rightarrowtail P \to B \twoheadrightarrow \Delta S$$

with $P = A/\mu$ projective as a torsion-free quotient of two cohomologically trivial modules (cf. [Se, IV, 5]). By means of Swan's generation theorem (cf. [Gk, p. 43]) we now construct a $\mathbb{Z}G$-lattice $Y_S$ which is related to $U_S$, and measure the deviation of the two in terms of $\Omega$ and $\mu$.

Denote by $L$ the image of the map $P \to B$. Then $L$ is a lattice. The rank $t$ of $B$, and so of $P$, is $\geq 2$, since $\Delta S$ cannot be cyclic over $\mathbb{Z}G$ according to our assumptions on $S$. For the same reason, with $\chi_{U_S}$ denoting the character of $U_S$, $\chi_{U_S} - \operatorname{ind}_1^G(1)$ contains every irreducible character of $G$ (cf. [RW1, Prop. 1b]) with $\bar{U} = U_S$). So the assumptions made in Swan's generation theorem are fulfilled and we arrive at some surjection $(\mathbb{Z}G)^t \twoheadrightarrow L$, the kernel of which we denote by $Y_S$.

**Observation 1.** *$Y_S$ is unique up to isomorphism.*

**Observation 2.** *$Y_S$ and $U_S$ are locally isomorphic.*

This is ensured by Schanuel's lemma and the theorem of Swan by which finitely generated projective $\mathbb{Z}G$-modules are locally free, in particular, $\mathbb{Z}_p \otimes_\mathbb{Z} P \simeq (\mathbb{Z}_pG)^t$ for all rational primes $p$. Indeed, by Schanuel's lemma the kernal $\tilde{Y}_S$ arising from a second surjection $(\mathbb{Z}G)^t \twoheadrightarrow L$ satisfies $Y_S \oplus (\mathbb{Z}G)^t \simeq \tilde{Y}_S \oplus (\mathbb{Z}G)^t$; moreover, because of Swan's theorem, $(\mathbb{Z}_p \otimes_\mathbb{Z} Y_S) \oplus (\mathbb{Z}_pG)^t \simeq (\mathbb{Z}_p \otimes_\mathbb{Z} U_S) \oplus (\mathbb{Z}_pG)^t$. In the first case we use the cancellation theorem (cf. [Sw, p. 175]) in order to get rid of $(\mathbb{Z}G)^t$ [RW1, Prop. 2c)]. In the second case we can already cancel $(\mathbb{Z}_pG)^t$ on account of the Krull-Remak-Schmidt theorem (cf. [CR1, p. 767]).

Putting things together we conclude that $U_S \oplus B \simeq U_S \oplus (\mathbb{Z}G)^t \simeq Y_S \oplus P$. In $K_0(\mathbb{Z}G)$ the difference $[P] - [B]$ is given by $\Omega$ and $\mu$, if the Tate sequence is built to $\tau_S$:

$$[P] - [B] = [A] - [\mu] - [B] = \Omega - [\mu];$$

remember that $\mu$ is cohomologically trivial and thus induces the element $[\mu] \in K_0(\mathbb{Z}G)$.

We claim that $U_S$ is fully detemined by $Y_S$, $\Omega$ and $\mu$. Note that only $\Omega$ depends on the Tate class $\tau_S$; $Y_S$, however, does not by the above.

In order to verify the claim, let $X$ be any $\mathbb{Z}G$-lattice with $X \oplus B' \simeq Y_S \oplus P'$ for projective modules $B'$, $P'$ having the same rank and satisfying $[P'] - [B'] = \Omega - [\mu]$. The equality $[B] + [P'] = [B'] + [P]$ in $K_0(\mathbb{Z}G)$ amounts to an isomorphism $B \oplus P' \oplus Z \simeq B' \oplus P \oplus Z \overset{\text{def}}{=} P_1$ for some free module $Z$. Therefore $U_S \oplus P_1 \simeq X \oplus P_1$, i.e., $U_S \simeq X$ by the cancellation property.

The situation becomes especially nice when $G = \langle g_0 \rangle$ is cyclic. We then just tensor

$$\mathbb{Z} \overset{\hat{G}}{\rightarrowtail} \mathbb{Z}G \overset{g_0-1}{\longrightarrow} \mathbb{Z}G \overset{\text{aug}}{\twoheadrightarrow} \mathbb{Z}$$

with $\Delta S$ and end up with $Y_S = \Delta S$.

## 3b. Small K

In this subsection, which is based on [Du], $K$ is restricted to be a cyclic extension of $k = \mathbb{Q}$ of odd prime degree $l$. The set $S$, however, can be arbitrary.

The peculiarity of the situation is the possibility of classifying all $\mathbb{Z}G$-lattices by means of the Diederichsen-Reiner theorem [CR0, pp. 508–513].[7] To recall it we need some notation first.

---

[7] At the same time one might object to using the theorem at all, because it is not possible to give a list of the $\mathbb{Z}G$-lattices for arbitrary groups $G$, and so the approach is limited.

Fix a primitive $l$-th root $\zeta_l$ of unity as well as a generator $g$ of $G$ and regard the $l$-th cyclotomic field $\mathbb{Q}(\zeta_l)$ as a $G$-module by $gx = \zeta_l x$ for $x \in \mathbb{Q}(\zeta_l)$. Each pair $\mathfrak{a}, a$ consisting of an ideal $\mathfrak{a}$ in $\mathbb{Q}(\zeta_l)$ (so a finitely generated $\mathbb{Z}\,G$-submodule of $\mathbb{Q}(\zeta_l)$) and an element $a \in \mathfrak{a}$ gives rise to a $\mathbb{Z}G$-module $\mathfrak{a} \oplus \mathbb{Z}$ with $g$-action $g(x, n) = (\zeta_l x + na, n)$, $x \in \mathfrak{a}, n \in \mathbb{Z}$; it will be denoted by $(\mathfrak{a}, a)$.

**Theorem (Diederichsen-Reiner).** *Let $M$ be a finitely generated $\mathbb{Z}G$-lattice. Then*

> *either* $M \simeq (\mathbb{Z}G)^{r_3} \oplus \mathbb{Z}[\zeta_l]^{r_2-1} \oplus \mathfrak{m} \oplus \mathbb{Z}^{r_1}$ *with integers* $r_3 \geq 0, r_2 \geq 1, r_1 \geq 0$ *and an ideal* $\mathfrak{m}$ *of* $\mathbb{Q}(\zeta_l)$,

> *or* $M \simeq (\mathbb{Z}G)^{r_3-1} \oplus (\mathfrak{m}, m) \oplus \mathbb{Z}^{r_1}$ *with integers* $r_3 \geq 1, r_1 \geq 0$ *and a pair* $(\mathfrak{m}, m)$ *such that* $m \notin (1 - \zeta_l)\mathfrak{m}$.

*The isomorphism type of $M$ is fully determined by the non-negative integers $r_1, r_2{}^8, r_3$ and the ideal class $\bar{\mathfrak{m}}$ of $\mathfrak{m}$ (which will be referred to as the Steinitz class of $M$). Moreover, if $_{\hat{G}}M = \{m \in M : \hat{G}(m) = 0\}$ then*

> $_{\hat{G}}M$ *is a $\mathbb{Z}[\zeta_l]$-module having Steinitz class $\bar{\mathfrak{m}}$ (cf. [Re, p. 49]) and rank $r_2 + r_3$,*

> $r_1 + r_3$ *is the $\mathbb{Z}$-rank of $M/_{\hat{G}}M$,*

> $r_3 = \dim_{F_l}(g-1)M/(\zeta_l - 1)_{\hat{G}}M$.

**Corollary.** $r_1 = \dim_{F_l} H^0(G, M), r_2 = \dim_{F_l} H^1(G, M)$.

Indeed, $H^0(G, M) = M^G/\hat{G}M$, $H^1(G, M) =_{\hat{G}}M/(g-1)M$; so the corollary readily follows from the decomposition of $M$ given in the above theorem.

We now turn to $U_S$ and discuss its invariants $r_1, r_2, r_3, \bar{u}_S$. To that end let $v_1, \ldots, v_t$ be the rational primes which ramify in $K$. They generate $l$-powers of $G$-invariant prime ideals $\mathfrak{v}_j$ in $K$, the ideal classes of which we denote by $\bar{\mathfrak{v}}_j$. The numbering of the $\mathfrak{v}_j$ shall be chosen in such a way that $\mathfrak{v}_1, \ldots, \mathfrak{v}_s \in S_f$ but $\mathfrak{v}_{s+1}, \ldots, \mathfrak{v}_t \notin S_f$.

As follows from the theory of genus fields, the group $cl/(g-1)cl$ is an $\mathbb{F}_l$-vector space of dimension $t - 1$ (cf. [Fr4, p. 26]), hence the same is true for $cl^G$ (the Herbrand quotient of $G$ with respect to $cl$ vanishes). This information is needed in case b) of the following theorem.

**Theorem.**

> a) $H^1(G, U_S) = 0$ *if* $s = t$.

> b) *Let the intersections of the subgroup of $cl$ generated by the primes in $S_f \setminus \{\mathfrak{v}_1, \ldots, \mathfrak{v}_s\}$ with $cl^G$ and with $\langle \bar{\mathfrak{v}}_1, \ldots, \bar{\mathfrak{v}}_s \rangle$ have $\mathbb{F}_l$-dimensions $d_1$ and $d_2$, respectively. Then*

---

${}^8 r_2 = 0$ means that the second case applies.

b1) $H^1(G, U_S) = (\mathbb{Z}/l)^{d_1-d_2}$ *if* $s \neq 0$ *and* $\bar{v}_1, \ldots, \bar{v}_s$ *are linearly dependent,*

b2) $H^1(G, U_S) = (\mathbb{Z}/l)^{d_1-d_2+1}$ *if* $s = 0$ *or if* $\bar{v}_1, \ldots, \bar{v}_s$ *are linearly indepen-*
*dent.*

The proof is based on the isomorphism $H^1(G, U_S) = H^1(G, E_S) \simeq P(S)^G/\phi$ $(\mathbb{Q}^\times)$ with $P(S)$ denoting the quotient of the group of principal ideals in $K$ by the subgroup $\{ao : a \in K^\times, w\mathfrak{p}(a) = 0 \text{ for } \mathfrak{p} \notin S_f\}$, and where $\phi$ is the natural map $K^\times \twoheadrightarrow P(S)$ having kernel $E_S$. For the actual computation of $P(S)^G/\phi(\mathbb{Q}^\times)$ see [Du, III]. – Observe that b2) applies to $S = S_\infty$, in which case $d_1 = d_2 = 0$.
  If $d_1 = d_2$, an immediate consequence is

in case b1): $U_S \simeq (\mathbb{Z}G)^d \oplus (u_S, u_S) \oplus \mathbb{Z}^{r-ld-1}$ with $d$ denoting the number
of rational primes below $S_f = \{\mathfrak{p}_1, \ldots, \mathfrak{p}_r\}$ which decompose in $K$;

in case b2): $U_S \simeq (\mathbb{Z}G)^d \oplus u_S \oplus \mathbb{Z}^{r-ld}$.

As before, $u_S$ is here an ideal in $\mathbb{Q}(\zeta_l)$ and $u_S \notin (1 - \zeta_l)u_S$. For $S = S_\infty$ we
obtain $U = U_{S_\infty} \simeq u = u_{S_\infty}$.
  We need to determine the ideal class of $u_S$. From the exact sequence $E \rightarrowtail E_S \rightarrow I_S \twoheadrightarrow cl(S)$, in which $I_S$ is the group of ideals in $K$ generated by the $\mathfrak{p}_i \in S_f$ and $cl(S)$ its image in $cl$ (so $cl/cl(S) = cl_S$), we get

$$U \rightarrowtail U_S \rightarrow I_S \twoheadrightarrow cl(S)$$

and furthermore, on passing to the Grothendieck group $G_0(\mathbb{Z}G)$ of finitely gener-
ated $\mathbb{Z}G$-modules and then projecting into the ideal class group of $\mathbb{Q}(\zeta_l)$ (cf. [Sw, p. 74]), the equation

$$(U_S) \cdot (cl(S)) = (U) \cdot (I_S).$$

Here $(M)$, for a $\mathbb{Z}G$-lattice $M$, is the class $\bar{m}$ appearing in the Diederichsen-Reiner
theorem; however, if $M$ is finite, it is the class of the $\mathbb{Z}[\zeta_l]$-order of the $\mathbb{Z}[\zeta_l]$-
module $(g - 1)M$ (cf. section 2c.).[9]
  As is readily seen, $I_S = (\mathbb{Z}S_f) = 1$. Putting everything together we arrive at

(*) $\qquad\qquad\qquad \bar{u}_S \cdot (cl_S)^{-1} = \bar{u} \cdot (cl)^{-1}.$

At this stage we bring $\Omega$ into play. As has been pointed out in section 2g., $\Omega = [A] - [B]$ vanishes in our case. Therefore, $1 = (\Omega) = (A) \cdot (B)^{-1}$ in the ideal class
group of $\mathbb{Q}(\zeta_l)$, into which $K_0(\mathbb{Z}G)$ is homomorphically mapped by sending the
projective module $M$ to its Steinitz class $\bar{m}$. From $U_S \rightarrowtail A \rightarrow B \twoheadrightarrow \Delta S$, with $S$

---

[9]Note that $H^{-1}(G, M)$ is an $l$-group and that consequently $\mathbb{Z}[\zeta_l]/(1 - \zeta_l)$ is the only type of
composition factor in $\hat{\phantom{x}}_G M/(g-1)M$. Therefore, $(M) = |(g-1)M|_{\mathbb{Z}[\zeta_l]} = |\hat{\phantom{x}}_G M|_{\mathbb{Z}[\zeta_1]}$. In particular,
we have $(cl(S)) = |cl(S)|_{\mathbb{Z}[\zeta_l]}$. This equally well applies to $cl$ and $cl_S$.

large, we obtain $1 = (\Omega) = \bar{u}_S \cdot (\Delta S)^{-1}$; however, $(\Delta S) = 1$, may $S$ be large or not. Consequently, and again for large $S$,

$$1 = (\Omega) = \bar{u}_S = \bar{u}_S(cl_S)^{-1} = \bar{u}(cl)^{-1},$$

i.e., $\bar{u} = (cl)$, from which $\bar{u}_S = (cl_S)$ results for arbitrary $S$ because of $(*)$ above. – We repeat

**Theorem.** *For each finite $G$-stable set $S \supset S_\infty$ we have $\bar{u}_S = (cl_S)$.*

**Corollary.** *The following are equivalent:*

$$\mathfrak{u} \simeq \mathbb{Z}[\zeta_l]; \, \bar{u} = 1; \, (cl) = 1; \, \exists \varepsilon \in U$$
$$\text{such that } \varepsilon, g(\varepsilon), \ldots, g^{l-2}(\varepsilon) \text{ is a } \mathbb{Z}\text{-basis of } U.$$

An $\varepsilon$ as above is called a Minkowski unit. Note that there exists a Minkowski unit whenever the ideal class group $cl$ of $K$ or the one of $\mathbb{Q}(\zeta_l)$ is trivial, so for sure if $l \leq 19$.

We close this subsection by displaying a unique ideal $\mathfrak{u}_1$ in the class $\bar{u}$ of $\mathfrak{u} \simeq U$. It is the inverse of an integral ideal and has the class number $h$ of $K$ as its norm.

To that end we look for a free sublattice $\tilde{\mathfrak{u}}$ in $\mathfrak{u}$ (so $\tilde{\mathfrak{u}}$ is a principal ideal) and write $\tilde{\mathfrak{u}} = \mathfrak{u}\mathfrak{a}$ with some integral ideal $\mathfrak{a}$ of $\mathbb{Q}(\zeta_l)$ (cf. [Re, p. 49]). Then $\bar{u} = \bar{\mathfrak{a}}^{-1}$ and we define $\mathfrak{u}_1 = \mathfrak{a}^{-1}$. With respect to its norm observe that the definition of the $\mathbb{Z}[\zeta_l]$-index yields $\mathfrak{a} = |\mathfrak{u}/\tilde{\mathfrak{u}}|_{\mathbb{Z}[\zeta_l]}$, so $N_{K/k}(\mathfrak{a}) = [\mathfrak{u} : \tilde{\mathfrak{u}}]_{\mathbb{Z}}$ (since $N_\mathfrak{q} = |\mathbb{Z}[\zeta_l]/\mathfrak{q}|$ for a prime ideal $\mathfrak{q}$ of $\mathbb{Z}[\zeta_l]$).

To carry out the just described programme we firstly identify $U$ with the norm 1 units in $E$ and so are allowed to regard $U$ itself as a $\mathbb{Z}[\zeta_l]$-module, i.e., $U = \mathfrak{u}$, namely because $\mathbb{Z}G/\hat{G} \simeq \mathbb{Z}[\zeta_l]$. The free submodule $\tilde{U}$ is obtained from the cyclotomic units. Let $f$ denote the conductor of $K$. By [Ha, p. 25],

$$\eta = \prod_{\substack{\pm a \in (\mathbb{Z}/f)^\times \\ a \bmod H}} \sqrt{\frac{(1 - \zeta_f^a)(1 - \zeta_f^{-a})}{(1 - \zeta_f^{za})(1 - \zeta_f^{-za})}} \in U.$$

Here, $\zeta_f$ denotes a primitive $f$-th root of unity, $H$ the Galois group of $\mathbb{Q}(\zeta_f)/K$ and $z \in (\mathbb{Z}/f)^\times$ is such that $\zeta_f \mapsto \zeta_f^z$ induces $g$ on $K$. Moreover,

$$\varphi : \Delta S_\infty \to U \leq E, g^i \infty \mapsto g^i \eta$$

is a $\mathbb{Z}G$-map with cokernel of order $h$; in particular it is injective. Above, $\infty$ is the infinite prime of $K$ which is induced by $K \subset \mathbb{Q}(\zeta_f) \hookrightarrow \mathbb{C}, \zeta_f \mapsto e^{2\pi i/f}$. Since $\mathbb{Z}S_\infty \simeq \mathbb{Z}G$ and $\hat{G}(g - 1) = 0$, we can view $\Delta S_\infty$ as a $\mathbb{Z}[\zeta_l]$-module which then is isomorphic to the principal ideal generated by $\zeta_l - 1$. Of course, $\tilde{U} = \varphi(\Delta S_\infty)$.

**Remark.** Observe that the map $\varphi$ is an explicit isogeny; cf. section 2a.

## 4. Idea of Proof of (SST)

In this section, which is a review of [RW2, RW3], we are going to sketch the proof of

(2) *Assume that $K$ is contained in a cyclotomic field $\mathbb{Q}(\zeta_m)$ in which 2 does not ramify. Then $A_\varphi(\check{\chi})\mathfrak{o}_F = \mathfrak{q}_\varphi(\chi)$.*

We begin by describing the strategy. First of all, in 4a., we generalize the concept of Tate sequences to small sets $S$, i.e., to such finite sets $S$ of primes of $K$ which are just $G$-stable and contain $S_\infty$. The new sequence looks like $E_S \rightarrowtail A \to B \twoheadrightarrow \nabla_S$ with cohomologically trivial $A$ and projective $B$ and a right end $\nabla_S$ having torsion $cl_S$ and incorporating $\Delta S$ as well as the ramified primes outside of $S$. Here, no restriction is imposed on the extension $K/k$.

We then, in 4b., extend the notion of an isogeny to include all $\mathbb{Z}G$-homomorphisms $\varphi$ with finite kernel and cokernel and mapping the right hand end $\nabla_S$ of the Tate sequence into $\tilde{E}_S \overset{\text{def}}{=} E_S \oplus (\mathbb{Z}G)^{r_S}$, where $r_S$ is the number of $G$-orbits of ramified primes which are not in $S$. Correspondingly, we define a new regulator, a new $A$-number and a new $q$-ideal. Whenever $S$ happens to be large, all new quantities coincide with the respective old ones. Large and small sets $S$ are brought together by means of the

**Theorem.** $\mathfrak{a}_\varphi(\chi) \overset{\text{def}}{=} A_\varphi(\check{\chi})\mathfrak{o}_F/\mathfrak{q}_\varphi(\chi)$ *is a rank 1 - $\mathfrak{o}_F$ - module in $\mathbb{C}$ which is independent of $\varphi$ and $S$.*

We may therefore restrict ourselves to only looking at $S = S_\infty$ and a suitable isogeny $\varphi$. Note, however, that the $\varphi$ is no longer a map from $\Delta S_\infty$ to $E$ and that only on this level we can expect to come by an explicit isogeny. Thus we are led to finding sort of a canonical way of getting a $\varphi : \nabla_\infty \to \tilde{E}$ from a given $\underline{\varphi} : \Delta S_\infty \to E$. A construction is given in 4d.

Having carried out these preparations, we turn to the special situation (2) and choose as $\varphi$ the Ramachandra map. It is derived from the Ramachandra units (cf. [Wa, p. 147]) and defined in 4e. The actual proof of $\mathfrak{a}_\varphi(\chi) = 1$ now emerges from a character theoretical reduction to local cases in which $\mathfrak{a}_\varphi(\chi)$ becomes a power of a prime ideal $\mathfrak{l}|l$ and $\chi$ is an $l$-adic character with $l \nmid |G/\ker \chi|$. In particular, we have arrived at a tame situation. Here, we can triply exploit number theory and obtain the claim from the validity of Leopoldt's conjecture, from a theorem of Fröhlich concerning Galois Gauß sums and resolvents, and from the Main Conjecture of Iwasawa theory.

### 4a. Tate Sequences for Small S

If $S$ is no longer large, we have to modify the construction given in 2d.. Indeed, for a $\mathfrak{p} \notin S_f$ the local unit group $U_\mathfrak{p}$ need not be cohomologically trivial anymore, since $\mathfrak{p}$ may be a ramified prime. However, starting from $K_\mathfrak{p}^\times \rightarrowtail V_\mathfrak{p} \twoheadrightarrow \Delta G_\mathfrak{p}$, with cohomologically trivial $V_{\mathfrak{p}}$, we can build the push-out along the valuation $K_{\mathfrak{p}}^{\times} \twoheadrightarrow \mathbb{Z}$ and arrive at

$$
\begin{array}{ccc}
U_{\mathfrak{p}} & = & U_{\mathfrak{p}} \\
\updownarrow & & \updownarrow \\
K_{\mathfrak{p}}^{\times} \rightarrowtail & V_{\mathfrak{p}} & \twoheadrightarrow \; \Delta G_{\mathfrak{p}} \\
\downarrow & \downarrow & \| \\
\mathbb{Z} \;\; \rightarrowtail & W_{\mathfrak{p}} & \twoheadrightarrow \; \Delta G_{\mathfrak{p}},
\end{array}
$$

which not only yields $U_{\mathfrak{p}} \rightarrowtail V_{\mathfrak{p}} \twoheadrightarrow W_{\mathfrak{p}}$ but also defines the inertia lattice $W_{\mathfrak{p}}$. It has been introduced in [GW1] together with the following commutative diagram with canonical outer vertical maps and the obvious lower row:

$$
\begin{array}{ccccc}
W_{\mathfrak{p}} & \rightarrowtail & \mathbb{Z}G_{\mathfrak{p}} \oplus \mathbb{Z}G_{\mathfrak{p}} & \twoheadrightarrow & W_{\mathfrak{p}}^{0} \; = \; \mathrm{Hom}(W_{\mathfrak{p}}, \mathbb{Z}) \\
\downarrow & & \downarrow & & \downarrow \\
\Delta G_{\mathfrak{p}} & \rightarrowtail & \mathbb{Z}G_{\mathfrak{p}} & \twoheadrightarrow & \mathbb{Z}.
\end{array}
$$

Note that $W_{\mathfrak{p}} = \mathbb{Z}G_{\mathfrak{p}}$, if $\mathfrak{p}$ is non-ramified.

Now, rather than merely placing the trivial sequences $U_{\mathfrak{p}} \rightarrowtail U_{\mathfrak{p}} \twoheadrightarrow 0$ for $\mathfrak{p} \notin S$ beside the sequences $K_{\mathfrak{p}}^{\times} \rightarrowtail V_{\mathfrak{p}} \twoheadrightarrow \Delta G_{\mathfrak{p}}$ for $\mathfrak{p} \in S$, we first choose a sufficiently large set $S' \supset S$ and then use

$$
\begin{array}{ccccccc}
U_{\mathfrak{p}} & \rightarrowtail & U_{\mathfrak{p}} & \twoheadrightarrow & 0 & \text{for} & \mathfrak{p} \notin S' \\
U_{\mathfrak{p}} & \rightarrowtail & V_{\mathfrak{p}} & \twoheadrightarrow & W_{\mathfrak{p}} & \text{for} & \mathfrak{p} \in S' \backslash S \\
K_{\mathfrak{p}}^{\times} & \rightarrowtail & V_{\mathfrak{p}} & \twoheadrightarrow & \Delta G_{\mathfrak{p}} & \text{for} & \mathfrak{p} \in S.
\end{array}
$$

Then, as in 2*d*., we get

$$
\begin{array}{ccc}
J_S & \rightarrowtail \;\; V_{S'} \;\; \twoheadrightarrow & \left[ \underset{S/G}{\oplus} \mathrm{ind}_{G_{\mathfrak{p}}}^{G} \Delta G_{\mathfrak{p}} \oplus \underset{(S'\backslash S)/G}{\bigoplus} \mathrm{ind}_{G_{\mathfrak{p}}}^{G} W_{\mathfrak{p}} \right] \overset{\mathrm{def}}{=} R \\
\downarrow & & \downarrow \\
C_K & \rightarrowtail \;\; \mathfrak{V} \;\; \twoheadrightarrow & \Delta G
\end{array}
$$

Because $S'$ has been chosen sufficiently large, the right vertical map is surjective. The left one has cokernel $cl_S$. Again, the diagram can be filled in with a surjective $\theta : V_{S'} \twoheadrightarrow \mathfrak{V}$. Thus $A = \ker \theta$ is cohomologically trivial. Applying the snake lemma to the diagram yields $E_S \rightarrowtail A \rightarrow B' \twoheadrightarrow cl_S$ with $B'$ denoting the kernel of the right vertical map.

We now go back to the second diagram in this subsection and combine its top row for the ramified $\mathfrak{p} \in S' \backslash S$ and the canonical sequences $\Delta G_{\mathfrak{p}} \rightarrowtail \mathbb{Z}G_{\mathfrak{p}} \twoheadrightarrow \mathbb{Z}$

for the unramified $p \in S'$ in order to perform an obvious construction yielding

$$
\begin{array}{ccc}
R & \rightarrowtail & N & \twoheadrightarrow & M \\
\downarrow & & \downarrow & & \downarrow \\
\Delta G & \rightarrowtail & \mathbb{Z}G & \twoheadrightarrow & \mathbb{Z}
\end{array}
$$

with

$$
N = \bigoplus_{\substack{S'/G \\ p \text{ non-ramified}}} \text{ind}_{G_p}^G \mathbb{Z}G_p \oplus \bigoplus_{\substack{(S'\backslash S)/G \\ p \text{ ramified}}} \text{ind}_{G_p}^G (\mathbb{Z}G_p)^2
$$

$$
\text{and } M = \bigoplus_{S/G} \text{ind}_{G_p}^G \mathbb{Z} \oplus \bigoplus_{\substack{(S'\backslash S)/G \\ p \text{ ramified}}} \text{ind}_{G_p}^G W_p^0 .
$$

Taking kernels leads to $B' \rightarrowtail B \twoheadrightarrow \bar{\nabla}_S$ and to the definition of $\bar{\nabla}_S$. Observe that $B$ is projective. Define $\nabla_S$ to be the push-out

$$
\begin{array}{ccc}
B' & \rightarrowtail & B & \twoheadrightarrow & \bar{\nabla}_S \\
\downarrow & & \downarrow & & \| \\
cl_S & \rightarrowtail & \nabla_S & \twoheadrightarrow & \bar{\nabla}_S .
\end{array}
$$

This together with $E_S \rightarrowtail A \to B' \twoheadrightarrow cl_S$ gives the new Tate sequence

$$
E_S \rightarrowtail A \to B \twoheadrightarrow \nabla_S .
$$

With respect to its naturalness and to all details we refer to [RW2]. For the sake of completeness we just want to add that

if $S$ is large, then $\nabla_S = \Delta S$ and the sequence belongs to Tate's canonical class $\tau_S$;

$[A] - [B] + r_S[\mathbb{Z}G]$ is Chinburg's $\Omega$-invariant;

the extension class of $cl_S \rightarrowtail \nabla_S \twoheadrightarrow \bar{\nabla}_S$ in $\text{Ext}_G^1(\bar{\nabla}_S, cl_S)$ can be described in terms of class field theory.

## 4b. Adjusting the Data

Since the notion of the $q$-index $q_\varphi(\chi)$, as a function on characters, requires a Tate sequence by which the domains of definition and values of $\varphi$ are linked, we redefine the concept of an isogeny. From now on this expression will be reserved for a $G$-homomorphism

$$
\nabla_S \xrightarrow{\varphi} E_S \oplus (\mathbb{Z}G)^{r_S} = \tilde{E}_S
$$

having finite kernel and cokernel. Note here that $\mathbb{Q} \otimes_{\mathbb{Z}} \nabla_S = \mathbb{Q} \otimes_{\mathbb{Z}} \bar{\nabla}_S = \mathbb{Q} \otimes_{\mathbb{Z}} \Delta S \oplus \mathbb{Q} \otimes_{\mathbb{Z}} (\mathbb{Z}G)^{r_S}$, where, as before, $r_S$ is the number of $G$-orbits of the ramified primes $p \notin S$. In particular, such isogenies exist and, moreover, nothing has changed for large $S$. The Tate sequence having ends $\tilde{E}_S$, $\nabla_S$ is, of course, $\tilde{E}_S \rightarrowtail A \oplus (\mathbb{Z}G)^{rs} \to B \twoheadrightarrow \nabla_S$, with the identity map on $(\mathbb{Z}G)^{rs}$. On replacing the old triple $\varphi : \Delta S \to E_S$ by the new $\varphi : \nabla_S \to \tilde{E}_S$ in the definition of $q_\varphi(\chi)$ in 2c. we obtain the general $q$-ideal, which again is an element in $\mathrm{Hom}_\Gamma^*(R(G), I_F)$.

We next turn to the $A$-number $A_\varphi(\chi)$. It is almost again a quotient of a certain regulator and an $L$-value, but we have to divide by an additional quantity:

$$A_\varphi(\chi) = R_\varphi(\chi)/\left(c_S(\chi)\prod_p e_p \log(N\mathfrak{p})^{\chi(1)}\right).$$

Here, $p$ runs through a set of $G$-representatives of the orbits of the ramified primes not in $S$ and $e_p$ is the ramification index of $p$.

We are left with defining $R_\varphi(\chi)$. To that end we need a generalized Dirichlet map

$$\tilde{\lambda} : \mathbb{R}\otimes_\mathbb{Z}\tilde{E}_S \to \mathbb{R}\otimes_\mathbb{Z}\nabla_S$$

which the isogeny $\varphi : \nabla_S \to \tilde{E}_S$ has to be compared with. As it turns out, we may take any $\tilde{\lambda}$ making the diagram below commute

$$
\begin{array}{ccccc}
\mathbb{R}\otimes_\mathbb{Z}E_S & \rightarrowtail & \mathbb{R}\otimes_\mathbb{Z}\tilde{E}_S & \twoheadrightarrow & (\mathbb{R}G)^{rs} \\
\downarrow\lambda & & \downarrow\tilde{\lambda} & & \downarrow\tilde{\rho} \\
\mathbb{R}\otimes_\mathbb{Z}\Delta S & \rightarrowtail & \mathbb{R}\otimes\bar{\nabla}_S & \twoheadrightarrow & \mathbb{R}\otimes^0 W_S
\end{array}
$$

and get an unambiguous $R_\varphi(\chi) \stackrel{\mathrm{def}}{=} \det(\tilde{\lambda}\varphi \mid \mathrm{Hom}_{CG}(\check{V}_\chi, \mathbb{C}\otimes_\mathbb{Z}\nabla_S))$. Of course, in the diagram $\lambda$ is the Dirichlet map from 2a. and $^0W_S = \oplus_p \mathrm{ind}_{G_p}^G W_p^0$ (see 4a). The right vertical map $\tilde{\rho}$ has to be picked carefully in order to make things work out later. To describe it, we first have to add some more information on the local inertia lattice $W_p$. In fact, $W_p$ can be viewed as a submodule of $\Delta G \oplus \mathbb{Z}\bar{G}_p$, where $\bar{G}_p = G_p/I_p$ is the Galois group of the residue field extension at $p$. If $f_p$ denotes its degree and $\phi_p$ its Frobenius automorphism, then $W_p$ is the free $\mathbb{Z}$-module on the basis

$$w_g = (g - 1, 1 + \phi_p + \cdots + \phi_p^{a(g)-1})$$

with $g \in G$ and $\bar{g} = \phi_p^{a(g)}$, $1 \le a(g) \le f_p$ [GW1]. Let now $\rho_p \in W_p^0$ be the map which takes $w_g$ to $1, 1 + e_p, e_p$ according to $g = 1, 1 \neq g$ & $\bar{g} = 1, \bar{g} \neq 1$. These $\rho_p$ determine $\tilde{\rho}$ by means of

$$\mathbb{R}\otimes_\mathbb{Z}(\mathbb{Z}G)^{rs} = \mathbb{R}\otimes_\mathbb{Z}(\oplus_p \mathrm{ind}_{G_p}^G \mathbb{Z}G_p) \to \mathbb{R}\otimes_\mathbb{Z}(\oplus_p \mathrm{ind}_{G_p}^G W_p^0),$$
$$1_p \mapsto \log(N_p)\otimes\rho_p.$$

Regarding the details compare [RW3, §3] (but observe that ours $\lambda$ is $-\lambda$ there).

## 4c. $a_\varphi(\chi)$

The independence of $a_\varphi(\chi)$ from $\varphi$ can already be traced back to Chinburg's work [Ch1]; we are not going to go into this (see [RW3, §4]). It is the independence from $S$ which really needs to be discussed. Here is the idea. Write $a_S(\chi)$ rather than $a_\varphi(\chi)$. It suffices to show that $a_S(\chi) = a_{S'}(\chi)$ for each sufficiently large set $S'$ containing $S$. This is done in three steps.

(i) Start out with $S$ and $\varphi$ and enlarge $S$ to $S_1$ by adjoining $G$-orbits of primes of absolute degree 1 which generate the $S$-class group $cl_S$. The degree condition can be realized because of Tchebotarev's density theorem; it is required in order to have $\mathbb{Z}[S_1 \backslash S]$ free over $\mathbb{Z}G$ which guarantees the existence of a $\varphi_1$ making

$$
\begin{array}{ccccc}
\bar{\nabla}_S & \rightarrowtail & \nabla_1 & \twoheadrightarrow & \mathbb{Z}[S_1 \backslash S] \\
\downarrow \varphi & & \downarrow \varphi_1 & & \downarrow h_S \\
\tilde{E}_S & \rightarrowtail & \tilde{E}_1 & \twoheadrightarrow & \tilde{E}_1/\tilde{E}_S
\end{array}
$$

commute. Here, we have chosen in advance an isogeny $\varphi : \nabla_S \to \tilde{E}_S$ which is trivial on $cl_S = \mathrm{tor}(\nabla_S)$. Moreover, the top row above is induced by $\mathbb{Z}S \subset \mathbb{Z}S_1$ and $^0W_S = {}^0W_1$. The right vertical map is due to $E_1/E_S \rightarrowtail \mathbb{Z}[S_1 \backslash S] \twoheadrightarrow cl_S, E_1 \ni u_1 \mapsto \sum_{\mathfrak{p} \in S_1 \backslash S} w_{\mathfrak{p}}(u_1)\mathfrak{p}$, so $h_S \cdot \mathbb{Z}[S_1 \backslash S] \subset E_1/E_S = \tilde{E}_1/\tilde{E}_S$.

A careful computation now shows that $q_{\varphi_1}(\chi)/q_\varphi(\chi)$ and the $\mathfrak{o}_F$-ideal generated by $A_{\varphi_1}(\check\chi)/A_\varphi(\check\chi)$ coincide [RW3, §5].

(ii) Assume that $S$ is such that $cl_S = 0$ and enlarge $S$ to $S_2$ by adjoining all ramified primes. In this case one starts out from a $\varphi_2 : \nabla_2 \to \tilde{E}_2$ and obtains a commutative diagram

$$
\begin{array}{ccc}
\nabla_S & \twoheadrightarrow & \nabla_2 = \Delta S_2 \\
\downarrow \varphi & & \downarrow n\varphi_2 \\
\tilde{E}_S & \twoheadrightarrow & \tilde{E}_2 = E_2
\end{array}
$$

with $\varphi$ arising from $n\varphi_2$ for some natural number $n$. Indeed, there is a canonical map $\delta$ between the kernels of the horizontal maps (see [RW3, (4) on p. 527]),

$$
\delta : \bigoplus_{(S_1 \backslash S)/G} \mathrm{ind}_{G_\mathfrak{p}}^G (\Delta G_\mathfrak{p})^0 \to \bigoplus_{(S_1 \backslash S)/G} \mathrm{ind}_{G_\mathfrak{p}}^G \Delta G_\mathfrak{p},
$$

induced by $\kappa_\mathfrak{p} \xrightarrow{\delta_\mathfrak{p}} |G_\mathfrak{p}| - \hat{G}_\mathfrak{p}$ where $\kappa_\mathfrak{p}(g - 1) = 0$ or $-1$ according to $g = 1$ or $g \neq 1$. Observe that $(\Delta G_\mathfrak{p})^0 = \mathbb{Z}G_\mathfrak{p} \cdot \kappa_\mathfrak{p}$. Hence, scaling $\varphi_2$ and $\delta$ by some $n \in \mathbb{N}$ yields a $\varphi$. Again $q_{\varphi_2}(\chi)/q_\varphi(\chi)$ and $(A_{\varphi_2}(\check\chi)/A_\varphi(\check\chi))\mathfrak{o}_F$ agree [RW3, §6].

(iii) It remains to study the case of a large set $S$ which gets enlarged to $S_3 = S \cup G\mathfrak{p}$ for some prime $\mathfrak{p}$. The discussion of this case is pretty straightforward (cf. [Ch1; RW3, §7]).

Since $a_\varphi(\chi)$ is independent of $\varphi$ and $S$ we prefer to write more suggestively $a_{K/k}(\chi)$ in future. The following observations can readily be checked

$a_{K/k}$ is additive, $a_{K/k}(1) = o_F$, $a_-$ respects induction and inflation of characters.

As a consequence, and by Brauer's induction theorem (cf. [CR1, p. 381]), the study of $a_{K/k}(\chi)$ is reduced to that of $a_{K'/k'}(\chi')$ where $K'/k'$ runs through the cyclic subextensions of $K/k$ and $\chi'$ through the correspoding characters.

If we happen to be in a situation to which Stark's conjecture applies – so $a_{K/k}(\chi) \subset F$ – we can do better and go locally:

Fix a prime $l$ and let $C_l$ denote a fixed completion of an algebraic closure of the $l$-adics $Q_l$. Regard $F$ as a subfield of $C_l$ and set $F_l = F \cdot Q_l$. Then there is a natural $1 - 1$ correspondence $\chi \leftrightarrow \chi_l$ between $F$-characters $\chi$ and $F_l$-characters $\chi_l$ of $G$. Define

$$a_{K/k}^{(l)}(\chi_l) = a_{K/k}(\chi)o_l$$

where $o_l$ is the ring of integers in $F_l$. Now we have

$a_{K/k}(\chi) = 1$, *if, for all $l$,* $a_{K'/k'}^{(l)}(\chi_l) = 1$ *whenever $\chi_l$ is a non-trivial $F_l$-character of a cyclic subextension $K'/k'$ of $K/k$ of degree prime to $l$.*

The result is taken from [We, §14]; it is kind of an induction theorem with respect to the maximal $l$-unramified subextension in $F$. In the case that $K$ is contained in a cyclotomic field $Q(\zeta_m)$ we may even restrict $K'/k'$ above to the *cyclic subextensions of $K/Q$ of degree prime to $l$ such that $k'/Q$ is a cyclic $l$-extension.*

## 4d. $\underline{\varphi}$ and $\varphi$

We assume to be given an embedding $\underline{\varphi} : \Delta S \rightarrowtail E_S$ and want to associate to it an isogeny $\varphi : \nabla_S \rightarrow \tilde{E}_S$. To that end we first require $\varphi(cl_S) = 0$ so that we are reduced to finding a map $\bar{\varphi} : \bar{\nabla}_S \rightarrow \tilde{E}_S$.

The existence of maps $\sigma_\infty$ and $\sigma_p$ in the following diagrams is due to $|G|$ and $|G_p|$ annihilating the $G$-, respectively $G_p$-, cohomology; there is, in fact, an explicit construction [RW3, p. 532]. The map $\delta_p$ has been defined in the previous subsection and induces $\varphi_p$:

$$
\begin{array}{ccccc}
\Delta S & \rightarrowtail & \mathbb{Z}S & \twoheadrightarrow & \mathbb{Z} \\
\downarrow \cdot |G| & & \downarrow \sigma_\infty & & \| \\
\Delta S & \rightarrowtail & \Delta S \oplus \mathbb{Z} & \rightarrow & \mathbb{Z} \\
\downarrow \underline{\varphi} & & \downarrow \varphi \oplus 1 & & \| \\
E_S & \rightarrowtail & E_S \oplus \mathbb{Z} & \twoheadrightarrow & \mathbb{Z}
\end{array}
\qquad
\begin{array}{ccccc}
(\Delta G_p)^0 & \rightarrowtail & W_p^0 & \twoheadrightarrow & \mathbb{Z} \\
\downarrow \cdot |G_p| & & \downarrow \sigma_p & & \| \\
(\Delta G_p)^0 & \rightarrowtail & (\Delta G_p)^0 \oplus \mathbb{Z} & \twoheadrightarrow & \mathbb{Z} \\
\downarrow \delta_p & & \downarrow \varphi_p & & \downarrow \cdot |G_p| \\
\Delta G_p & \rightarrowtail & \mathbb{Z}G_p & \twoheadrightarrow & \mathbb{Z}.
\end{array}
$$

Putting things together we first obtain the vertical map in the centre below and then the definition of $\bar{\varphi}$:

$$
\begin{array}{ccccc}
\bar{\nabla}_S & \rightarrowtail & \mathbb{Z}S \oplus \bigoplus \mathrm{ind}_{G_p}^G W_p^0 & \twoheadrightarrow & \mathbb{Z} \\
\downarrow \bar{\varphi} & & \downarrow & & \| \\
\tilde{E}_S & \rightarrowtail & E_S \oplus \mathbb{Z} \oplus (\mathbb{Z}G)^{rs} & \twoheadrightarrow & \mathbb{Z}.
\end{array}
$$

Here, the sum ranges over the $G$-orbits of ramified primes not in $S$. The construction of $\varphi$ from $\underline{\varphi}$ is justified by the following result which is to be seen in connection with the local reduction at the end of the last subsection where the group order has become a unit in $F_l$.

*Assume that $G$ is abelian and set $\mathfrak{o}' = \mathfrak{o}_F\left[\frac{1}{|G|}\right]$. Then*

$$
a_{K/k}(\chi) \cdot \mathfrak{o}' \;=\; \frac{\det(\lambda\underline{\varphi}|\mathrm{Hom}_{\mathbb{C}G}(V_\chi, \mathbb{C}\otimes_\mathbb{Z} \Delta S))}{c_S(\check{\chi})} \cdot \frac{|(\mathfrak{o}'\otimes_\mathbb{Z} cl_S)^\chi|_{\mathfrak{o}'}}{|(\mathfrak{o}'\otimes_\mathbb{Z} \mathrm{coker}\varphi)^\chi|_{\mathfrak{o}'}}
$$

*for every irreducible $\chi$.*

Above, $X^\chi = e_\chi X$ is the $\chi$-eigenspace of the $\mathfrak{o}'G$-module $X$; $e_\chi = \frac{1}{|G|}\sum_{g\in G} \chi(g^{-1})g \in \mathfrak{o}'G$. The result is proved by computing both, $q_\varphi(\chi)$ and $A_\varphi(\check{\chi})$, in terms of $\underline{\varphi}$ [RW3, Prop. 10].

## 4e. The Ramachandra Map

From the assumption of (2) we know $K \subset \mathbb{Q}(\zeta_m)$. Following [Wa, p. 147], we write $m = \prod_{i=1}^s p_i^{e_i}$ and $m_I = \prod_{i\in I} p_i^{e_i}$ for each subset $I \subsetneqq \{1, \dots, s\}$, and set

$$
\xi_K = N_{\mathbb{Q}(\zeta_m)/K}\left(\prod_I (1 - \zeta_m^{m_I})(1 - \zeta_m^{-m_I})\right).
$$

Let $\infty$ be the infinite prime of $K$ arising from $K \subset \mathbb{Q}(\zeta_m) \subset \mathbb{C}$, $\zeta_m = e^{2\pi i/m}$. Then there is a $G$-homomorphism $\mathbb{Z}S_\infty \to K^\times$, $\infty \mapsto \xi_K$ which, when restricted to $\Delta S_\infty$, will be our $\underline{\varphi}$. Observe that there is only one orbit of infinite primes in $\mathbb{Q}(\zeta_m)$ with respect to the Galois group of $\mathbb{Q}(\zeta_m)/\mathbb{Q}$ and, moreover, that $\xi_K^{g-1}$ is, in fact, a (real) unit in $K$, if $g \in G$. The calculations in [Wa, pp. 152/153] can be refined (see [RW3, §10]) in order to arrive at the value, say $Y$, of the first factor in the product $a_\varphi(\chi)\mathfrak{o}'$ displayed at the end of the last subsection. As $Y \neq 0$, $\underline{\varphi}$ is injective.

## 4f. Number Theoretical Means

Taking everything into account we have reached a situation where, for a given prime $l$,

$\phantom{x}$ $l \nmid [K:k]$, $[k:\mathbb{Q}]$ *is an $l$-power,* $\chi = \chi_l$ *is faithful and non-trivial,* $S = S_\infty$, $\varphi$ *is induced by the Ramachandra $\underline{\varphi}$.*

In order to check whether $a^{(l)}(\chi) = \mathfrak{o}_l$ we are left with studying

$$
|(\mathfrak{o}_l \otimes_\mathbb{Z} cl)^\chi|_{\mathfrak{o}_l} / |(\mathfrak{o}_l \otimes_\mathbb{Z} \mathrm{coker}\,\underline{\varphi})^\chi|_{\mathfrak{o}_l}.
$$

For this, let us concentrate on just one case and assume that $k$ is real and $\chi(c) = 1$ for the restriction $c$ of complex conjugation to $K$.

We start out from the canonical embedding $E \longmapsto \prod_{\mathfrak{L}/l} U_{\mathfrak{L}}$, where $U_{\mathfrak{L}}$ is the group of units in the localization $K_{\mathfrak{L}}$. Due to the Leopoldt conjecture, which holds true for absolutely abelian number fields $K$, it induces an injection

$$\alpha : \mathbb{Z}_l \otimes_{\mathbb{Z}} E \longmapsto \prod_{\mathfrak{L}/l} U_{\mathfrak{L}}^1$$

into the product of the principal units. The cokernel of $\alpha$ is known to be canonically isomorphic to the Galois group $G(M/L)$ where $L$ and $M$ are the maximal abelian $l$-extensions of $K$ which are unramified, respectively unramified outside $l$. In particular, $G(L/K) = \mathbb{Z}_l \otimes_{\mathbb{Z}} cl$. Hence, the quotient above equals

$$|G(M/K)^{\chi}|/|(\text{coker } \alpha\underline{\varphi})^{\chi}|$$

where, from now on, we omit the reference to $\mathfrak{o}_l$.

The denominator turns out to be the product of the $l$-adic $L$-value $L_l(1, \chi)$ and $-Y/2$ with the $Y$ from the previous subsection. This is seen as follows:

We compose $\alpha\underline{\varphi}$ with the $l$-adic logarithm $\log = \prod_{\mathfrak{L}/l} \log_{\mathfrak{L}}$ and are led to work out lattice indices on the space $F_l \otimes_{\mathbb{Q}_l} \prod_{\mathfrak{L}/l} K_{\mathfrak{L}}$; essentially

$$\left| \left( \prod_{\mathfrak{L}/l} \mathfrak{o}_{\mathfrak{L}} \right)^{\chi} : (\text{im } \log \alpha\underline{\varphi})^{\chi} \right| .$$

It is the ideal generated by $\det(\mathfrak{a}_{i,\mathfrak{p}})$ with the matrix $\mathfrak{a}_{i,\mathfrak{p}}$ fitting into an equation

$$(s(\log \alpha\underline{\varphi})e_{\chi}(1 \otimes \mathfrak{p}))_{\mathfrak{p},s} = (\mathfrak{a}_{i,\mathfrak{p}})_{\mathfrak{p},i}(sv_i)_{i,s},$$

where $\mathfrak{p}$ runs through $S_{\infty}/G$ and the elements $v_1, \ldots, v_{[k:\mathbb{Q}]}$ constitute an $\mathfrak{o}_l$-basis of $\prod_{\mathfrak{L}/l} \mathfrak{o}_{\mathfrak{L}}$ and where the $s$ form a set of orbit representatives of all $\mathbb{Q}$-embeddings of $K$ into $\mathbb{C}_l$ with respect to the (right) $G$-action.

The determinant of the matrix on the left can be dealt with by means of the $l$-adic class number formula; remember here that we are allowed to neglect $l$-units in $\mathbb{C}_l$.

The determinant of the matrix on the far right is, again up to $l$-units, $\prod_{\mathfrak{L}/l} d_{\mathfrak{L}}^{1/2} \tau_l$ $(\chi_{\mathfrak{L}})$ where $d_{\mathfrak{L}}$ is the discriminant of $k_{\mathfrak{L}}/\mathbb{Q}_l$ and $\chi_{\mathfrak{L}}$ the restriction of $\chi$ to the decomposition group of $\mathfrak{L}$. The proof of this result exploits a theorem of Fröhlich relating norm resolvents and Galois Gauß sums; it is by no means straightforward.

It remains to compute $|G(M/K)^{\chi}|$. But this quantity can be read off the Main Conjecture of Iwasawa theory – at least when 2 does not ramify in $K$:

$$|G(M/K)^{\chi}| = 2^{-[k:\mathbb{Q}]} L_l(1, \chi) \mathfrak{o}_l.$$

For the details and references to the literature see [RW3, §§11–13].

## 5. The Refined Root Number Conjecture

In this section we discuss a new invariant $\Omega_\varphi$ which, as has already been said, is an element in the Grothendieck group $K_0 T(\mathbb{Z}G)$ of finite $\mathbb{Z}G$-modules having finite projective dimension and which is attached to the Tate class $\tau_S$ and to an isogeny $\varphi : \Delta S \rightarrowtail E_S$. The set $S$ is always large.

The invariant has been introduced in [GRW] together with a refined root number conjecture stated as (RRN) in subsection $5a$.. We will content ourselves with recalling its definition and recollecting its main properties (in fact, we do not see a way which would enable us to illuminate the ideas behind the respective computations without actually carrying them out). However, we will point out why and where it has become necessary to replace the usual Dirichlet map $\mathbb{R} \otimes E_S \mapsto \mathbb{R} \otimes \Delta S, u \mapsto \sum_{\mathfrak{p} \in S} \log|u|_\mathfrak{p} \mathfrak{p}$ by its negative $\lambda$, cf. section $2a$. Furthermore, in $5b$., we sketch the proof of

($1'$) *the image of $\Omega_\varphi$ in $K_0 T(\Lambda_{max})$ is represented by* $q_\varphi \in \mathrm{Hom}_\Gamma^*(R(G), I_F)$,

which sharpens Chinburg's result (1) mentioned in section $2d$.. The above statement refers to the natural isomorphism $\mathrm{Hom}_\Gamma^*(R(G), I_F) \simeq K_0 T(\Lambda_{max})$ resulting from the Fröhlich Hom language for a maximal order $\Lambda_{max} \subset \mathbb{Q} G$ containing $\mathbb{Z}G$. Indeed, for any order $\mathbb{Z}G \subset \Lambda \subset \mathbb{Q} G$, the construction of $2e$. of a representing homomorphism of an element $[P_1] - [P_2] \in Cl(\Lambda)$, when applied to projective $\Lambda$-modules $P_1, P_2$ fitting into a short exact sequence $P_1 \rightarrowtail P_2 \twoheadrightarrow C$ with finite $C$, leaves less room for choice because of the natural identification $\mathbb{Q} \otimes_\mathbb{Z} P_1 = \mathbb{Q} \otimes_\mathbb{Z} P_2$. As a matter of fact, the factor $\mathrm{Hom}_\Gamma^+(R(G), F^\times)$ occuring in the denominator on the left hand side of

$$\mathrm{Hom}_\Gamma^+(R(G), J_F)/\mathrm{Hom}_\Gamma^+(R(G), F^\times) \, \mathrm{Det}\, U(\Lambda) \simeq Cl(\Lambda)$$

disappears when the right hand side gets replaced by $K_0 T(\Lambda)$.

We close the section by looking, in $5c$., at a first series of examples where the refined conjecture (RRN) holds true (cf. [RW4]).

### 5a. Definition and Properties of $\Omega_\varphi$

There is a straightforward construction of embeddings $\tilde{\varphi} : B \rightarrow A$ if a Tate sequence $E_S \rightarrowtail A \rightarrow B \twoheadrightarrow \Delta S$ and an isogeny $\varphi : \Delta S \rightarrowtail E_S$ is given. To describe it, split the Tate sequence into two short exact sequences $E_S \rightarrowtail A \twoheadrightarrow L$, $L \rightarrowtail B \twoheadrightarrow \Delta S$, thereby introducing the lattice $L$. Choose two monomorphisms $\alpha, \beta$ of $L$ which induce the zero map in cohomology (as, for example, $\alpha = \beta = $ multiplication by $|G|$). Then the pull-back and push-out along $\alpha$ and $\beta$ of $E_S \rightarrowtail A \twoheadrightarrow L$ and $L \rightarrowtail B \twoheadrightarrow \Delta S$, respectively, induce split sequences

$$
\begin{array}{ccccccccc}
E_S & \rightarrowtail & L \oplus E_S & \twoheadrightarrow & L & \quad & L & \rightarrowtail & B & \twoheadrightarrow & \Delta S \\
\| & & \downarrow{\tilde{\alpha}} & & \downarrow{\alpha} & \downarrow{\beta} & & & \downarrow{\tilde{\beta}} & & \| \\
E_S & \rightarrowtail & A & \twoheadrightarrow & L & \quad & L & \rightarrowtail & L \oplus \Delta S & \twoheadrightarrow & \Delta S
\end{array}
$$

and yield maps $\tilde{\alpha}$, $\tilde{\beta}$ as shown. We set

$$\tilde{\varphi} : B \xrightarrow{\tilde{\beta}} L \oplus \Delta S \xrightarrow{1 \oplus \varphi} L \oplus E_S \xrightarrow{\tilde{\alpha}} A.$$

Of course, all these $\tilde{\varphi}$ are injective and have finite and cohomologically trivial cokernel. So we arrive at elements [coker $\tilde{\varphi}$] $\in K_0 T(\mathbb{Z}G)$ which obviously map to $\Omega$ when passing from $K_0 T(\mathbb{Z}G)$ to $K_0(\mathbb{Z}G)$; however, there is no such canonical preimage of $\Omega$. Hence, the first question to be asked is how [coker $\tilde{\varphi}$] varies with the choices of $\alpha$, $\beta$, $\tilde{\alpha}$, $\tilde{\beta}$ when the Tate sequence is kept fixed; the second question is how it changes if only the Tate class $\tau_S$ is kept fixed. Surprisingly enough, it needs not much of an adjustment of [coker $\tilde{\varphi}$] in order to end up with a unique element in $K_0 T(\mathbb{Z}G)$, which then will be denoted by $\Omega_\varphi$:

The composite map $\alpha\beta$ induces an automorphism of $\mathbb{Q} \otimes_\mathbb{Z} L$ and so, as $\mathbb{Q} \otimes_\mathbb{Z} L$ is a projective $\mathbb{Q} G$-module, the element $[\mathbb{Q} \otimes_\mathbb{Z} L, 1 \otimes \alpha\beta]$ in $K_1(\mathbb{Q} G)$ (cf. [CR2, p. 62]).[10] Applying to it the differential $\partial$ of the localization sequence in $K$-theory (cf. [CR2, p. 65]),

$$K_1(\mathbb{Z}G) \to K_1(\mathbb{Q} G) \xrightarrow{\partial} K_0 T(\mathbb{Z}G) \to K_0(\mathbb{Z}G) \to K_0(\mathbb{Q} G),$$

we obtain $\partial(L, \alpha\beta)$, which vanishes when passing on to $K_0(\mathbb{Z}G)$. In particular,

$$\Omega_\varphi \overset{\text{def}}{=} [\text{coker } \tilde{\varphi}] - \partial(L, \alpha\beta)$$

is a preimage of $\Omega$.

**Theorem 1.**

 *1) $\Omega_\varphi$ only depends on the Tate class $\tau_S$ and on $\varphi$.*

 *2) If $S' = S \cup G\mathfrak{p}$ and if $\varphi$, $\varphi'$ fit into a commutative diagram*

$$\begin{array}{ccccc}
\Delta S & \hookrightarrow & \Delta S' & \twoheadrightarrow & \mathbb{Z}[S'\backslash S] \\
\downarrow \varphi & & \downarrow \varphi' & & \downarrow \cdot n \\
E_S & \hookrightarrow & E_{S'} & \twoheadrightarrow & \mathbb{Z}[S'\backslash S],
\end{array}$$

 *then $\Omega_{\varphi'} - \Omega_\varphi = \partial(\mathbb{Q} G, \eta) + 2\partial(\text{ind}_{G\mathfrak{p}}^G \mathbb{Z}, |G|)$, where*

$$\eta(1) = |G_\mathfrak{p}|^{-2}(|G| - \hat{G}_\mathfrak{p}) \sum_{i=0}^{|G_\mathfrak{p}|-1} i\phi_\mathfrak{p}^i + n|G|^{-2}\hat{G}_\mathfrak{p}.$$

 *3a) If $H$ is a subgroup of $G$, then $\text{res}_H^G \Omega_\varphi = \Omega_{\text{res}_H^G \varphi}$ where $\text{res}_H^G$ is the restriction map from $G$ to $H$ (i.e., view $G$-modules and -maps as $H$-modules and -maps, respectively).*

---

[10]Elements $[\mathbb{Q} \otimes_\mathbb{Z} X, 1 \otimes \gamma] \in K_1(\mathbb{Q}G)$, where $X$ is a $\mathbb{Z}G$-module and $\gamma : X \to X$ has finite kernel and cokernel, are abbreviated by $[X, \gamma]$ in the following.

3b) If $N$ is a normal subgroup of $G$ with fixed field $\bar{K}$ and $\bar{G} = G/N = \mathrm{Gal}(\bar{K}/k)$, if the set $\bar{S}$ of primes of $\bar{K}$ below $S$ is large for $\bar{K}/k$, and if $\varphi, \bar{\varphi}$ fit into the commutative square

$$
\begin{array}{ccc}
\Delta S & \twoheadrightarrow & \Delta \bar{S} \\
\downarrow \varphi & & \downarrow \bar{\varphi} \\
E_S & \xrightarrow{\hat{N}} & E_{\bar{S}},
\end{array}
$$

then $\mathrm{defl}^G_{\bar{G}} \Omega_\varphi = \Omega_{\bar{\varphi}}$. Here, $\mathrm{defl}^G_{\bar{G}}$ denotes the deflation map $K_0 T(\mathbb{Z}G) \to K_0 T(\mathbb{Z}\bar{G})$, $[C] \mapsto [\mathbb{Z}\bar{G} \otimes_{\mathbb{Z}G} C = C_N]$.

Regarding the diagrams in 2) and 3b) of the theorem some remarks are in place: First of all, there is no difficulty in keeping track of how $\Omega_\varphi$ changes with $\varphi$. Hence, we may as well work with $\varphi$ that are adapted to the special situation. This being said, note that a diagram with vertical maps $\varphi$, $\varphi'$, $n \in N$, as is shown in 2), can easily be derived from some auxiliary $\varphi_0 : \Delta S \rightarrowtail E_S$ by tensoring

$$
\begin{array}{ccc}
\Delta S & \rightarrowtail & \Delta S' & \twoheadrightarrow & \mathbb{Z}[S' \backslash S] \\
\updownarrow \varphi_0 & & & & \\
E_S & \rightarrowtail & E_{S'} & \twoheadrightarrow & \mathbb{Z}[S' \backslash S]
\end{array}
$$

with $\mathbb{Q}$ and so arriving at split horizontal sequences and vertical maps $\varphi_0$, $\varphi_0 \oplus 1$, $1$. Now we go back to the integral level.

The square in 3) results from a given $\varphi$ and from the norm map $\hat{N} : K^\times \to \bar{K}^\times$ on observing

$$
\Delta S \xrightarrow{\hat{N}} (\Delta S)_N \xrightarrow{\cap \bar{K}} \Delta \bar{S} \text{ and } E_S \xrightarrow{\hat{N}} E_S^N = E_{\bar{S}}.
$$

The left sequence above requires $\bar{S}$ to be large.

The precise statements given in Theorem 1 invite to look at the dependence of $A_\varphi$ on $\varphi$ and on $S$ and its behaviour with respect to restriction and inflation. The resulting information is best collected in terms of $\omega_\varphi \in K_0 T(\mathbb{Z}G)$

which, by definition, is such that $\Omega_\varphi + \omega_\varphi$ has representing homomorphism $[\chi \mapsto A_\varphi(\check{\chi})/W_{K/k}(\check{\chi})]$ in $\mathrm{Hom}^+_\Gamma(R(G), J_F)$.

## Theorem 2.

i) $\omega_\varphi$ is independent of the choice of $\varphi : \Delta S \rightarrowtail E_S$ and of $S$, whence is an invariant of $K/k$, denoted by $\omega(K/k)$.

ii) $\mathrm{res}^G_H \omega(K/k) = \omega(K/k')$, if $k'$ is the fixed field of the subgroup $H$ of $G$.

$\mathrm{defl}^G_{\bar{G}} \omega(K/k) = \omega(\bar{K}/k)$, if $\bar{K}$ is the fixed field of the normal subgroup $N$ of $G$ with $G/N = \bar{G}$.

It is the invariance of $\omega_\varphi$ with respect to enlarging $S$ that forces us to pick our choice $\lambda$ of a Dirichlet map. In fact, when replacing $\lambda$ by $-\lambda$ in the definition of $A_\varphi$ (or rather $R_\varphi$) we no longer get $A_\varphi$ changed in the same way as $\Omega_\varphi$ if, as before, an orbit $G\mathfrak{p}$ is added to $S$ and the earlier maps $\varphi$, $\varphi'$ and $n$ are used. It is only for that reason that the formula in 2) of Theorem 1 is stated; the derivation of the analogous formula for $A_\varphi$ is pretty straightforward (cf. [We, Prop. 86) (with $\lambda$)]).

Theorem 2 right away gives rise to our refined root number conjecture:

**Conjecture (RRN).** $\omega(K/k) = 0$.

Due to the splitting $K_0 T(\mathbb{Z}G) = \oplus_l K_0 T(\mathbb{Z}_l G)$, where $l$ runs through the rational primes (cf. [CR2, p. 221]), this amounts to

$$(\text{RRN})_l \quad \omega^{(l)}(K/k) = 0 \text{ for all } l,$$

where $\omega^{(l)}(K/k)$ denotes the component at $l$ of $\omega(K/k)$.

Of course, (RRN) implies (RN).

Taking pattern from the argumentation in [Fr1, pp. 86–88] one can reduce the verification of $(\text{RRN})_l$ to certain special extension $K/k$:

**Theorem 3.** $\omega^{(l)}(K/k) = 0$ if

a) $\omega^{(l)}(K'/k')$ is a torsion element in $K_0 T(\mathbb{Z}G')$ for every intermediate Galois extension $K'/k'$ of $K/k$ with group $G'$, and moreover if

b) $\omega^{(l)}(K'/k') = 0$ for all intermediate extensions $K'/k'$ having an $l$-elementary Galois group $G'$, i.e., $G'$ is a direct product of a cyclic group and an $l$-group.

For all details in this subsection and for even finer reductions see [GRW].

## 5b. Proof of $(1')$

Without going into computational details we now sketch the proof of

$(1')$ $\Omega_\varphi \bmod DT(\mathbb{Z}G)$ has representing homomorphism $q_\varphi$.

Recall that $DT(\mathbb{Z}G)$ is the kernel of $K_0 T(\mathbb{Z}G) \to K_0 T(\Lambda_{\max})$; it is independent of the special choice of $\Lambda_{\max}$ (indeed, as can be readily seen from the Hom language, $DT(\mathbb{Z}G)$ is the torsion subgroup of $K_0(\mathbb{Z}G)$).

In passing, let us metnion that if $K$ happens to be contained in a cyclotomic field $\mathbb{Q}(\zeta_m)$ in which 2 does not ramify, then we have result (2) at our disposal, i.e., (SST) is valid and so $q_\varphi(\chi) = A_\varphi(\check{\chi}) \circ_F$. Consequently, $\omega^{(l)}(K/k) \in DT(\mathbb{Z}_l G)$ for all rational primes $l$ (remember that $W_{K/k}(\check{\chi})$ is trivial at the finite primes) and, as a direct result, (RRN) is true modulo $DT(\mathbb{Z}G)$. Equally well, condition a) in Theorem 3 is satisfied.

The model of what follows is section 6 in [GRW].

First of all, we observe that if $\tilde{\varphi} : B \rightarrowtail A$ is a map as in 5a., then $[\text{coker } \tilde{\varphi}]$ is represented by

$$\chi \mapsto |\text{Hom}_{o_F G}(M_\chi, o_F \otimes_{\mathbf{Z}} \text{coker } \tilde{\varphi})|_{o_F}$$

in $\text{Hom}_\Gamma^*(R(G), I_F) = K_0 T(\Lambda_{\max})$. Here, $M_\chi$ is, as before, an $o_F G$-lattice spanning the $FG$-module $V_\chi$ with character $\chi$.

The above assertion just reflects the connection between determinants and module indices. Namely, if $P_1$ is a projective $\mathbf{Z}G$-module mapping onto $A$ and if $P_2$ is defined by means of the pull-back

$$
\begin{array}{ccccc}
P_2 & \rightarrowtail & P_1 & \twoheadrightarrow & C \\
\downarrow & & \downarrow\downarrow & & \| \\
B & \overset{\tilde{\varphi}}{\rightarrowtail} & A & \twoheadrightarrow & \text{coker } \tilde{\varphi},
\end{array}
$$

then, by 2e., $[C] \in K_0 T(\mathbf{Z}G)$ is represented by $\chi \mapsto \det(\gamma|V_\chi)^t$, where $\gamma$ is the idèle having component 1 at the primes $p \nmid |C|$ and component $\gamma_p \in GL(t, \mathbf{Q}_p G)$ at $p \mid |C|$ with $\gamma_p(\mathbf{Z}_p \otimes_{\mathbf{Z}} P_1) = \mathbf{Z}_p \otimes_{\mathbf{Z}} P_2$. Above, $t$ is the rank of $P_1$. Since we are only interested in the ideal content of $\det(\gamma|V_\chi)$, we do not bother about the infinite primes. Now, $\gamma_p(\mathbf{Z}_p \otimes_{\mathbf{Z}} P_1) = \mathbf{Z}_p \otimes_{\mathbf{Z}} P_2 \subset \mathbf{Z}_p \otimes_{\mathbf{Z}} P_1$ and the factor is $\mathbf{Z}_p \otimes_{\mathbf{Z}} C$. Hence $\det \gamma_p$ generates the ideal $|\mathbf{Z}_p \otimes_{\mathbf{Z}} C|_{\mathbf{Z}_p}$. Of course, the argument is compatible with taking $\chi$-eigenspaces first. So it remains to observe that

$$\det(\gamma|V_\chi) = \det(\gamma|\text{Hom}_{FG}(V_\chi, FG)) = |\text{Hom}_{o_F G}(M_\chi, o_F \otimes_{\mathbf{Z}} C)|_{o_F};$$

cf. [GRW, Lemmas A1, A2].

We construct $\tilde{\varphi}$ from $\varphi$ and $\alpha = \beta = $ multiplication by $|G|$ on $L$. Thus, $\Omega_\varphi = [\text{coker } \tilde{\varphi}] - 2\partial(L, |G|)$. From

$$\det : K_1(\mathbf{Q}\,G) \to \text{Hom}_\Gamma(R(G), F^\times) \text{ taking } [L, |G|] \text{ to}$$
$$[\chi \mapsto \det(|G| |\text{Hom}_{FG}(V_\chi, F \otimes_{\mathbf{Z}} L))]$$

we read off the representing homomorphism for $\partial(L, |G|)$ in $\text{Hom}_\Gamma^*(R(G), J_F)$:

$$\chi \mapsto |\text{Hom}_{o_F G}(M_\chi, o_F \otimes_{\mathbf{Z}} L)/|G|\text{Hom}_{o_F G}(M_\chi, o_F \otimes_{\mathbf{Z}} L)|_{o_F}$$

(cf. [GRW, (5.1) and Appendix A]).

We next observe that $|\text{Hom}_{o_F G}(M_\chi, o_F \otimes_{\mathbf{Z}} \text{coker } \tilde{\varphi})|_{o_F} = \mathfrak{q}_\chi(\tilde{\varphi}_M)$, if $\tilde{\varphi}_M$ is attached to $\tilde{\varphi}$ in the same way as $\varphi_M$ to $\varphi$. This follows from $B \overset{\tilde{\varphi}}{\rightarrowtail} A \twoheadrightarrow \text{coker } \tilde{\varphi}$ with all three modules, $B$, $A$ and coker $\tilde{\varphi}$, being cohomologically trivial. It reduces our proof to verifying

$$\mathfrak{q}_\chi(\tilde{\varphi}_M) = \mathfrak{q}_\chi(\varphi) \cdot \partial(L, |G|)^2.$$

Before turning to this we would like to ease the notation by abbreviating

$$\text{Hom}_{o_F}(M_\chi, o_F \otimes_{\mathbf{Z}} -) \text{ by } \text{Hom}(M, -), \quad |-|_{o_F} \text{ by } |-| \text{ and } \mathfrak{q}_\chi \text{ by } \mathfrak{q}.$$

We also attach to a homomorphism $h : X \to Y$ of $\mathbb{Z}G$-modules with finite kernel and cokernel the new maps

$$h_M : \text{Hom}(M, X)_G \xrightarrow{\hat{G}} \text{Hom}(M, X)^G \xrightarrow{h} \text{Hom}(M, Y)^G,$$

$$\bar{h}_M : \text{Hom}(M, X)^G \xrightarrow{h} \text{Hom}(M, Y)^G,$$

$$\underline{h}_M : \text{Hom}(M, X)_G \xrightarrow{h} \text{Hom}(M, Y)_G.$$

Note that for $h = \varphi$ or $h = \tilde{\varphi}$ we get back $h_M = \varphi_M, h_M = \tilde{\varphi}_M$, respectively. Define

$$q(h) = |\text{coker } h|/|\text{ker } h| \text{ and correspondingly with } h \text{ replaced by } h_M, \bar{h}_M, \underline{h}_M.$$

The basic property of $q(h)$ is its multiplicativity

$$q(h_1 h_2) = q(h_1)q(h_2)$$

for maps $Z_1 \xrightarrow{h_1} Z \xrightarrow{h_2} Z_2$ of $\mathbb{Z}G$-modules with finite kernel and cokernel [Ta2, p. 58]. As a consequence, $\tilde{\varphi} : B \xrightarrow{\tilde{\beta}} L \oplus \Delta S \xrightarrow{1 \oplus \varphi} L \oplus E_S \xrightarrow{\tilde{\alpha}} A$ yields $q(\tilde{\varphi}_M) = q(\underline{\tilde{\beta}}_M)q((1 \oplus \varphi)_M)q(\tilde{\bar{\alpha}}_M)$. Now,

$$q((1 \oplus \varphi)_M) = q(1_M)q(\varphi_M) \text{ and}$$
$$q(1_M) = |H^0(G, \text{Hom}(M, L)|/|H^{-1}(G, \text{Hom}(M, L))|.$$

Furthermore, as $\tilde{\beta}$ is obtained from $L \xrightarrow{|G|} L$, we have $B \xrightarrow{\tilde{\beta}} L \oplus \Delta S \twoheadrightarrow L/|G|L$. Apply $\text{Hom}(M, -)$ and get

$$q(\underline{\tilde{\beta}}_M) = |(\text{Hom}(M, L)/|G|\text{Hom}(M, L))_G| \cdot |H^{-2}(G, \text{Hom}(M, L \oplus \Delta))|/$$
$$|H^{-2}(G, \text{Hom}(M, L/|G|L)))|.$$

Similarly,

$$q(\tilde{\bar{\alpha}}_M) = |\text{Hom}(M, L)^G/|G|\text{Hom}(M, L)^G|/|H^1(G, \text{Hom}(M, L \oplus E_S))|.$$

Putting things together we arrive at a formula for $q(\tilde{\varphi}_M)/q(\varphi_M)$ in terms of $o_F$-orders of certain cohomology groups. By means of

$$H^{-2}(G, \text{Hom}(M, L \oplus \Delta S)) = H^{-2}(G, \text{Hom}(M, L)) \oplus H^{-1}(G, \text{Hom}(M, L))$$

and an analogous formula for $H^{-1}(G, \text{Hom}(M, L \oplus E_S))$, which both are due to the Tate sequence $L \rightarrowtail B \twoheadrightarrow \Delta S$, $E_S \rightarrowtail A \twoheadrightarrow L$, the quotient $q(\tilde{\varphi}_M)/q(\varphi_M)$ looks like

$$\frac{|\text{Hom}(M, L/|G|L)^G| \cdot |\text{Hom}(M, L/|G|L)_G| \cdot |H^{-2}(G, \text{Hom}(M, L))|}{|H^1(G, \text{Hom}(M, L))| \cdot |H^{-2}(G, \text{Hom}(M, L/|G|L))|}.$$

Using $L \xrightarrow{|G|} L \twoheadrightarrow L/|G|L$ and the derived short exact sequences with respect to $\mathrm{Hom}(M, -)^G$ and $\mathrm{Hom}(M, -)_G$, the above expression gets changed into

$$\frac{|\mathrm{Hom}(M, L)^G/|G|\mathrm{Hom}(M, L)^G| \cdot |\mathrm{Hom}(M, L)_G/|G|\mathrm{Hom}(M, L)_G|}{|H^{-1}(G, \mathrm{Hom}(M, L))|}.$$

We finally look at

$$H^{-1}(G, \mathrm{Hom}(M, L)) \rightarrowtail \mathrm{Hom}(M, L)_G \twoheadrightarrow \hat{G}\mathrm{Hom}(M, L),$$

$$\hat{G}\mathrm{Hom}(M, L) \rightarrowtail \mathrm{Hom}(M, L)^G \twoheadrightarrow H^0(G, \mathrm{Hom}(M, L))$$

and apply $\cdot|G|$ to each entry. The snake lemma with respect to the resulting kernel and cokernel sequences yields

$$H^{-1}(G, \mathrm{Hom}(M, L)) \rightarrowtail \mathrm{Hom}(M, L)_G/|G|\mathrm{Hom}(M, L)_G$$

$$\twoheadrightarrow \hat{G}\mathrm{Hom}(M, L)/|G|(\hat{G}\mathrm{Hom}(M, L))$$

as well as

$$H^0(G, \mathrm{Hom}(M, L)) \rightarrowtail \hat{G}\mathrm{Hom}(M, L)/|G|(\hat{G}\mathrm{Hom}(M, L)) \rightarrow$$

$$\rightarrow \mathrm{Hom}(M, L)^G/|G|\mathrm{Hom}(M, L)^G$$

$$\twoheadrightarrow H^0(G, \mathrm{Hom}(M, L)).$$

Thus,

$$\frac{\mathfrak{q}(\tilde{\varphi}_M)}{\mathfrak{q}(\varphi_M)} = |\mathrm{Hom}(M, L)^G/|G|\mathrm{Hom}(M, L)^G|$$

$$\cdot |\hat{G}\mathrm{Hom}(M, L)/|G|(\hat{G}\mathrm{Hom}(M, L))|$$

$$= |\mathrm{Hom}(M, L)^G/|G|\mathrm{Hom}(M, L)^G|^2.$$

## 5c. An Example

We turn to the discussion of a first series of examples with respect to (RRN); cf. [RW4]. Fix three odd pairwise distinct primes $l$, $p_1$, $p_2$ subject to the condition that $p_1$ is not an $l$-th power modulo $p_2$. Let $K/\mathbb{Q}$ be a cyclic extension of degree $l$ in which precisely $p_1$ and $p_2$ ramify. Then $K \subset \mathbb{Q}(\zeta_{p_1 p_2})$ and $l | p_i - 1, i = 1, 2$. Moreover, the Galois group $G$ of $K/\mathbb{Q}$ is generated by the Frobenius automorphism $g_0$ of any inert (unramified) prime $p_0$.

From genus theory (cf. [Fr4, p. 26]) we see that the $l$-part of the ideal class group $cl$ of $K$ satisfies $(\mathbb{Z}_l \otimes_{\mathbb{Z}} cl_G = \mathbb{Z}/l$. Nakayama's lemma thus yields $\mathbb{Z}_l \otimes_{\mathbb{Z}} cl = \mathbb{Z}_l G/\langle \hat{G}, (g_0 - 1)^{h+1} \rangle$ with some $h \geq 0$; observe that $\hat{G}$ annihilates $cl$ and that $\mathbb{Z}_l G/\hat{G} \simeq \mathbb{Z}_l[\zeta_l]$ via $g_0 \mapsto \zeta_l$. Hence, $\mathbb{Z}_l \otimes_{\mathbb{Z}} cl$ is a cyclic $\mathbb{Z}_l G$-module. We can choose a prime ideal $\mathfrak{P}$ of $K$ as a generator. Of course, the rational prime $p \in \mathfrak{P}$ splits in $K$.

Let $\mathfrak{p}_i | p_i$ be the primes in $K$ dividing $p_i (i = 1, 2)$ and set

$$\mathfrak{P}^{b_i(g_0-1)^h} = \mathfrak{p}_i(\beta_i)$$

with suitable elements $\beta_i \in K^\times$ and integers $b_i$. It is possible since the $\mathfrak{p}_i$ are fixed by $G$ and so their ideal classes lie in

$$(\mathbb{Z}_l \otimes_{\mathbb{Z}} cl)^G = (\zeta_l - 1)^h \mathbb{Z}_l[\zeta_l]/(\zeta_l - 1)^{h+1} \mathbb{Z}_l[\zeta_l].$$

Note that not both, $\mathfrak{p}_1$ and $\mathfrak{p}_2$, can be principal (cf. [Fr4, p. 26]); say $\mathfrak{p}_2$ is not, so $l \nmid b_2$.

Choose $S$ to be the set of primes

$$G \cdot \infty \cup G \cdot \mathfrak{P} \cup \{\mathfrak{p}_0, \mathfrak{p}_1, \mathfrak{p}_2\} \cup \ldots$$

with $\infty$ denoting a fixed infinite prime, $\mathfrak{p}_0 | p$, and the dots indicating orbits of primes of $K$ which generate the part of $cl$ that is prime to $l$. We do not worry about them in what follows. In fact, as has been pointed out at the beginning of section 5, in order to prove (RRN) for $K/\mathbb{Q}$ it suffices to prove (RRN)$_l$ for our prime $l$.

We proceed by defining a map $\varphi : \Delta S \to E_S$ by

$$\infty^g - \infty \mapsto \xi_K^{g-1} = u_K^{g-1}$$

$$\mathfrak{p}_0 - \infty \mapsto \xi_K^{-1}$$

$$\mathfrak{p}^g - \mathfrak{p}_0 \mapsto \begin{cases} \beta_1^g & \text{if } \mathfrak{p} = \mathfrak{P} \\ p^{x_i} \mathfrak{p}_0^{-1} & \text{if } \mathfrak{p} = \mathfrak{p}_i (i = 1, 2) \end{cases},$$

where $g \in G$ and[11]

$$\xi_K = N_{\mathbb{Q}(\zeta_{p_1 p_2})/K}(1 - \zeta_{p_1})(1 - \zeta_{p_1}^{-1})(1 - \zeta_{p_2})(1 - \zeta_{p_2}^{-1})$$

$$= p_1^{2m_2} p_2^{2m_1} u_K$$

gives the definition of the global unit $u_K \in E$ (we have abbreviated $m_i = \frac{p_i - 1}{l}$, $i = 1, 2$) and

$c_i x_i \equiv d_i + 1 \bmod l$ with $g_0^{c_i}$ and $g_0^{d_i}$ denoting the images of $p$ and $p_0$, respectively, under the Artin reciprocity maps at $p_i$. Because $c_1 + c_2 \equiv 0 \bmod l$ and $1 + d_1 + d_2 \equiv 0 \bmod l$, by the reciprocity law (cf. [CF, p. 170]), the congruences for $x_i$ are solvable.

From the definition of $\varphi$ we obtain the commutative diagram with exact rows

$$
\begin{array}{ccccc}
\Delta S_\infty & \rightarrowtail & \Delta S & \overset{\mu}{\twoheadrightarrow} & \mathbb{Z} S_f \\
\downarrow \varphi_\infty & & \downarrow \varphi & & \downarrow \tilde{\varphi} \\
E & \rightarrowtail & E_S & \twoheadrightarrow & E_S/E
\end{array}
, \qquad
\mu(\mathfrak{p} - \mathfrak{p}_\infty) = \begin{cases} \mathfrak{p}, & \text{if } \mathfrak{p} \in S_f \\ 0, & \text{if } \mathfrak{p} \notin S_f, \end{cases}
$$

----
[11] compare 4e.

which is to be read together with

$$
\begin{array}{ccccc}
\mathbb{C}\otimes_\mathbb{Z} E & \rightarrowtail & \mathbb{C}\otimes_\mathbb{Z} E_S & \twoheadrightarrow & \mathbb{C}\otimes_\mathbb{Z} E_S/E \\
\downarrow{\lambda_\infty} & & \downarrow{\lambda} & & \downarrow{\tilde{\lambda}} \\
\mathbb{C}\otimes_\mathbb{Z} \Delta S_\infty & \rightarrowtail & \mathbb{C}\otimes_\mathbb{Z} \Delta S & \twoheadrightarrow & CS_f
\end{array}
$$

in order to arrive at the factorization

$$
A_\varphi(\check{\chi}) \;=\; \frac{\det(\lambda_\infty \varphi_\infty \mid \mathrm{Hom}_{CG}(V_\chi, \mathbb{C}\otimes_\mathbb{Z} \Delta S))}{c_{S\infty}(\check{\chi})}
$$

$$
\cdot \det(\tilde{\lambda}\tilde{\varphi} \mid \mathrm{Hom}_{CG}(V_\chi, CS_f)) \cdot \frac{c_{S\infty}(\check{\chi})}{c_S(\check{\chi})}
$$

of which the outer factors have already been calculated in [RW3, Proposition 12 (with $-\lambda$) and Lemma 7]. The computation of the middle one comes as a direct consequence from knowing $\tilde{\varphi}$ explicitly. In fact, we end up with

i) $A_\varphi(\check{\chi}) \neq 0$, *so $\varphi$ is injective*

ii) $X \overset{\mathrm{def}}{=}$ coker $\varphi$ *is cohomologically trivial*

iii) $(RRN)_l$ *is valid if, and only if, $[\mathbb{Z}_l \otimes_\mathbb{Z} X]$ in $K_0T(\mathbb{Z}_l G)$ is represented by*

$$
\chi \mapsto \begin{cases} -4lm_1(x_2 - x_1), & \chi = 1 \\ 4b_1\chi(g_0)(\chi(g_0) - 1)^{h+1}, & \chi \neq 1. \end{cases}
$$

Remember that $b_1$ occurs in the relation $\mathfrak{P}^{b_1(g_0 - 1)^h} = \mathfrak{p}_1(\beta_1)$. The above map is regarded as an element in $\mathrm{Hom}_{\Gamma_1}(R(G), F_1^\times)$ with $F_1 = \mathbb{Q}_l(\zeta)_l)$ and $\Gamma_1 = \mathrm{Gal}(\mathbb{Q}_l(\zeta_l)/\mathbb{Q}_l)$.

The proofs of i), ii) and iii) are omitted; see [RW4] instead. It should be remarked, however, that the map $\varphi$ has been designed in order to arrive at a cohomologically trivial cokernel $X$ which is closely related to $\Omega_\varphi$, and that it is the Tate sequence to $\tau_S$ together with the fact that $G$ is cyclic which enable us to do so.

A direct consequence is, firstly, the exactness of $(\Delta S)^G \rightarrowtail (E_S)^G \twoheadrightarrow X^G$ (since $X^G = \hat{G}X$) and, secondly, that $(\mathbb{Z}_l \otimes_\mathbb{Z} X)^G$, and so $\mathbb{Z}_l \otimes_\mathbb{Z} X$ itself, is $\mathbb{Z}_l G$-cyclic, namely generated by $p_2$ and $\beta_2$, respectively. Hence there is a presentation $\mathbb{Z}_l G \overset{1 \mapsto \alpha}{\rightarrowtail} \mathbb{Z}_l G \twoheadrightarrow \mathbb{Z}_l \otimes_\mathbb{Z} X$ and $\alpha \in \mathbb{Z}_l G$ determines the relation $\beta_2^\alpha \equiv 1$ in $\mathbb{Z}_l \otimes_\mathbb{Z} X$. Moreover, $[\chi \mapsto \chi(\alpha)]$ is a representing homomorphism for $\mathbb{Z}_l \otimes_\mathbb{Z} X \in K_0T(\mathbb{Z}_l G)$, which modulo $DT(\mathbb{Z}_l G)$ is known, i.e., $\chi(\alpha)$ differs from the $\chi$-component displayed in iii) above by a unit (since they both generate the same ideal). From the fibre diagram

$$
\begin{array}{ccc}
\mathbb{Z}_l G & \overset{g_0 \mapsto \zeta_l}{\longrightarrow} & \mathbb{Z}_l[\zeta_l] \\
\downarrow{g_0 \mapsto 1} & & \downarrow \\
\mathbb{Z}_l & \longrightarrow & \mathbb{F}_l
\end{array}
$$

we therefore see that $\alpha$ and $-4\hat{G}m_1(x_2 - x_1) + 4b_1g_0(g_0 - 1)^{h+1}$ differ by some $u_1 \in \mathbb{Z}_l[\zeta_l]^\times$ and $u_2 \in \mathbb{Z}_l^\times$ in the respective fibres. Since $\alpha$ can be changed by a unit in $\mathbb{Z}_l G$, we may assume $u_1 = 1$ and $u_2 = j, 1 \le j \le l - 1$. Of course, (RRN)$_l$ predicts $j = 1$. In any case,

$$\beta_2^{\hat{G}m_1(x_1 - x_2) + jb_1(g_0 - 1)^{h+1}} \equiv 1$$

is the relation for the generator $\beta_2$ of $\mathbb{Z}_l \otimes_{\mathbb{Z}} X$.

Define $\varepsilon = (\beta_1^{-b_2}\beta_2^{b_1})^{g_0-1}$. It is a global unit because $\beta_1^{-b_2}\beta_2^{b_1}$ generates the $G$-invariant ideal $\mathfrak{p}_1^{b_2}\mathfrak{p}_2^{-b_1}$. From knowing $(\mathbb{Z}_l \otimes_{\mathbb{Z}} X)^G$ we readily infer

$$\beta_1 \equiv 1, \beta_2^{b_1(g_0-1)} \equiv \varepsilon, u_K \equiv p_2^{-2m_1},$$

with all congruences referring to the quotient $X$ of $E_S$, and so obtain

$$u_K \equiv \varepsilon^{2c_2 j(g_0-1)^h}.$$

We can, in fact, improve on this congruence. Namely,

$$\mathbb{Z}_l G/(\hat{G}m_1(x_1 - x_2) + jb_1(g_0 - 1)^{h+1}) = \mathbb{Z}_l \otimes_{\mathbb{Z}} X$$

yields $(\mathbb{Z}_l \otimes_{\mathbb{Z}} X)^G = \mathbb{Z}_l/lm_1$. In particular, $u_K \not\equiv 1$ and so $\varepsilon^{(g_0-1)^h} \not\equiv 1$; however, $u_K^{g_0-1} \equiv 1$ forces $\varepsilon^{(g_0-1)^{h+1}} \equiv 1$. Combining we get

$$\mathbb{Z}_l G/\langle \hat{G}, (g_0 - 1)^{h+1} \rangle \rightarrowtail \mathbb{Z}_l \otimes_{\mathbb{Z}} X_\infty$$

by $1 \mapsto \varepsilon$ and with $X_\infty = E/\mathrm{im}\, \varphi_\infty \subset X$. The above injection is actually an isomorphism, as follows from the Ramachandra index formula (cf. [RW3, §11]). Therefore, $\mathbb{Z}_l \otimes_{\mathbb{Z}} E = \varepsilon^{\mathbb{Z}_l G}$, namely because $\mathbb{Z}_l \otimes_{\mathbb{Z}} \mathrm{im}\, \varphi_\infty = u_K^{\mathbb{Z}_l \otimes_{\mathbb{Z}} \Delta G} = \xi_K^{\mathbb{Z}_l \otimes_{\mathbb{Z}} \Delta G}$, and $u_K = \varepsilon^{z(g_0-1)^h}$ with $z \in \mathbb{Z}_l G$ such that $\chi(z) \equiv 2c_2 j \bmod \zeta_l - 1$ (if $\chi(g_0) = \zeta_l$).

We have reached the stage where we need to know that $p_1$ is not an $l$-th power modulo $p_2$.[12] By definition of $u_K$ we have $u_k \equiv p_1^{2m_2}\bmod \mathfrak{p}_2$, whence $u_K \not\equiv 1 \bmod \mathfrak{p}_2$ and $\varepsilon^{z(g_0-1)^h} \not\equiv 1 \bmod \mathfrak{p}_2$. This implies $h = 0$, since $\mathfrak{p}_2$ is fixed by $g_0$, and enables us to verify $j = 1$ by reducing everything modulo $\mathfrak{p}_2$.

In order to determine $\varepsilon \bmod \mathfrak{p}_2$ we use the local symbol for the extension $K_{\mathfrak{p}_2}/\mathbb{Q}_{p_2}$ (cf. [Se, XIV]); observe that $\zeta_l \in \mathbb{Q}_{p_2}$ and that $K_{\mathfrak{p}_2}$ is a tamely ramified extension of degree $l$ over $\mathbb{Q}_{p_2}$. It follows[13] $\varepsilon \equiv p_0^{m_2 b_1 d_2^{-1}} \bmod \mathfrak{p}_2$ and $u_K \equiv \varepsilon^{2c_2 j} \bmod \mathfrak{p}_2$. Applying the norm $\hat{G}$ to $\mathfrak{P}^{b_1} = \mathfrak{p}_1(\beta_1)$ we obtain $p^{b_1} = p_1 \beta_1^{\hat{G}}$ and thus that $p_1$ and $p_0^{c_2 b_1 d_2^{-1}}$ have the same image under the reciprocity map for $K_{\mathfrak{p}_2}/\mathbb{Q}_{p_2}$, i.e.,

$$p_1 = p_0^{c_2 b_1 d_2^{-1}} y^{\hat{G}}, y \text{ a unit in } K_{\mathfrak{p}_2}.$$

---

[12] Actually, we do not have to know this. However, without the assumption it takes quite an effort to conclude the proof: see [RW4].

[13] The exponents are read in $\mathbb{F}_l$.

Hence, $p_1^{2m_2} \equiv p_0^{2m_2c_2b_1d_2^{-1}} (y^{lm_2})^2 \equiv p_0^{2m_2c_2b_1d_2^{-1}}$ mod $p_2$.

Taking everything into account we have

$$p_0^{2m_2c_2b_1d_2^{-1}} \equiv p_0^{2m_2c_2b_1d_2^{-1}j} \text{ mod } p_2,$$

so $j = 1$ (if we only choose our inert prime $p_0$ to be a primitive root modulo $p_2$).

## 6. Conclusion

We finish by adding a few remarks regarding an extension of the framework of this paper. In the past years a series of conjectures concerning $L$-values has emerged from a much more general point of view, originating from arithmetic algebraic geometry. To give a rough account of the basic matter already requires a great deal of time; however, within the scope of this survey the choice between entering into a detailed description and merely referring to the literature must be in favour of the latter, so we have decided not to recapitulate any definition at all.

Work of Beilinson, Bloch and Deligne (cf. [Bel, 2; Bs; De]) on the $L$-function $L(h^i(X), s)$ of a smooth projective variety $X$ defined over an algebraic number field $K$ has led to conjectures about the order $r$ at a zero $s = m$ of $L(h^i(X), s)$ as well as about the actual value

$$L^*(h^i(X), m) = \lim_{s \to m} (s - m)^r L(h^i(X), s)$$

upto a rational factor $\neq 0$. If $X$ happens to be an abelian variety, the Birch and Swinnerton-Dyer conjecture predicts the exact value $L^*(h^1(X), 1)$. Investigations by Bloch and Kato [BK] into $L$-functions attached to motives of negative weight and by Fontaine and Perrin-Riou [FR] with respect to motives with commutative coefficients aim at getting a good prediction of the exact value in general. Recently, an even more general conjecture, stated by Burns in [Bu2] and comprising the earlier mentioned conjectures in the commutative case, has built a bridge towards Galois module theory by, on the one hand, finding supporting evidence for it from results there and, on the other hand, yielding new impulses to tackle open questions here. In this vein also a generalized Main Conjecture of Iwasawa theory has to be mentioned, which has been formulated by Kato [Ka] (see also [BF1, 2, 3]) and which Burns [Bu3] relates to the conjectures (ST), (SST), (RN) and (RRN). It is in that context that the discussion in 5c again deserves attention because the examples studied there seem to be the first ones confirming this generalized Iwasawa Main Conjecture. In fact, the equation $\varepsilon^{z(g_0-1)^h} = u_K$ which is derived in 5c with a modulo $\langle \hat{G}, g_0-1 \rangle$ uniquely determined $z \in \mathbb{Z}_l G$ not only reflects information that goes beyond the one arising from the known Main Conjecture of Iwasawa theory, but also appears to provide new congruences for $l$-adic $L$-values (cf. [RW4]).

## References

[Bel]   Beilinson, A., Higher regulators and values of L-functions. *J. Sov. Math.* **30**, 2036–2070, 1985.

[Be2]    Beilinson, A., Height pairing between algebraic cycles. *Contemp. Math.* **67**, 1–24, 1987.

[Bw]     Bley, W., Elliptic Curves and Module Structure Over Hopf Orders and The Conjecture of Chinburg-Stark for Abelian Extensions of a Quadratic Imaginary Field. Habilitation thesis, Augsburg, December 1997.

[Bs]     Bloch, S., A note on height pairing, Tamagawa numbers and the Birch and Swinnerton-Dyer conjecture. *Invent. Math.* **58**, 65–76, 1980.

[BK]     Bloch, S., and Kato, K., L-functions and Tamagawa numbers of motives. "*The Grothendieck Festschrift*", 1, *Prog. Math.* **86**, 333–400, Birkhäuser Verlag, 1990.

[Br]     Brown, K.S., Cohomology of Groups. Springer TGM **87**, 1982.

[Bu1]    Burns, D., On multiplicative Galois structure invariants. *Amer. J. Math.* **117**, 875–903, 1995.

[Bu2]    ——————, Iwasawa theory and p-adic Hodge theory over non-commutative algebras, I, II. Preprints (1997; King's College London).

[Bu3]    ——————, Equivariant Tamagawa Numbers and Galois Module Theory, I. *Preprint* (1997; King's College London).

[BF1]    Burns, D., Flach, M., Motivic L-functions and Galois Module Structures. *Math. Ann.* **305**, 65–102, 1996.

[BF2]    ——————, On Galois structure invariants associated to Tate motives. *Preprint* (1997; King's College London and Caltech).

[BF3]    ——————, Equivariant Tamagawa Numbers of Motives. *Preprint* (1997; King's College London and Caltech).

[Ch1]    Chinburg, T., On the Galois structure of algebraic integers and S-units. *Invent. Math.* **74**, 321–349, 1983.

[Ch2]    ——————, Exact sequences and Galois module structure. *Ann. Math.* **121**, 351–376, 1985.

[Ch3]    ——————, The analytic theory of multiplicative Galois structure. *Mem. AMS* **77**, 1989.

[CF]     Cassels, J.W.S. and Fröhlich, A., Algebraic Number Theory. Brighton-Proceedings (1965), Academic Press, 1967.

[CR0]    Curtis, C.W., and Reiner, I., Representation Theory of Finite Groups and Associative Algebras. John Wiley & Sons (1962).

[CR1]    ——————, Methods of Representation Theory, vol. 1. John Wiley & Sons, 1981.

[CR2]    ——————, Methods of Representation Theory, vol. 2. John Wiley & Sons, 1987.

[De]     Deligne, P., Valeurs de fonctions L et périodes d'intégrales. *Proc. Symp. pure Math.* **33**, 313–346, 1979.

[Du]     Dubois, I., Structure galoisienne des S-unités et unités d'une extension cyclique de degré premier, *Thesis*, Bordeaux (1997).

[FP]     Fontaine, J-M. and Perrin-Riou, B., Autour des conjectures de Bloch et Kato: cohomologie galoisienne et valeurs de fonctions L. *Proc. Symp. pure Math.* **55**, 599–706, 1994.

[Fr1]    Fröhlich, A., Galois Module Structure of Algebraic Integers. Springer-Verlag, 1983.

[Fr2]    ——————, Classgroups and Hermitian Modules. Birkhäuser Verlag, PM **48**, 1984.

[Fr3]     Fröhlich, A., Galois module structure. In: *Algebraic Number Fields, Durham-Proc.* (1975) (*Ed.* Fröhlich) A., Academic Press, 1977.

[Fr4]     ————, Central extensions, Galois groups, and ideal class groups of number fields. *Contemp. Math.* **24**, AMS, 1983.

[Gc]      Greither, C., The structure of some minus class groups, and Chinburg's third conjecture for abelian fields. *Preprint* (1997; Universitè Laval et CICMA, Canada).

[Gk]      Gruenberg, K.W., Relation Modules of Finite Groups. *Reg. Conf. Sr. Math.* **25**, AMS, 1976.

[GRW]     Gruenberg, K.W., Ritter, J. and Weiss, A., A Local Approach to Chinburg's Root Number Conjecture. To appear in Proc. LMS.

[GW1]     Gruenberg, K.W. and Weiss, A., Galois invariants for units. *Proc. LMS* **70**, 264–284, 1995.

[GW2]     ————, Galois invariants for S-units. *Am. J. Math.* **119**, 953–983, 1997.

[Ha]      Hasse, H., Über die Klassenzahl Abelscher Zahlkörper. Springer Verlag (Nachdruck der Ausgabe des Akademie-Verlags von 1952).

[Ka]      Kato, K., Iwasawa theory and p-adic Hodge theory. *Kodai Math. J.* **16**, 1–31, 1993.

[La]      Lang, S., Algebraic Number Theory. Addison-Wesley, 1970.

[MW]      Mazur, B. and Wiles, A., Class fields of abelian extensions of $\mathbb{Q}$. *Invent. Math.* **76**, 179–330, 1984.

[Re]      Reiner, I., Maximal Orders. Academic Press (1975).

[RW1]     Ritter, J. and Weiss, A., On the local Galois structure of S-units. In: *Algebra and Number Theory, Proc. Conf. Inst. experiment. Math.* University of Essen, Germany, December 2–4, 1992 (*Eds.* G. Frey, J. Ritter), de Gruyter Proc. Math. 229–245, 1994.

[RW2]     ————, A Tate sequence for global units. *Comp. Math.* **102**, 147–178, 1996.

[RW3]     ————, Cohomology of units and L-values at zero. *J. AMS* **10**, 513–552, 1997.

[RW4]     ————, *The Lifted Root Number Conjecture for some cyclic extensions of* $\mathbb{Q}$. Submitted to Acta Arithmetica.

[Se]      Serre, J.P., Corps Locaux. Hermann, Paris, 1968.

[Sw]      Swan, R.G., K-Theory of Finite Group and Orders (Notes by E.G. Evans). *Springer LNM* **149**, 1970.

[Ta1]     Tate, J., The cohomology groups of tori in finite Galois extensions of numer fields. *Nagoya Math. J.* **27**, 709–719, 1966.

[Ta2]     ————, Les Conjectures de Stark sur les Fonctions L d'Artin en $s = 0$. *Prog. Math.* **47**, Birkhäuser, 1984.

[Wa]      Washington, L., Introduction to Cyclotomic Fields. Springer-Verlag (1982).

[We]      Weiss, A., Multiplicative Galois Module Structure. Fields Institute Monographs **5**, AMS (1996).

[Wi1]     Wiles, A., The Iwasawa Conjecture for Totally Real Fields. *Ann. Math.* **131**, 493–540, 1990.

[Wi2]     ————, On a conjecture of Brumer. *Ann. Math.* **131**, 555–565, 1990.

Institut für Mathematik, Universität Augsburg, D-86135 Augsburg, Germany

# A Survey of Groups in Which Normality or Permutability is a Transitive Relation

*Derek J.S. Robinson*

## 1. Introduction

In group theory it is a familiar observation that normality of subgroups is not in general a transitive relation, i.e., $H \lhd K \lhd G$ need not imply that $H \lhd G$. The smallest group exhibiting this phenomenon is $\text{Dih}(8)$, the dihedral group of order 8.

**Definition.** A group $G$ is said to be a *T-group* if

$$H \lhd K \lhd G \text{ implies that } H \lhd G.$$

Equivalently, one could require that every subnormal subgroup of $G$ be normal. Examples of $T$-groups that quickly come to mind are: abelian groups, simple groups, the symmetric groups $S_n, n \neq 4$.

The theory of $T$-groups really began with a paper of Dedekind published in 1897. In this work Dedekind investigated the structure of finite groups in which every subgroup is normal. Subsequently, Baer (1933) extended Dedekind's work to infinite groups. The definitive result is:

**Theorem 1.1.** *A group $G$ has all its subgroups normal if and only if $G$ is either abelian or hamiltonian, i.e., $G = Q \times E \times O$ where $Q$ is a quaternion group of order 8, $E$ is an elementary abelian 2-group and $O$ is an abelian group with all its elements of odd order.*

The groups appearing in (1.1) are now generally called *Dedekind groups*.

### Motivation

The motivation for Dedekind's work came from algebraic number theory. Suppose that $K$ is an algebraic number field with Galois group $G = \text{Gal}(K/\mathbb{Q})$. The familiar Galois correspondence between subfields of $K$ and subgroups of $G$ shows that:

(i) every subfield of $K$ is normal over $\mathbb{Q}$ if and only if $G$ is a Dedekind group;

(ii) normality of subfields of $K$ is transitive, (i.e., if we have subfields $A \subseteq B$ with $A$ normal over $\mathbb{Q}$ and $B$ normal over $A$, then $B$ is normal over $\mathbb{Q}$), if and only if $G$ is a $T$-group.

## Permutable Subgroups

A subgroup $H$ of a group $G$ is said to be *permutable* (or sometimes *quasinormal*) in $G$ if $HK = KH$ for all subgroups $K$ of $G$. Since every normal subgroup is permutable, we should regard permutability as a rather weak form of normality. We will write

$$H \, per \, G$$

to indicate that $H$ is a permutable subgroup of the group $G$.

The connection between permutability and normality is quite strong for finite groups, as the following well-known result shows.

**Theorem 1.2. (Ore, 1939)** *In a finite group $G$ every permutable subgroup $H$ is subnormal in $G$.*

**Proof.** It is enough to show that if $H$ is a *maximal* permutable subgroup of $G$, then $H \triangleleft G$. Suppose this is false, so that $H \neq H^g$ for some $g \in G$. Then $HH^g$ is a permutable subgroup properly containing $H$. Therefore, $G = HH^g$ and $g \in HH^g$. However, this implies that $g \in H$ and $H^g = H$.

A permutable subgroup of an infinite group need not be subnormal, although it must be ascendant by a result of Stonehewer (1972).

**Definition.** A group $G$ will be called a *PT*-group if permutability is transitive in $G$, i.e., if $H \, per \, K \, per \, G$ implies that $H \, per \, G$.

For finite groups this definition can be reformulated in terms of subnormality as follows:

**Lemma 1.3.** *Let $G$ be a finite group. Then $G$ is a PT-group if and only if every subnormal subgroup of $G$ is permutable.*

**Proof.** If $G$ is a *PT*-group and $H$ is subnormal in $G$, then certainly $H$ is permutable. Conversely, if $H \, per \, K \, per \, G$, then $H$ is subnormal in $G$ by (1.2); hence $H \, per \, G$.

**Corollary 1.4.** *A finite $T$-group is a PT-group.*

However, in general, a $T$-group need not be a *PT*-group, as the following example shows (Menegazzo, 1968). Let $G = \langle x \rangle \ltimes A$ where $A \simeq 3^\infty$ and $a^x = a^5$ for $a \in A$. Then every subnormal subgroup of $G$ contains $A$ or is contained in it, so $G$ is a $T$-group. But it is not a *PT*-group because, if $a_1 \in A$ has order 3, then $\langle x^2 a_1 \rangle$ *per* $\langle x^2, A \rangle \triangleleft G$, but $\langle x^2 a_1 \rangle$ is not permutable in $G$.

## Iwasawa Groups

We will call a group an *Iwasawa group* if every subgroup is permutable. These groups form a wider class than the class of Dedekind groups. A finite Iwasawa group has every subgroup subnormal by (1.2), so it is nilpotent. Thus to find all finite Iwasawa groups one has only to determine the Iwasawa groups of prime power order; these turn out to be just the finite $p$-groups with modular subgroup lattice (Schmidt, 1994, p. 55).

The definitive result is due to Iwasawa (1941).

**Theorem 1.5.** *A finite group is an Iwasawa group if and only if it is a direct product $G = G_1 \times \cdots \times G_k$ where either $G_i$ is a Dedekind group or $G_i = \langle x_i \rangle A_i$ is a $p_i$-group with $A_i$ abelian and $a^{x_i} = a^{1+p_i^{s_i}}$, $(a \in A_i)$, where $s_i > 0$ and $s_i > 1$ if $p_i = 2$. Here the $p_i$ are distinct primes.*

For an excellent account of this theory and for the structure of infinite Iwasawa groups see Schmidt (1994), (2.3, 2.4), and also Iwasawa (1943).

In the next section, we shall discuss what is known about the structure of soluble $T$-groups and $PT$-groups. Until the recent classification of finite simple groups (CFSG) little or nothing was known about insoluble $T$-and $PT$-groups. In later sections, we will show how CFSG allows us to draw some remarkably strong conclusions about the structure of finite $T$-groups and finite $PT$-groups.

## 2. Soluble Groups

The first article to focus on non-Dedekind $T$-groups was published in 1942 by Best and Taussky (1942). Their main result is:

**Theorem 2.1.** *A finite group with cyclic Sylow subgroups is a $T$-group.*

Thus, in particular, every group of square-free order is a $T$-group. Now the structure of groups with cyclic Sylow subgroups had previously been determined by Hölder; in particular such groups are soluble. From this structure it is not hard to see that these groups are $T$-groups.

The first characterization of finite soluble $T$-groups was given by Zacher (1952), who found necessary and sufficient conditions on the Sylow structure. The definitive result on the structure of finite soluble $T$-groups was published by Gaschütz (1957).

**Theorem 2.2.** *A finite group $G$ is a soluble $T$-group if and only if it has the form $G = Q \rtimes L$, (a semidirect product), where $Q$ is a Dedekind group, $L$ is abelian of odd order, $|Q|$ and $|L|$ are relatively prime, and each $x$ in $Q$ induces a **power automorphism** in $L$, (i.e., $a^x = a^n$ for all $a \in L$, where $n$ is an integer).*

Gaschütz's theorem provides us with not only a lucid description of the group structure, but also a means of constructing all finite soluble $T$-groups. Moreover, is not hard to decide when two of the constructed groups are isomorphic. Even today Gaschütz's theorem remains a prime example of that rarity in group theory, a perfect classification theorem.

**Definition.** A $\bar{T}$-*group* is a group all of whose subgroups are $T$-groups. This is a much narrower class of groups then $T$. One sees easily from (2.2) that every finite soluble $T$-group is a $\bar{T}$-group. On the other hand, the infinite dihedral 2-group, $G = \langle x \rangle \ltimes A$, where $A \simeq 2^{\infty}$, $x^2 = 1$ and $a^x = a^{-1}$, $(a \in A)$, is an infinite soluble $T$-group that is not a $\bar{T}$-group.

Finite $\bar{T}$-groups turn out to be soluble, so that *for finite groups $\bar{T}$ is equivalent to soluble $T$*. The underlying result here is due to Peng (1969) and Robinson (1968):

**Theorem 2.3.** *If $G$ is a finite group, the following conditions are equivalent:*

(i) *$G$ is a soluble $T$-group;*

(ii) *$G$ is a $T$-group;*

(iii) *for each prime $p$, if $P$ is a Sylow $p$-subgroup of $G$, then every subgroup of $P$ is normal in $N_G(P)$.*

Minimal non-$T$-groups, i.e., non-$T$-groups whose proper subgroups are $T$, are studied in Robinson (1969); as a consequence of the results one obtains:

**Theorem 2.4.** *A finite group is a $\bar{T}$-group if and only if every 2-generator subgroup of order divisible by at most two primes is a $T$-group.*

Soluble non-$T$-groups whose proper quotients are $T$-groups are determined in Robinson (1973).

Finally, finite groups generated by subnormal $T$-subgroups are studied in Cossey (1995).

The study of finite soluble $PT$-groups was initiated by Zacher (1964), who proved a result very similar to Gaschütz's.

**Theorem 2.5.** *A finite group $G$ is a soluble $PT$-group if and only if it has the form $G = Q \ltimes L$, where $Q$ is an Iwasawa group, $L$ is abelian of odd order, $|Q|$ and $|L|$ are relatively prime and each $x$ in $Q$ induces a power automorphism in $L$.*

**Corollary 2.6.** *A finite soluble $PT$-group $G$ is metabelian.*

For we have $G = Q \ltimes L$ and $G' = Q'[L, Q]$. Now $[L, Q'] = 1$ since power automorphisms commute; also $Q'$ is abelian by (1.5). Thus $G'$ is abelian.

Note also the consequence of (2.5) that subgroups of finite soluble $PT$-groups are $PT$-groups. In fact all soluble $PT$-groups are metabelian, as may be read off from results of Menegazzo (1968).

Recently a characterization of soluble $PT$-groups similar to (2.3) has been given by Brewster, Beidleman and the author.

*Infinite Soluble Groups*

Infinite soluble $T$-groups were first considered in Robinson (1964), where it was shown that the following holds:

**Theorem 2.7.** *A finitely generated soluble T-group is either finite or abelian.*

It turns out to be convenient to divide non-abelian soluble $T$-groups into three sub-classes:

    ($a$) periodic soluble $T$-groups;

    ($b$) soluble $T$-groups $G$ for which $C_G(G')$ is non-periodic;

    ($c$) non-periodic soluble $T$-groups $G$ for which $C_G(G')$ is periodic.

The theory is most satisfying in case ($a$), where there is an analogue of Gaschütz's theorem, and case ($b$), where a complete set of invariants for $G$ can be given.

A similar study of infinite soluble $PT$-groups has been carried out in two papers of Menegazzo (1968, 1970).

For a detailed account of the theories of infinite soluble $T$-groups and infinite soluble $PT$-groups the reader is referred to the works cited.

## 3. Insoluble Groups

In the next four sections we shall indicate how the classification of finite simple groups impinges on the theory of finite insoluble $T$-groups and $PT$-groups. Generally proofs will not be given and details will appear elsewhere. *From now on it is understood that all groups are finite.*

According to the CFSG every finite (non-abelian) simple group falls into one of the three classes:

    ($a$) the alternating groups $A_n, n \geq 5$;

    ($b$) the simple groups of Lie type;

    ($c$) the 26 sporadic simple groups.

In addition a great deal of information about finite simple groups is now available. For example, the Schur multiplicator $M(S)$ and the universal covering group $\hat{S}$ of all finite simple groups $S$ are known. Recall that these form a central extension $M(S) \rightarrowtail \hat{S} \twoheadrightarrow S$. It is noteworthy that if $S$ is not of projective or unitary type, then $M(S)$ is a $\{2, 3\}$-group. As a reference for these facts we cite Griess (1980).

Also important for our purposes is knowledge of the outer automorphism group $\text{Out}(S) = \text{Aut}(S)/\text{Inn}(S)$. This group is always soluble, (with derived length $\leq 3$), thus verifying the *Schreier conjecture*. In fact $\text{Out}(S)$ has very small order if $S$ is of type ($a$) or ($c$); when $S$ is of Lie type, there is a well-understood decomposition of an automorphism into inner, diagonal, field and graph automorphisms (see Carter, 1992; Steinberg, 1960). Thus the structure of $\text{Out}(S)$ can be considered known.

## Necessary Conditions

Let us now analyze the structure of an insoluble *PT*-group $G$. First we record an easy lemma which shows the role of power automorphisms, (already indicated by (2.5)).

**Lemma 3.1.** *Let $G$ be a PT-group and let $N$ be a normal p-subgroup of $G$. Then $p'$-elements of $G$ induce power automorphisms in $N$. Consequently $G/C_G(N)$ is an extension of an abelian $p'$-group by a p-group, and hence is soluble.*

To see this, let $a \in N$ and note that $\langle a \rangle$ *per* $G$ since $\langle a \rangle$ is subnormal. Thus for any $p'$-element $x$ we have $\langle x, a \rangle = \langle x \rangle \langle a \rangle$, and $a^{\langle x \rangle} = a^{\langle x \rangle} \cap (\langle x \rangle \langle a \rangle) = (a^{\langle x \rangle} \cap \langle x \rangle) \langle a \rangle = \langle a \rangle$ since $a^{\langle x \rangle}$ is a $p$-group. Hence, $a^x = a^m$ for some integer $m$.

Now consider a *PT*-group $G$ and write $D$ for the limit of the derived series of $G$. Then

$$D = G''$$

by (2.6), and $D = D'$, i.e., $D$ is a perfect group. The structure of

$$Q := G/D$$

is known from (2.5), so we can assume that $D \neq 1$.

Let $E$ denote the soluble radical, (i.e., the maximum soluble normal subgroup), of $D$, and write $C_i = C_D(E^{(i)}/E^{(i+1)})$, where $E^{(i)}$ is a term of the derived series of $E$. Then $D/C_i$ is soluble by (3.1). But $D$ is perfect, so $D = C_i$ and $D$ centralizes each $E^{(i)}/E^{(i+1)}$. This means that $E$ coincides with the hypercentre of $D$. But, since $D$ is perfect, the hypercentre equals the centre $Z(D)$. Therefore, $E = Z(D)$; we will write

$$Z = Z(D).$$

We now have the central extension $Z \rightarrowtail D \twoheadrightarrow R$ where $R = D/Z$. Since $D = D'$, this group $D$ an image of the universal cover of $R$, and there is a commutative diagram

$$\begin{array}{ccccc} M(R) & \rightarrowtail & \hat{R} & \twoheadrightarrow & R \\ \downarrow & & \downarrow & & \| \\ Z & \rightarrowtail & D & \twoheadrightarrow & R \end{array}$$

(For background on universal covers see Stammbach, 1973).

The next step is to identify $R$; it is at this point that the truth of the Schreier Conjecture is important.

Since $R$ is a semisimple group, (i.e., there are no non-trivial abelian normal subgroups), it has a unique largest normal subgroup $X/Z$ which is a direct product of simple groups $U_i/Z$. Now $U_i$ is subnormal in $G$, so $U_i$ *per* $G$. If $g \in G$, then $U_i^{\langle g \rangle} = U_i^{\langle g \rangle} \cap (U_i \langle g \rangle) = U_i(U_i^{\langle g \rangle} \cap \langle g \rangle)$, and so $U_i^{\langle g \rangle}/U_i$ is cyclic. But $U_i^{\langle g \rangle}/Z$ is certainly perfect, so $U_i^{\langle g \rangle} = U_i$ and $U_i \triangleleft G$.

By the last paragraph $D$ induces a group of automorphisms in each $U_i/Z$. Also $\text{Out}(U_i/Z)$ is soluble, by the Schreier Conjecture, and $D$ is a perfect group. Taken together, these statements imply that $D = XC_D(X/Z)$. But $C_D(X/Z) = Z$, so $D = X$. Therefore, $R$ is a direct product of simple groups $R_i = U_i/Z$,

$$R = R_1 \times R_2 \times \cdots \times R_k.$$

By now it is clear that we know a good deal about the structure of $G$. Indeed $Z$ is an image of $M(R)$ and $M(R) \simeq M(R_1) \oplus \cdots \oplus M(R_k)$, while $M(R_i)$ is known. Also $G/C_G(D)D$ is isomorphic with a subgroup of $\text{Out}(R_1) \times \cdots \times \text{Out}(R_k)$.

Further arguments enable us to establish:

**Theorem 3.2.** *Let $G$ be a PT-group with $D = G''$; let $Z = Z(D)$ and let $S$ denote the soluble radical of $G$. Then*

(i) *$Z$ is the soluble radical of $D$ and $R = D/Z$ is a direct product of simple groups $R_i = U_i/Z$ with $U_i \triangleleft G$;*

(ii) *$D$ is an image of the universal cover $\hat{R}$ and $Z$ is an image of $M(R) \simeq M(R_1) \oplus \cdots \oplus M(R_k)$; also $Q := G/D$ is a soluble PT-group;*

(iii) *$S = C_G(D) = C_G(D/Z)$;*

(iv) *$SD/D$ equals the kernel of the coupling of $D \rightarrowtail G \twoheadrightarrow Q$, so that $G/SD$ is isomorphic with a subgroup of $\underset{i=1}{\overset{k}{\text{Dr}}} \text{Out}(R_i)$; and*

(v) *$p'$-elements of $G$ induce power automorphisms in $Z_p$ for all primes $p$.*

Here the *coupling* of the extension $D \rightarrowtail G \twoheadrightarrow Q$ is the homomorphism $\chi : Q \to \text{Out}(D)$ that arises from conjugating in $D$ by elements of $G$.

Of course (3.2) also applies to $T$-groups; note that $(v)$ can then be strengthened since *all* elements of $G$ induce power automorphisms in $Z$. Also $Q$ is, of course, a soluble $T$-group.

## 4. Sufficient Conditions for PT

Although the structural information provided in (3.2) is quite detailed, it does not imply that the group $G$ is a *PT*-group. For example, consider the central product

$$G = SL_2(5) \underset{Z_2}{Y} \text{Dih}(8)$$

of order 480. Here $D = SL_2(5)$, $S = \text{Dih}(8)$ and $|Z| = 2$. But $G$ is not a *PT*-group because $\text{Dih}(8)$ is not. What we need are further restrictions on normal $p$-subgroups of $G$.

Let $N$ be a normal $p$-subgroup of a group $G$ satisfying the conclusions of (3.2). Then essentially we understand the action of $p'$-elements of $G$ on $ND/D$; indeed such elements induce power automorphisms. Also $N \cap D = N \cap Z$, and again $p'$-elements induce power automorphisms. The problem is to ensure that these actions "fit together" in a harmonious manner. A similar problem arises with the action of $p$-elements.

## Conditions $N_p$ and $P_p$

Suppose that $G$ is a group and $p$ is a prime. We shall define two conditions $N_p$ and $P_p$ on $G$ as follows:

(a) the group $G$ satisfies *the condition* $N_p$ if, whenever $N \triangleleft G$ with $N$ soluble, $p'$-elements of $G$ induce power automorphisms in $O_p(G/N)$;

(b) the group $G$ satisfies *the condition* $P_p$ if, whenever $N \triangleleft G$ with $N$ soluble, all subgroups of $O_p(G/N)$ are permutable in a Sylow $p$-subgroup of $G/N$.

Notice that $P_p$ and $N_p$ are valid in any *PT*-group $G$ and its quotients for all primes $p$.

Using these technical conditions we can formulate a criterion for a group to be a *PT*-group.

**Theorem 4.1.** *Let $G$ be a group with a perfect normal subgroup $D$ such that*

(i) $Q := G/D$ *is a soluble PT-group;*

(ii) *if $Z := Z(D)$, then $D/Z = U_1/Z \times \cdots \times U_k/Z$ where $U_i/Z$ is simple and $U_i \triangleleft G$;*

(iii) $G/U'_{i_1} \cdots U'_{i_k}$ *has $P_p$ for all $p \in \pi(Z)$ and $N_p$ for all $p \in \pi(D)$ (where $1 \leq i_j \leq k$).*

*Then $G$ is a PT-group.*

Of course all the conditions given are also necessary for *PT*, so (4.1) provides necessary and sufficient conditions for the property *PT*.

A more accessible set of necessary and sufficient conditions for a group to be a *PT*-group are furnished by:

**Theorem 4.2.** *A group $G$ is a PT-group if and only if it has a perfect normal subgroup $D$ such that $G/D$ is soluble, $D/Z(D)$ is a direct product of $G$-invariant simple groups, and all quotients of $G$ satisfy $N_p$ and $P_p$ for all $p$.*

As part of the proof of this result one shows that: *a soluble group which satisfies $N_p$ and $P_p$ for all $p$ is a PT-group.*

## 5. Sufficient Conditions for *T*.

To obtain sufficient conditions for a group to have *T*, it is necessary to replace the two conditions $P_p$ and $N_p$ by a single stronger one.

## Condition $T_p$

Let $G$ be a group and $p$ a prime. Then $G$ satisfies *the condition $T_p$* if, whenever $N \triangleleft G$ with $N$ soluble, and $H/N$ is a $G$-invariant $p$-group with nilpotent class $\leq 2$, elements of $G$ induce power automorphisms in $H/N$.

The reason why one can restrict to factors with nilpotent class $\leq 2$ in this result is that Dedekind groups, unlike Iwasawa groups, have class $\leq 2$. Notice that $T_p$ holds in any $T$-group and its quotients for all primes $p$.

We can now formulate a criterion for $T$.

**Theorem 5.1.** *Let $G$ be a group with a perfect normal subgroup $D$ such that:*

*(i) $Q := G/D$ is a soluble $T$-group;*

*(ii) if $Z = Z(D)$, then $D/Z = U_1/Z \times \cdots \times U_k/Z$ where $U_i/Z$ is simple and $U_i \triangleleft G$;*

*(iii) $G/U'_{i_1} \ldots U'_{i_k}$ has $T_p$ for all $p \in \pi(Z)$, (where $1 \leq i_j \leq k$).*

*Then $G$ is a $T$-group.*

Again there is a simpler form of the criterion.

**Theorem 5.2.** *A group $G$ is a $T$-group if and only if it has a perfect normal subgroup $D$ such that $G/D$ is soluble, $D/Z(D)$ is a direct product of $G$-invariant simple groups, and $G$ induces power automorphisms in every $G$-invariant $p$-factor of nilpotent class $\leq 2$.*

Next suppose now that $G$ is a $PT$-group with $D = G''$. Notice that (5.2) can be applied to the group $D$ since the power automorphism condition is satisfied; hence, $D$ is a $T$-group. Thus we have an interesting connection between $PT$-groups and $T$-groups.

**Theorem 5.3.** *If $G$ is a $PT$-group, then $G''$ is a $T$-group.*

Finally, we mention that the methods of sections 4 and 5 can be applied to give necessary and sufficient conditions for a direct product of finite $PT$-groups ($T$-groups) to be a $PT$-group ($T$-group).

## 6. Constructing T-Groups

A good test of a classification theorem in group theory is the extent to which it tells us how to construct the relevant groups. Consider, for example, the problem of constructing insoluble $T$-groups. To begin one should choose some simple groups $R_1, \ldots, R_k$. Write $R = R_1 \times \cdots \times R_k$, so that $M(R) \simeq M(R_1) \oplus \cdots \oplus M(R_k)$. Now choose a subgroup $V$ of $M(R)$, form the covering group $\hat{R}$ of $R$, and define

$$D := \hat{R}/V, \quad Z := M(R)/V.$$

Thus $Z \rightarrowtail D \twoheadrightarrow R$ is a stem cover of $R$.

The next step is to choose a soluble $T$-group $Q$—these are all known by Gaschütz's theorem. To form an extension of $D$ by $Q$, one must first choose a homomorphism

$$\chi : Q \to \text{Out}(D) = N_{\text{Out}R}(V)$$

as the coupling. In the present case we must ensure that $\text{Im}(\chi) \le N_{\text{Dr}_i(\text{Out}R_i)}(V)$, so that each $R_i$ will be $G$-invariant.

There is a further requirement since not every such $\chi$ gives rise to an extension. Recall from extension theory that $\chi$ determines an element $\hat{\chi}$ of $H^3(Q, Z_\chi)$, the *obstruction*, where $Z_\chi$ is $Z$ with the $Q$-module structure dictated by $\chi$. It is well-known that the condition for $\chi$ to give rise to an extension of $D$ by $Q$ is that the obstruction should vanish, i.e., $\hat{\chi} = 0$.

Once $\chi$ has been so chosen, we form an extension $D \rightarrowtail G \twoheadrightarrow Q$ by settling on a cohomology class $\psi \in H^2(Q, Z_\chi)$. Then there is the condition $T_p$ to be verified for each $G/U'_{i_1} \cdots U'_{i_k}$ where $U_i/Z = R_i$. Think of this as a kind of compatibility condition between $\chi$ and $\psi$. At the end of this procedure we will then have a $T$-group

$$G = G(R_1, \ldots, R_k | V, Q | \chi, \psi) :$$

and of course every $T$-group has this form.

It will be clear to the reader that the awkward points in this constructional method are the verifications that $\hat{\chi} = 0$ and that $T_p$ holds. However in some cases this is not so hard. For example, when $R$ is simple and $H^3(Q, Z_x) = 0$, then $\hat{\chi} = 0$. Also $T_p$ need only be verified for $G$, which should not be hard since $Z$ is a small group.

# References

Baer, R. Situation der Untergruppen und Struktur der Gruppe, *Sitz.-Ber. Heidelberg. Akad. Wiss.* **2**, 12–17, 1933.

Best, E. and Taussky, O. A class of groups, *Proc. Roy. Irish Acad. Sect A* **47**, 55–62, 1942.

Carter, R.W. *Simple Groups of Lie Type*, Wiley-Interscience, New York, 1972..

Cossey, J. Finite groups generated by subnormal $T$-subgroups, *Glasgow Math. J.* **37**, 363–371, 1995.

Dedekind, R. Über Gruppen, deren sämmtliche Teiler Normalteiler sind, *Math. Ann.* **48**, 548–561, 1897.

Gaschütz, W. Gruppen, in denen das Normalteilersein transitiv ist, *J. reine angew. Math.* **198**, 87–92, 1957.

Griess, R.L. Schur multipliers of the known finite simple groups II, *Proc. Symp. Pure Math.* **37**, 279–282, 1980.

Iwasawa, K. Über die endlichen Gruppen und die Verbände ihrer Untergruppen, *J. Fac. Sci. Imp. Univ. Tokyo* **I.4**, 171–199, 1941.

————, On the structure of infinite M-groups, *Jap. J. Math.* **18**, 709–728, 1943.

Menegazzo, F. Gruppi nei quali la relazione di quasi-normalità è transitiva I, II, *Rend. Sem. Mat. Univ. Padova* **40**, 347–361, 1968; *ibid.* **42**, 389–399, 1970.

Ore, O. Contributions to the theory of groups of finite order, *Duke Math* **5**, 431–460, 1939.

Peng, T.A. Finite groups with pronormal subgroups, *Proc. Amer. Math. Soc.* **20**, 232–234, 1969.

Robinson, D.J.S. Groups in which normality is a transitive relation, *Proc. Camb. Phil. Soc.* **60**, 21–38, 1964.

————, A note on finite groups in which normality is transitive, *Proc. Amer. Math. Soc.* **19**, 933–937, 1968.

————, Groups which are minimal with respect to normality being intransitive, *Pacific J. Math.* **31**, 777–785, 1969.

————, Groups whose homomorphic images have a transitive normality relation, *Trans. Amer. Math. Soc.* **176**, 181–213, 1973.

————, *A Course in the Theory of Groups*, Second Edition, Springer, New York, 1996.

Schmidt, R. *Subgroup Lattices of Groups*, De Gruyter, Berlin, 1994.

Stammbach, U. *Homology in Group Theory, LNM.*, **359**, Springer, Berlin, 1973.

Steinberg, R. Automorphisms of linear groups, *Canad. J. Math.* **12**, 606–615, 1960.

Stonehewer, S.E. Permutable subgroups of infinite groups, *Math. Z.* **125**, 1–16, 1972.

Zacher, G. Caratterizzazione dei t-gruppi risolubili, *Ricerche Mat.* **1**, 287–294, 1952.

————, I gruppi risolubili in cui i sottogruppi di composizione coincidono con i sottogruppi quasi-normali, *Lincei Rend Sc. Fis. Mat. Nat.* **37**, 150–154, 1964.

Department of Mathematics, University of Illinois, 1409, West Green Street, Urbana, IL 61801, USA

# The Structure of Some Group Rings*

*Klaus W. Roggenkamp*

## 1. Blocks and Defect Groups

For details of block theory we refer to [CR; 82, 6].

Let $G$ be a finite group and let $R$ be a complete Dedekind domain with maximal ideal p and field of fractions $K$, a local number field, and residue field $\mathfrak{k}$ of characteristic $p > 0$.

In general, the group ring $RG := \{\sum_{g \in G} r_g \cdot g : r_g \in R\}$ will decompose into a direct sum of rings. The indecomposable ring direct summands of $RG$ are called BLOCKS or more generally $p$-BLOCKS of $RG$. Such a block is uniquely determined – note that the Krull-Schmidt theorem holds, since $R$ complete – by a primitive central – note that these are unique – idempotent $e = e(B)$ with $B = RG \cdot e$, called the BLOCK IDEMPOTENT.

Since the representation theory of $RG$ is determined by the representation theory of the blocks, these blocks are the main ingredients of $p$-adic representation theory. For a block $B$ we denote by $_B latt$ the category of LEFT $B$-LATTICES, i.e., left $B$-modules which are $R$-free of finite rank. The aim is the classification – if possible – of the indecomposable $B$-lattices.

An important tool of constructing $RG$-lattices is the process of INDUCTION. Let $H$ be a subgroup of $G$ and $M \in {}_{RG} latt$ be indecomposable, then there exists a – unique up to conjugation – $p$-subgroup $V$ of $G$, called the VERTEX of $M$ and an indecomposable $RV$-module $S$ such that $M$ is a direct summand of $S \uparrow_V^G :=$ $RG \otimes_{RH} S$; i.e., it is a direct summand of an induced module from $V$. The module $S$ is unique up to conjugation and is called the SOURCE OF $M$.

**Definition 1.1.** Let $B$ be a block, then there exists a minimal – unique up to conjugation – p-group $D = D(B)$, the DEFECT GROUP OF $B$, such that every indecomposable $B$-module is a direct summand of a module induced from $D$; equivalently, the defect group of $B$ is the vertex $D$ of $B$ as $R(G \times G^{op})$-module – this vertex then can be identified naturally with a subgroup of $G$.

*This research was partially supported by the Deutsche Forschungsgemeinschaft.

**Example 1.2.**

1. There is a unique block of $RG$ containing the trivial representation; it is called the PRINCIPAL BLOCK $B_0$; its defect group is the Sylow $p$-subgroup of $G$.

2. If the Sylow $p$-subgroup is normal, then every block has the Sylow $p$-subgroup as vertex, since the defect group of every block is the intersection of two Sylow $p$-subgroups.

3. A block $B$ of DEFECT ZERO is a block whose defect group is trivial; i.e., every indecomposable lattice is induced from the trivial group and hence projective. It even turns out that $B \simeq Mat(n, S)$, where $S$ is an unramified extension of $R$; i.e., $rad(S) = rad(R) \cdot S$. This means that $B$ is an indecomposable separable $R$-order.

4. For example, $\mathbb{Z}_3 S_3$ is a single block, but $\mathbb{Z}_2 S_3 = Mat(2, \mathbb{Z}_2) \oplus \mathbb{Z}_2 C_2$ is the direct sum of a block of defect zero and the group ring of $C_2$, the cyclic group of order 2.

## 2. Blocks with Normal Defect Group

We keep the notation from above. Let $B$ be a block of $RG$ with NORMAL DEFECT GROUP $D$ and let $e$ be the block idempotent of $B$.

$$(1) \qquad\qquad 0 \longrightarrow I_R(D) \cdot G \longrightarrow RG \xrightarrow{\epsilon_D} RG/D \longrightarrow 0,$$

where $\epsilon_D$ is induced by the residue map $G \longrightarrow G/D$. We then get the commutative diagram with exact rows and columns:

$$
\begin{array}{ccccccccc}
& & 0 & & 0 & & 0 & & \\
& & \downarrow & & \downarrow & & \downarrow & & \\
0 & \longrightarrow & 0 & \longrightarrow & 0 & \longrightarrow & |D| \cdot \bar{B} & \longrightarrow & 0 \\
& & \downarrow & & \downarrow & & \downarrow & & \\
0 & \longrightarrow & I_R(D) \cdot B & \longrightarrow & B & \longrightarrow & \bar{B} & \longrightarrow & 0. \\
& & \downarrow & & \downarrow & & \downarrow & & \\
0 & \longrightarrow & I_R(D) \cdot B & \longrightarrow & \Lambda_B & \longrightarrow & \bar{B}/(|D| \cdot \bar{B}) & \longrightarrow & 0 \\
& & \downarrow & & \downarrow & & \downarrow & & \\
& & 0 & & 0 & & 0 & &
\end{array}
$$

(2)

Since the middle sequence splits off the exact sequence (1), the ring $\bar{B}$ is an $R$-lattice and hence it is a block of $RG/D$.

**Claim 2.1.** $\bar{B}$ *is a direct sum of blocks of defect zero of $RG/D$ and hence it is a separable R-order.*

**Proof.** We shall be using the follows facts:-

1. Since $D$ is normal, we have $D = O_p(G)$, the maximal normal $p$-subgroup; in fact, $D$ is the intersection of two Sylow $p$-subgroup.

2. The vertex of every $B$-lattice is contained in $D$ - note that $D$ is normal.

3. A block all of whose lattices are projective is a block of defect zero.

Let $M$ be an indecomposable $\bar{B}$-lattice. If we view it as $B$-lattice, its source is the trivial $D$-module, and thus, since $M$ is a $\bar{G} := G/O_p(G)$-module, it is a projective $\bar{B}$-module. Hence $\bar{B}$ is a direct sum of blocks of defect zero of $RG/D$. **q. e. d.**

**Note 2.2.** The indecomposable $\bar{B}$-lattices have maximal vertex. Then also the syzygies of these lattices have maximal vertex, and hence under the Green correspondence (cf. Section 4) go over into lattices with maximal vertices.

Let us now ASSUME that in the above setup the defect group $D$ is cyclic. If $D$ is generated by $c$ of order $p^n$, then $I_R(D) \cdot B = (c - 1) \cdot B = B \cdot (c - 1)$ is a two-sided principal ideal, isomorphic to $\Lambda_B$. Let us for the moment replace $B$ by its basic order; i.e., we apply a Morita equivalence to $B$ to obtain an order $B_1$ in which each indecomposable projective module occurs with multiplicity one. This does not change the property that $I_R(D) \cdot B$ is a principal ideal for $\Lambda_B$. Then $\bar{B} = \prod_{i=1}^s S_i$ and so $\bar{B}/|D| \cdot \bar{B}$ is a product of uniserial algebras. Moreover, surely $(c - 1) \cdot B \subset rad(\Lambda_B)$, since $D$ is a $p$-group.

Hence we can apply the following result:

**Lemma 2.3. ([Ro; 92])** *Let $\Lambda$ be a connected – i.e., indecomposable as a ring – and basic R-order in a separable K-algebra A; i.e., $\Lambda$ is R-free of finite rank, spanning A. Assume that $\Lambda$ has a two-sided ideal J such that*

1. $J_\Lambda \simeq \Lambda$, *as left and as right module.*

2. $\Lambda/J = \prod_{i=1}^s U_i$ *is a product of uniserial algebras, then all the algebras $U_i$ are isomorphic, say $U_i \simeq U_0$. Moreover, $\Lambda$ is isomorphic to the order*

$$(3) \qquad T(\Omega, \omega, n) := \begin{pmatrix} \Omega & \Omega & \Omega & \cdots & \Omega & \Omega \\ \omega & \Omega & \Omega & \cdots & \Omega & \Omega \\ \omega & \omega & \Omega & \cdots & \Omega & \Omega \\ \cdots & \cdots & \cdots & \ddots & \cdots & \cdots \\ \omega & \omega & \omega & \cdots & \Omega & \Omega \\ \wp & \omega & \omega & \cdots & \omega & \Omega \end{pmatrix}$$

where $\omega = \alpha \cdot \Omega = \Omega \cdot \alpha$ *is a two-sided principal ideal of the local R-order $\Omega$; moreover, $\Omega/\omega = U_0$ is uniserial. We call the order $T(\Omega, \omega, n)$ a* TRIANGULAR ORDER. *The ideal $\rho$ generated by*

(4)
$$\rho_0 := \begin{pmatrix} 0 & 1 & 0 & \cdots & 0 & 0 \\ 0 & 0 & 1 & \cdots & 0 & 0 \\ 0 & 0 & 0 & \cdots & 0 & 0 \\ \vdots & \vdots & \vdots & \ddots & \vdots & \vdots \\ 0 & 0 & 0 & \cdots & 0 & 1 \\ \omega_0 & 0 & 0 & \cdots & 0 & 0 \end{pmatrix}_n$$

*is two-sided, i.e., $T \cdot \rho_0 = \rho_0 \cdot T$, and so $\rho$ is isomorphic to $T$ as left and as right module; not as bimodule though.*

*Moreover, conjugation with $\rho_0$ induces an automorphism which cyclicly permutes the indecomposable projective direct summands of $T$. The quotient $T/\rho$*

$$T/\rho = \prod_1^n U_0.$$

*is a product finite local uniserial algebras.*

IN OUR CASE, $\Lambda_B = T, \rho = J$ and $U_0$ *is an indecomposable ring direct summand of $\bar{B}$.*

THE DETERMINATION OF $\Omega$ AND $\omega$ FOR BLOCKS WITH CYCLIC NORMAL DEFECT CAN NOT BE DONE IN GENERAL, as will be shown in Section 3. We shall now use Diagram 2 and Lemma 2.3 to describe the block $B$. Before we can do so, we have to define a certain class of orders. In the above pull-back (cf. Diagram 2) the order $\bar{B} = \prod_1^s S_i$ maps onto $\prod_1^s U_i$. Moreover, we have an isomorphism, $\bar{\phi}_i : U_i \longrightarrow U_0$.

**Definition 2.4.** Let us be given two local orders $\Omega_i$ for $i = 1, 2$ with principal ideals $\omega_i$ such that there are epimorphisms $\alpha_i : \Omega_i \longrightarrow U_0$ for a local $R$-torsion $R$-finitely generated (as module) ring $U_0$ with $Ker(\alpha_i) = \omega_i$, then we shall denote the pull-back as follows:

$$\begin{array}{ccc} \Omega_i \xrightarrow{\ (\alpha_1, \alpha_2)\ } \Omega_2 & \longrightarrow & \Omega_1 \\ \downarrow & & \downarrow {\scriptstyle \alpha_1} \\ \Omega_2 & \xrightarrow{\ \alpha_2\ } & U_0 \end{array}$$

Let us now be given a tree $T = (V, E)$ with vertex set $V = \{v_1, \ldots, v_\nu\}$ and edges $E = \{e_1, \ldots, e_\mu\}$, which is embedded into the plane. BECAUSE OF THIS EMBED-DING, EACH VERTEX $v$ HAS LOCAL EDGES, NUMBERED CLOCKWISE, $(\epsilon_1^v, \ldots, \epsilon_{v_v}^v)$. Given an edge from the vertex $v$ to the vertex $w$, this edge meets $v$ at the local edge

$\epsilon_i^v$ and $w$ at the local edge $\epsilon_j^w$. We shall write $e_{\epsilon_i^v, \epsilon_j^w}$ for this edge. (Note that this uses heavily that $T$ is embedded into the plane.)

We now construct an $R$-order to $T$, given the following data:

**Data 2.5.**

1. A finite uniserial local $R$-algebra $U$.

2. For a vertex $v$ with valency $v_v$ the following data are given:

   (a) An order $\Omega_v$ with principal prime ideal $\omega_v$.

   (b) An $R$-order homomorphism $\alpha^v : \Omega_v \longrightarrow U$, which has kernel $\omega_v$.

   (c) The $R$-order $T_v = T_{\Omega_v, \omega_v, v_v}$ as in Equation 3.

   (d) We number the quotients modulo the ideal $\rho_v$ (cf. Equation 4) according to the numbering of the local edges of $v$:

$$ T_v / \rho_v \xrightarrow{(\alpha_1^v, \alpha_2^v, \ldots, \alpha_{v_v}^v)} \prod_{i=1}^{v_v} U_i^v, $$

   where each of the orders are equal: $U_i^v = U$.

**Definition 2.6.** The GREEN ORDER $\mathcal{G}$ constructed from the above tree and from the data 2.5 is defined as a sub-order of $T = \prod_{v \in V} T_v$. The only difference between $T$ and $\mathcal{G}$ lies in the diagonal entries: We link certain diagonal entries in $T$:

$$ T_v(i, i) \times T_w(j, j) \text{ is changed to } \Omega_v \xrightarrow[(\alpha_i^v, \alpha_j^w)]{} \Omega_w \text{ in } \mathcal{G} $$

according to Definition 2.4.

**Remark 2.7.** Instead of a tree we can also construct a GREEN-ORDER TO A GRAPH, WHICH IS LOCALLY EMBEDDED INTO THE PLANE according to the above recipe; i.e., the local edges at a vertex are numbered cyclicly – this comes from the local embedding into the plane.

We now return to blocks with cyclic normal defect group $D = \langle c : c^{p^n} \rangle$.

**Lemma 2.8.** *Let $B$ be a block of $RG$ with cyclic normal defect group $D$. Then $B$ is Morita equivalent to a Green order with Brauer tree a star. The* EXCEPTIONAL VERTEX *is in the center and the orders at the leaves*[1] *are separable and are all isomorphic; i.e., the exceptional vertex has the form:*

$$ (5) \qquad T(\Omega, \omega, n) = \begin{pmatrix} \Omega & \Omega & \Omega & \cdots & \Omega & \Omega \\ \omega & \Omega & \Omega & \cdots & \Omega & \Omega \\ \omega & \omega & \Omega & \cdots & \Omega & \Omega \\ \cdots & \cdots & \cdots & \ddots & \cdots & \cdots \\ \omega & \omega & \omega & \cdots & \Omega & \Omega \\ \omega & \omega & \omega & \cdots & \omega & \Omega \end{pmatrix}_n $$

---

[1]A LEAF is a vertex at the end of a branch; i.e., a vertex with only one edge.

where $\omega = \alpha \cdot \Omega = \Omega \cdot \alpha$ is a two-sided principal ideal of the local $R$-order $\Omega$. Contrary to the other vertices, the exceptional vertex involves in general ramification (cf. Section 3).

**Proof.** If $D$ is cyclic, generated by $c$, then $I_R(D) \cdot B = (c-1) \cdot B$ is a two-sided principal ideal, isomorphic to $\Lambda_B$. The diagram 2 then shows that $B$ is a Green order with Brauer tree a star, the exceptional vertex in the center and separable orders (unramified extensions) are at the leaves (cf. [Ro; 92]).          **q. e. d.**

**Note 2.9.** Let $T(\Omega, \omega, n)$ be a triangular order. Then the different indecomposable projective lattices are obtained as the orbit $\{^{\omega^j} P\}$ from one indecomposable projective lattice $P$ under the conjugation operation with $\rho_0$. In particular, if we have a block with cyclic defect $D = \langle c \rangle$, then the number of indecomposable projective lattices in $\Lambda_B$ is the orbit under $c - 1$ of a fixed indecomposable projective lattice.

**Notation 2.10.** Let the defect group $D = \langle c \rangle$ of $B$ be normal cyclic of order $p^n$ and put $H = G/C_G(D)$.

1. $H = C_q \times C_{p^\nu} \subseteq C_{p-1} \times C_{p^{n-1}} = Gal(\mathbb{Z}_p[\theta_{p^n}]/\mathbb{Z}_p)$, where the latter are the $p$-adic integers in the $p^n$-th cyclotomic field. $C_m$ denotes the cyclic group of order $m$ and $\theta_\nu$ stands for a primitive $\nu$-th root of unity.

2. Denote by $H'$ the kernel of the map $G \longrightarrow C_q \subseteq C_{p-1}$, where $C_q$ is the $p'$-part of $H$.

3. Let $\{\tilde{P}_1, \ldots, \tilde{P}_e\}$ be a complete set of non isomorphic indecomposable $B$-modules, and denote by $\{P_1, \ldots, P_e\}$ the projections of $\{\tilde{P}_i\}$ into $\Lambda_B$.

4. Note that all the $\{P_i\}$ are non isomorphic and constitute a complete set of indecomposable projective $\Lambda_B$-lattices.

**Claim 2.11.** *The $RG$-modules $\{P_i\}_{1 \leq i \leq e}$ are all isomorphic when restricted to $H'$. Consequently, $e \leq q$. Moreover, $e \mid q$.*

**Proof.** Let $P$ be a Sylow $p$-subgroup of $G$, then all the modules $\tilde{P}_i$ are isomorphic when restricted to $P$; but then so are the projections $P_i$ into $\Lambda_B$.

According to the structure of Green orders, the non isomorphic indecomposable projective $\Lambda_B$-modules are of the form

$$(c-1)^i \cdot P_1, \text{ where } 0 \leq i \leq e - 1.$$

Obviously, these modules are all isomorphic when restricted to $C_G(D)$.

In general we can not conclude from these observations, that the modules $P_i$ are isomorphic, when restricted to $H'$. Let $\gamma \in H$ be such that it generates

$C_{p^\nu}$ modulo $C_G(D)$. We then have $\gamma \cdot (c-1) = (c \cdot c^{p^\nu} - 1) \cdot \gamma$. Moreover, $(c^i - 1) = (c-1) \cdot (\sum_{j=0}^{i-1} c^j)$. Thus $(c^i - 1) \cdot P_1 \subseteq (c-1) \cdot P_1$. Now

$$(c \cdot c^{p^\nu} - 1) \cdot P_1 = (c \cdot (c^{p^\nu} - 1) + (c-1)) \cdot P_1 \subseteq (c-1) \cdot P_1.$$

But $(c^{p^\nu} - 1) = (c-1)^{p^\nu} + p \cdot x$ for some element $x \in P_1$. In particular, $p \cdot x \in p \cdot P_1 \cap (c-1) \cdot P_1$, unless $p \cdot P_1 = (c-1) \cdot P_1$, in which case $e = 1$.

We thus may assume $p \cdot P_1 \cap (c-1) \cdot P_1 \neq (c-1) \cdot P_1$. Then the structure of $\Lambda_B$ (cf. Equation 5) shows that

$$p \cdot P_1 \cap (c-1) \cdot P_1 \subseteq \omega \cdot P_1 = (c-1)^e \cdot P_1.$$

From the above observations we conclude:

$$\gamma \cdot (c-1) \cdot y = (c-1) \cdot \gamma \cdot y \text{ modulo } p^\epsilon \text{ for } y \in P_1,$$

where $\epsilon = \min(e, \nu)$.

We thus have an $H'$-homomorphism $\phi : P_1 \longrightarrow (c-1) \cdot P_1/(c-1)^\epsilon \cdot P_1$. Since $(c-1) \cdot P_1$ also maps onto the latter, we conclude $P_1 \simeq (c-1) \cdot P_1$ as $RH$-module. Since there are only $q$ possibilities for $C_q$ to act, we conclude $e \leq q$. In fact, since $H'$ contains the Sylow $p$-subgroup of $G$, we deduce that $P_i$ is a direct summand of $P_i \downarrow H' \uparrow G$ and so $\oplus_{i=1}^e P_i$ is a direct summand of $P_1 \downarrow H' \uparrow G$. Counting ranks, it follows that $e \leq q$.

It remains to show that $e$ divides $q$. We note that $c$ acts on $\Lambda_B$ as multiplication with a primitive $|D|$-th root of unity. Thus we have on $\Lambda_B$:

$$(c^i - 1) = (c-1) \cdot \left( \sum_{j=0}^{i-1} c^j \right), \text{ which is a fundamental unit if } (i, p) = 1.$$

Thus for $(i, p) = 1$ we have $(c^i - 1) \cdot P_1 = (c-1) \cdot P_1$. This implies, that $(c-1)^q \cdot P_1 \simeq P_1$. In fact, $(c-1)^q \cdot P_1 \simeq \prod_{i_1}^q (c^{\beta^i} - 1) \cdot P_1 \simeq P_1$. Here $\beta$ generates $C_q$ modulo $C_G(D)$. Thus we have an action of $C_q$ on $\{P_i\}_{1 \leq i \leq e}$. Hence $e$ must divide $q$.   **q. e. d.**

## 3. Examples of Blocks with Normal Cyclic Defect

In this section we shall consider several typical examples of blocks with normal cyclic defect.

### 3.1. The Affine Groups

Let $G = G(p^n, q) = C_{p^n} \rtimes C_q = \langle a, b : a^{p^n}, b^q, {}^b a = a^r \rangle$ be the affine group of order $p^n \cdot q$ where $q \mid (p-1)$. Let $R = \mathbb{Z}_p$ be the $p$-adic integers. The group ring is then the principal block, and its structure is given in Lemma 2.8. We have to describe – in the notation of the above lemma – the ring $\bar{B}$, the ring $\Omega$ with the ideal $\langle \alpha \rangle$ and the amalgamation homomorphisms.

$\bar{B} = \mathbb{Z}C_q \simeq \prod_1^q \mathbb{Z}_p$, since $\mathbb{Z}_p$ contains the $q$-th roots of unity (note $q|(p-1)$). The matrix ring $\mathcal{T}(\Omega, \omega, n)$ thus has size $n = q$ over $\Omega$.

The integrated group ring of the cyclic group $C_{p^n}$ can be written as a pullback

$$
\begin{array}{ccc}
\mathbb{Z}C_{p^n} & \longrightarrow & \mathbb{Z} \\
\downarrow & & \alpha_{p^n} \downarrow \\
\Lambda_{p^n} & \xrightarrow{\alpha_0} & \mathbb{Z}/(p^n \cdot \mathbb{Z})
\end{array}
$$

(6)

Since $C_q$ is a subgroup of the automorphism group of $C_{p^n}$ it acts on $\mathbb{Z}C_{p^n}$, and hence also on $\Lambda_{p^n}$. We denote by $\Omega_{p^n}$ the fixed ring of $\Lambda_{p^n}$ under $C_q$.

The above data are GLOBAL. We indicate the COMPLETION of these data at $p$ by the subscript $(-)_p$.

The central "exceptional vertex" in the group ring $\mathbb{Z}_p G$ is thus the triangular order of size $q$ over $(\Omega_{p^n})_p$ with ideal $(\omega_{p^n})_p = \langle \prod_{i=1}^q (^{b^i}a - 1) \rangle_p$. The amalgamation maps are $\alpha_0^{C_q} : (\Omega_{p^n})_p \longrightarrow \mathbb{Z}/(p^n \cdot \mathbb{Z})$ and $(\alpha_{p^n})_p$. Note that here "taking fixed points" is exact: In fact, we have the exact sequence

$$
0 \longrightarrow (a-1) \cdot \Lambda_{p^n} \longrightarrow \Lambda_{p^n} \longrightarrow \mathbb{Z}/(p^n \cdot \mathbb{Z}) \longrightarrow 0.
$$

The fixed point functor is $Hom_{\mathbb{Z}G}(-, \mathbb{Z})$, which is left exact; however, as $C_q$-module under the conjugation action the module $(a-1) \cdot \Lambda_{p^n} \simeq \Lambda_{p^n}$ is free, since $C_q$ acts freely on $C_{p^n} \backslash \{1\}$. So taking fixed points exact on the above sequence. Completing at $p$ gives the desired result.

In general, the order $\Lambda_{p^n}$ is a quite complicated ring.

However, in case $n = 1$, we have $\Lambda_p = \mathbb{Z}_p[\theta_p]$, where $\theta_p$ is a primitive $p$-th root of unity. The ring $\Omega_p$ is then the fixed ring under $C_q$ and $\omega_p$ is the ideal above $p$ in $\Omega_p$.

The above structure of $\mathbb{Z}_p G(p^n, q)$ is also valid, if $\mathbb{Z}_p$ is replaced by an unramified extension $S'$ of $\mathbb{Z}_p$. Here we have described the $p$-ADIC STRUCTURE of the integral group ring $\mathbb{Z}G(p^n, q)$. Recently, H. Weber [We; 98] has described the GLOBAL STRUCTURE.

**Proposition 3.1. (Weber [We; 98])** *The integral group ring $\mathbb{Z}G(p^n, q)$ of the affine group $G(p^n, q)$ can be described via the commutative diagram with exact* *rows and columns:*

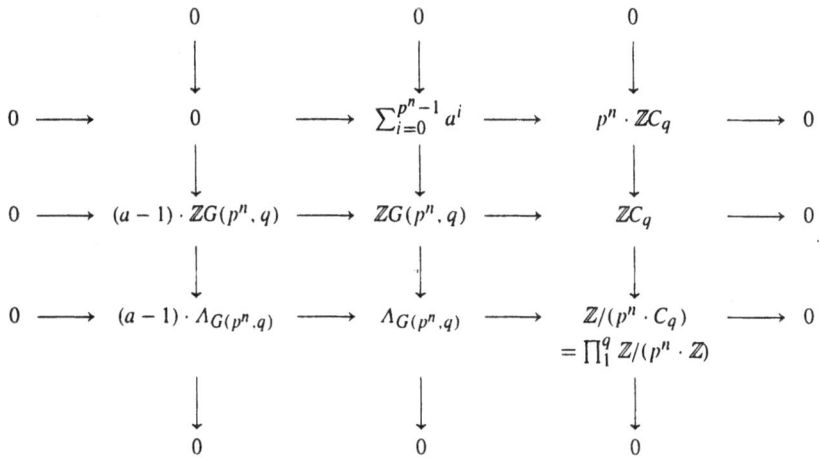

where $\Lambda_{G(p^n,q)}$ is the triangular order of size $q$ over the $\mathbb{Z}$-order $\Omega_{p^n}$ (cf. Equation 6 ff.).

This example can be used to construct NON PRINCIPAL BLOCKS with cyclic defect. Let $H$ be a finite group of order prime to $p$. Then the group ring $\mathbb{Z}_p H = \prod_{i=1}^{n} Mat(n_i, R_i)$ is the product of matrix rings over unramified extensions $R_i$ over $\mathbb{Z}_p$. Hence the group ring of $G(p^n, q) \times H$ decomposes into blocks

$$\mathbb{Z}_p(G(p^n, q) \times H) = \prod_{i=1}^{n} Mat(n_i, R_i G(p^n, q)).$$

### 3.2. Central Extensions of Affine Groups

Let $C_s = \langle c : c^s = 1 \rangle$ map onto $C_q$ with kernel $C_r = \langle c^q \rangle$ and assume that $(p, s) = 1$. We can then form the central extension

$$
\begin{array}{ccccccccc}
0 & \longrightarrow & C_{p^n} & \longrightarrow & G(p^n, q) & \longrightarrow & C_q & \longrightarrow & 0 \\
 & & \uparrow & & \uparrow & & \uparrow & & \\
0 & \longrightarrow & c_{p^n} & \longrightarrow & G(p^n, q, s) & \longrightarrow & C_s & \longrightarrow & 0
\end{array}
$$

(7)

The centre of $G(p^n, q, s)$ is $C_r$, and $C_{p^n} \times C_r$ is an abelian normal subgroup of $G(p^n, q, s)$. Moreover, $O_{p'}(G(p^n, q, s)) = C_r$.

The principal block of $\mathbb{Z}_p G(p^n, q, s)$ is $\mathbb{Z}_p G(p^n, q)$. The NON PRINCIPAL BLOCKS can have various structures, as we shall show next.

**Example 3.2.** Let us look at $G(17, 4, 8)$, whose order in $2 \times |G(17, 4)|$. Apart from the principal block, there is one additional block, coming from the exact sequence

$$0 \longrightarrow (c^4 - 1) \cdot \mathbb{Z}_{17} G(17, 4, 8) \rightarrow \mathbb{Z}_{17} G(17, 4, 8) \rightarrow \mathbb{Z}_{17} G(17, 4) \longrightarrow 0,$$

which is split exact, $(c^4 - 1)/2$ being a central idempotent in $\mathbb{Z}_{17}G(p^n, q, s)$. Since $C_{17}$ does not act trivially, $B_1 := (c^4 - 1)\mathbb{Z}_{17}G(17, 4, 8)$ is a block, i.e., it is indecomposable as ring. Since $\mathbb{Z}_{17}$ contains the primitive 8-th roots of unity, we conclude that $\mathbb{Z}_{17}C_8 \simeq \prod_1^8 \mathbb{Z}_{17}$. For $B_1$ we have the exact sequence

$$(8) \qquad 0 \longrightarrow (a - 1) \cdot B_1 \longrightarrow B_1 \longrightarrow \prod_1^4 \mathbb{Z}_{17} \longrightarrow 0,$$

and we conclude that as rings the principal block $B_0$ and the non principal block $B_1$ are isomorphic.

**Example 3.3.** Let $G = G(3, 2, 4)$ be the central extension of the symmetric group on three letters $S_3$; $G = \langle a, b : a^3, b^4, {}^b a = a^2 \rangle$. As in the Example 3.2 one argues that apart from the principal block, which is the group ring $\mathbb{Z}_3 S_3$, there is one other block $B_1$. Contrary to the previous example, $\mathbb{F}_3$ is not a splitting field for $C_4$. We have $\mathbb{Z}_3 C_4 = \mathbb{Z}_3^+ \times \mathbb{Z}_3^- \times \mathbb{Z}_3[i]$. As in Example 3.5 we have an exact sequence

$$0 \longrightarrow (a - 1) \cdot B_1 \longrightarrow B_1 \longrightarrow \mathbb{Z}_3[i] \longrightarrow 0,$$

which shows, that $B_1$ is indecomposable as module. Hence in this case $e = 1$ but $q = 2$ (cf. Claim 2.11). This sequence gives as usual rise to the pull-back description of $B_1$:

$$\begin{array}{ccc} B_1 & \longrightarrow & \mathbb{Z}_3[i] \\ \downarrow & & \downarrow \\ \Lambda & \overset{\alpha}{\longrightarrow} & \mathbb{F}_9 \end{array} \; ,$$

since $\mathbb{Z}_3[i]$ is unramified. Moreover, the kernel of $\alpha$ is $(a - 1) \cdot B_1$.

**Claim 3.4.** $\Lambda$ is the maximal order in a skew-field of degree 2 over $\mathbb{Z}_3$.

**Proof.** Since $Ker(\alpha) \simeq \Lambda$, we conclude that $\Lambda$ is hereditary and indecomposable as left module. Thus $\Lambda$ is a maximal order in a skew-field. **q. e. d.**

Let us write down the matrix representation of $G$ in $\Lambda$. We quote from Reiner [Re; 75] the structure of the maximal order in the skew-field of index 2 over $\mathbb{Z}_3$. Let $\rho$ be a primitive 8-th root of unity over $\mathbb{Z}_3$. Then $\rho$ is unramified over $\mathbb{Z}_3$ and $S := R[\rho]$ has degree 2 over $\mathbb{Z}_3$, since $\mathbb{F}_9$ contains a primitive 8-th root of unity. The Galois automorphism sends $\rho$ to $\rho^3$. The maximal order in the skew field is given as

$$(9) \qquad \Lambda := \begin{pmatrix} a + b \cdot \rho & \alpha + \beta \cdot \rho \\ 3 \cdot (\alpha + \beta \cdot \rho^3) & a + b \cdot \rho^3 \end{pmatrix}$$

with $a, b, \alpha, \beta \in S := \mathbb{Z}_3[\rho]$.

We can now find the group $G$ in $\Lambda$:

$$a \longrightarrow \begin{pmatrix} 2 & 1 \\ 3 & 2 \end{pmatrix} \text{ and } b \longrightarrow \begin{pmatrix} \rho^2 & 0 \\ 0 & \rho^3 \end{pmatrix}.$$

One sees that WITH THESE CENTRAL EXTENSIONS QUITE OFTEN SKEW-FIELDS OCCUR. This depends on the splitting behavior of $\mathbb{Z}_p C_s$. We conclude this section with yet another example:

**Example 3.5.** We consider the group $G := G(5, 2, 8)$. Then $\mathbb{Z}_5 C_8 = (\times^4 \mathbb{Z}_5) \times (\times^2 \mathbb{Z}_5[\rho])$, where $\rho$ is a primitive 8-th root of unity. Moreover, $\mathbb{Z}_5[\rho]/(5 \cdot \mathbb{Z}_5[\rho]) = \mathbb{F}_{25}$. The exact sequence

$$0 \longrightarrow (c^4 - 1) \cdot \mathbb{Z}_5 G \longrightarrow \mathbb{Z}_5 G \longrightarrow \mathbb{Z}_5 G(5, 2, 4) \longrightarrow 0.$$

splits. The group ring $\mathbb{Z}_5 G(5, 2, 4)$ has two blocks, $B_0$ and $B_1$. There are now TWO POSSIBILITIES:

**Case 1:** $B_3 := (c^4 - 1) \cdot \mathbb{Z}_5 G$ is a block. Then $\Lambda_{B_3}$ has rank 16 and is a triangular order or size 2 over a skew-field of index 2. The centre of $\Lambda_{B_3}$ is then necessarily trivial. On the other hand, $c^2 \cdot 1_{\Lambda_{B_3}}$ is a non-trivial central element. Thus this case cannot occur.

**Case 2:** $(c^4 - 1) \cdot \mathbb{Z}_5 G = B_3 \times B_4$ is the direct sum of two conjugate blocks, each of rank 8 over $\mathbb{Z}_5$. Then $\Lambda_{B_3}$ has centre $S := \mathbb{Z}_5[\rho]$ and is a maximal order in a skew-field of index two over $S$.

These examples show how delicate the description of blocks with cyclic normal defect can be. As far as I know THERE IS NO GENERAL THEORY TO DESCRIBE THESE BLOCKS. In each case one has to find new arguments to obtain their structure.

## 4. Green Correspondence

We shall now develop the theory which allows to describe blocks with cyclic defect from the knowledge of blocks with normal cyclic defect.

**Data 4.1.** Let $G$ be a finite group and $R$ the ring of integers in a local number field $K$. Let $B$ be a block of $RG$ with cyclic defect, say $D$ of order $p^n$. Let $H = N_G(D)$ be the normalizer of $D$.

We consider two functors,

1. the RESTRICTION FUNCTOR

$$\mathbb{F}_0 : {}_{RG}latt \longrightarrow {}_{RH}latt : N \mapsto N \downarrow_H^G; \text{ this is } N \text{ as } H\text{-module}$$

2. and the INDUCTION FUNCTOR

$$\mathbb{G}_0 : {}_{RH}latt \longrightarrow {}_{RG}latt : M \mapsto M \uparrow_H^G := RG \otimes_{RN} M.$$

Then there exists a unique block $b$ of $RH$, the BRAUER CORRESPONDENT of $B$, such that

1. for $N \in {}_B latt$ with vertex $D$, the module $\mathbb{F}_0(N)$ has a unique direct summand $\mathbb{F}(N) \in {}_b latt$ of vertex $D$.

2. for $M \in {}_b latt$ with vertex $D$, the module $\mathbb{G}_0(M)$ has a unique direct summand $\mathbb{G}(M) \in {}_B latt$ of vertex $D$.

3. The functors $\mathbb{F}$ and $\mathbb{G}$ are inverse equivalences between the lattices of vertex $D$ in $b$ and $B$ and homomorphisms modulo maps factoring *via* lattices of smaller vertex.

The above equivalence is called GREEN CORRESPONDENCE. For the induction $\mathbb{G}$ which is induced from $\mathbb{Z}G \otimes_{RH} -$ we can also use $B \otimes_b -$.

**Note 4.2.** The functor $Hom_R(-, R) : {}_{RG}latt \longrightarrow latt_{RG}; M \mapsto M^*$ is a DUALITY mapping projective lattices to projective lattices.

By using the anti-involution $g \mapsto g^{-1}$ on $G$, the dual $M^*$ of a left lattice $M$ can be viewed also as a left lattice, called the CONTRAGREDIENT.

Since the Green correspondence commutes with this duality, and since for the $b$-lattices corresponding to the leaves, their duals also correspond to leaves, the above statement also holds for the dual modules. The same holds for the contragredients.

We shall next derive some properties on $B$, which can be derived from the known structure of $b$ (cf. Section 2) by using the Green correspondence. We assume now that Data 4.1 are given. Recall that $M \in {}_b latt$ is said to be IRREDUCIBLE provided $K \otimes_R M$ is simple.

**Claim 4.3.** *Let* $M \in {}_b latt$ *be irreducible with vertex* $D$, *corresponding to a leaf of the Brauer-tree of* $b$, *which is by Claim 2.8 a star. Then* $\mathbb{G}(M)$ *is irreducible.*

**Proof.** We start with a

**Remark 4.4.** Given an exact sequence of $RG$-lattices,

$$0 \longrightarrow M' \longrightarrow M \longrightarrow M'' \longrightarrow 0$$

we obtain the long exact sequence

$$\cdots \longrightarrow Ext^1_{RG}(M'', -) \to Ext^1_{RG}(M, -) \to Ext^1_{RG}(M', -) \longrightarrow \cdots.$$

Consequently, we have for the vertices:

$$|vx(M)| \leq \max(|vx(M')|, |vx(M'')|).$$

WE NOW TURN TO THE PROOF OF CLAIM 4.3: Assume to the contrary, that $N := \mathbb{G}(M)$ is not irreducible; then there is an exact sequence of $B$-lattices

$$0 \longrightarrow N' \xrightarrow{\alpha} N \xrightarrow{\beta} N'' \longrightarrow 0.$$

Neither $\alpha$ nor $\beta$, can factorize via modules of smaller vertex, as shows an argument similar to that of Remark 4.4. Since $N$ has maximal vertex, one of $N'$ or $N''$ must have maximal vertex according to Remark 4.4. We note that $N$ is irreducible if and only if $N^* := Hom_R(N, R)$ is irreducible; moreover, the vertex does not change under the duality. Hence, applying the duality functor if necessary, we may assume that $N''$ has vertex $D$, and that $\beta$ does not factorize via a module of smaller vertex. We now apply the functor $\mathbb{F}$ and obtain a non zero morphism $\mathbb{F}M \xrightarrow{\mathbb{F}(\beta)} \mathbb{F}(N'')$ in $_b latt$ – modulo maps which factorize via lattices of smaller vertex. Since $M$ corresponds to the irreducible lattice of a leaf, the only non-zero morphisms are from $M$ to $M$. Thus $\mathbb{F}(N'') \simeq M$ and hence $N \simeq N''$, a contradiction. Consequently, $N$ is irreducible. **q. e. d.**

**Claim 4.5.** *Let $M$ be an indecomposable b-lattice corresponding to the exceptional vertex; i.e., it is the projection of an indecomposable projective b-lattice into the order of the exceptional vertex. Then $N := G(M)$ is a local B-lattice; i.e., it is simple modulo its radical.*

**Proof.** Assume there is a minimal projective resolution with $P$ and $Q$ non zero projective:

$$0 \longrightarrow L \longrightarrow P \oplus Q \xrightarrow{\phi} N \longrightarrow 0.$$

Since the Green correspondence is induced from induction, $L$ is the Green correspondent of the first syzygy of $M$, and hence is irreducible, since it corresponds to a leaf (cf. Claim 4.3).

Let $X := Im(\phi \downarrow P)$. Then we have the commutative diagram with exact rows and columns which we obtain from the "ker-coker-lemma":

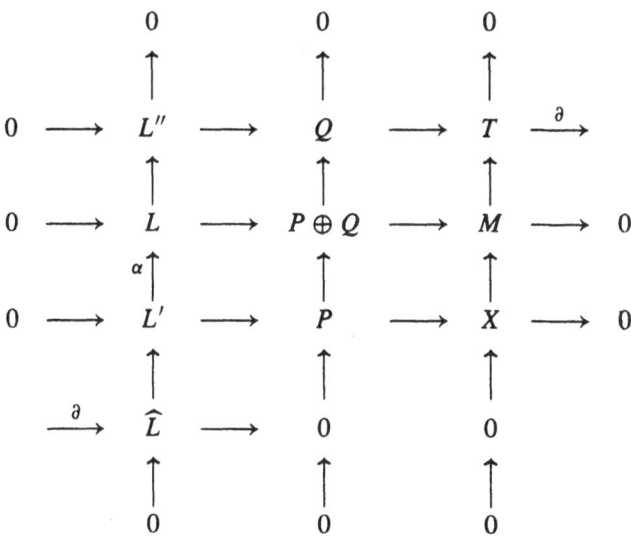

If $L'' \neq 0$, then it must be a lattice as submodule of $Q$. But then $L' = 0$ and so $X \simeq P$ is projective, a contradiction, since $T$ is a lattice.

So we may assume that $L'' = 0$, and we have an exact sequence of lattices

$$0 \longrightarrow Q \longrightarrow T \longrightarrow \widehat{L} \longrightarrow 0,$$

which is split since $\widehat{L}$ is a lattice. But then $Q$ is a direct summand of $M$, a contradiction.                                                                    **q. e. d.**

We are slowly working our way to show that $B$ is a Green order.

**Lemma 4.6.** *Assume that the star of b has n vertices, and let $M \in {}_b latt$ be a module corresponding to a leaf. If $N := \mathbb{G}(M)$ is its Green correspondent in ${}_B latt$, then N has a periodic projective resolution of period $v := 2 \cdot n$, in which all projectives occurring are indecomposable. In particular, all syzygies of N are local.*

**Proof.** Let

$$\mathbb{P}(M) : \cdots \longrightarrow P_i \xrightarrow{\phi_i} P_{i-1} \longrightarrow \cdots \longrightarrow P_0 \xrightarrow{\phi_0} M$$

be a periodic minimal resolution of $M$. Since $M$ corresponds to a leaf, the syzygies $\Omega_i(M)$ of $M$ belong to leaves for $i$ even, and they belong to the central (exceptional vertex) for $i$ odd.

We tensor the resolution $\mathbb{P}(M)$ with $B$ and split off the modules of smaller vertex to obtain a minimal projective resolution of $N$,

$$\mathbb{Q}(N) : \cdots \longrightarrow Q_i \xrightarrow{\psi_i} Q_{i-1} \longrightarrow \cdots \longrightarrow Q_0 \xrightarrow{\psi_0} N.$$

Note that this is possible since with $N$ the syzygies also have vertex $D$. Then the syzygies of $N$ are the Green correspondents of the syzygies of $M$, that is $\Omega_i(N) \simeq \mathbb{G}(\Omega_i(M))$. Because of Claim 4.5, the modules $\Omega_i(N)$ are local for even $i$, and so $Q_i$ is indecomposable projective for $i$ even. However, $M^*$ also corresponds to a leaf, since $M$ corresponds to a leaf, and so the dual $\mathbb{P}(M)^*$ of the projective resolution $\mathbb{P}(M)$ is also a minimal projective resolution of $M*$, which also corresponds to a leaf.

We now repeat the above procedure and conclude that $\mathbb{Q}(N)^*$ is also a projective resolution of $N^*$. But now, $Q^*_{v-1}$ maps onto $N^*$. Consequently, the modules $Q_i$ are indecomposable projective for $i$ odd, since $v$ is even.

This shows that all the modules $Q_i$ are indecomposable projective, and hence all syzygies of $N$ are local.                                                      **q. e. d.**

We conclude this section with a note on graphs:

**Note 4.7.** Let $G = (V, E)$ be a connected graph with $|V| =: v$ vertices and $|E| =: \epsilon$ edges. Then $G$ is a tree if and only if $v = \epsilon + 1$ Moreover, these latter condition are the minimal ones which can be realized for a connected graph. To see this one uses induction on the number of vertices. If $\Gamma$ is a tree then this is obvious. Conversely, assume that $\Gamma$ is not a tree, then it has a cycle. Omitting one edge, one still has a connected graph, and the result follows by the minimality for trees.

## 5. Stable Equivalence of Morita Type

For details we refer the reader to the paper of Linckelmann [Li; 96].

We start with pointing out the difference between a stable equivalence and a stable equivalence of Morita type.

Let $R = \mathbb{Z}_3[\theta_3]$ where $\theta_3$ is a primitive 3rd root of unity. Then we have a homomorphism $\alpha : R \longrightarrow \mathbb{F}_3$ and $\beta : \mathbb{Z}_3 \longrightarrow \mathbb{F}_3$, and the orders $R \overline{{}_{(\alpha,\alpha)}} R$ and $\mathbb{Z}_3 \overline{{}_{(\beta,\beta)}} \mathbb{Z}_3$ are stably equivalent; but they are not at all stably equivalent of Morita Type, since there centers are different (cf. below Note 5.4).

Now we are turning to a general stable equivalence of Morita type, between a Green order $\Lambda$, whose underlying tree is a star and an order $\Gamma$, which is stably equivalent to $\Lambda$ via the bimodule $B :=_\Gamma B_\Lambda$.

Let $\mathcal{P}_\Lambda$ be the projective resolution for $\Lambda$ associated to the walk around the Brauer tree and let $\mathcal{P}_\Gamma$ be the projective resolution for $\Gamma$ obtained *via* the stable equivalence of Morita type. We then have

$$B \otimes_\Lambda B^* \simeq \Gamma \oplus M \text{ for a projective } (\Gamma, \Gamma)\text{-bimodule } M \text{ and}$$

$$B^* \otimes_\Gamma B^* \simeq \Gamma \oplus N \text{ for a projective } (\Lambda, \Lambda)\text{-bimodule } N.$$

The aim of this section is to prove the following result.

**Theorem 5.1.** *Let $\Lambda$ be a Green order whose underlying graph is a star. Assume that the order $\Gamma$ is stably equivalent of Morita type to $\Lambda$. Then $\Gamma$ is a Green order whose underlying graph is a tree.*

If we apply this to blocks with cyclic defect, we obtain:

**Theorem 5.2.** *Let $B$ be a block of $\mathbb{Z}_p G$ with cyclic defect $D$ which is $TI$; i.e., $D \cap {}^x D = \{1\}$ for $x \in G \backslash N_G(D)$. Then $B$ is a Green order with underlying graph a tree. If $b$ is the Brauer correspondent to $B$, then $b$ and $B$ have the same centers.*

*Proof of Theorem 5.2:* Since the defect group is TI, the Green correspondence between $B$ and $b$ is a stable equivalence of Morita type.  **q. e. d.**

### Note 5.3.

1. For more details we refer the reader to the Ph.D. thesis of M. Kauer [Ka; 98].

2. Theorem 5.2 remains true for blocks $B$ without the "TI"-assumption, provided the defect group is cyclic. The argument is an induction where at each step one has a stable equivalence of Morita type [CR; 82, 86] or a Morita equivalence.

3. As we know, both $b$ and $B$ have an exceptional vertex. Except at the exceptional vertex, the orders are of the form $Mat(n_i, R_i)$ for unramified extensions $R_i$ over $\mathbb{Z}_p$. At the exceptional vertex we have the local orders $\Omega_B$ and $\Omega_b$, which do not need to be commutative. It is MY FEELING that the Green correspondence preserves $\Omega$; i.e., $\Omega_b \simeq O_B$. I CANNOT PROVE IT THOUGH.

Let us recall some facts from stable equivalences of Morita type:

**Note 5.4.**

1. The centers of $\Lambda$ and $\Gamma$ are the same. In fact, essentially

$$Hom_{(\Gamma, \Lambda)}(B, B) = Hom_\Lambda(B, B)^\Gamma$$

and vice versa. For details we refer to [Ri; 89, I], [Ri; 89,II].

2. $\Lambda$ and $\Gamma$ have isomorphic Grothendieck groups of the finite length torsion modules.

3. In particular they have the same number of indecomposable non isomorphic projective lattices, say $n$. To see this we note, that the above stable equivalence induces a stable equivalence of Morita type of the $\bar{\Lambda} := \Lambda/(p \cdot \Lambda)$ and $\bar{\Gamma} := \Gamma/(p \cdot \Gamma)$ modules. The Grothendieck groups are also generated by the non-projective modules, provided no simple $\bar{\Lambda}$-module is projective and the same for $\Gamma$. The images of the simple $\Gamma$-modules generate the Grothendieck-group of $\Gamma$-modules.

4. Since the Brauer tree of $b$ is a star, there is only one projective resolution of the lattice at a leaf. Hence, if $B$ is the Green order to a locally embedded graph, then there is also only one periodic resolution for $B$.

We shall now define another invariant of the stable equivalence class:

**Definition 5.5.** For an $R$-order $\Lambda$ denoted by

$$\mathcal{G}_1(\Lambda) = \{X \, a(\Lambda/(p \cdot \Lambda))\text{-module} : hd_\Lambda(X) \leq 1\}.$$

The projective $\bar{\Lambda} = \Lambda/(p \cdot \Lambda)$-modules always lie in there.

We point out that a stable equivalence of Morita type also induces a stable equivalence of Morita type between the module categories $_{\bar\Lambda}\text{mod}$ and $_{\bar\Gamma}\text{mod}$, which sends the non projective $\bar{\Lambda}$-modules in $\mathcal{G}_1(\Lambda)$ to the corresponding modules in $\mathcal{G}_1(\Gamma)$.

**Claim 5.6.** *Assume that $\bar{\Gamma} := \Gamma/\pi \cdot \Gamma$ a uniserial $\mathfrak{k} := R/\pi \cdot R$-algebra. If the graph G associated to the Green order $\Gamma$ is a tree then $\mathcal{G}_1(\Gamma)$ consists solely of projective $\bar{\Gamma}$-modules.*

Note that this is satisfied if $\Gamma$ is a block with cyclic normal defect.

**Note 5.7.**

1. If the graph of $\Gamma$ is a tree, then $\Gamma$ is derived equivalent and hence also stably equivalent to an order $\Lambda$ whose graph is a star $C$ (cf. [Ri; 89, II], [KöZi; 98]). Assume that $\bar{\Lambda}$ is uniserial. Then $\mathbb{G}(\mathcal{G}_1(\Lambda))$ consists solely of projective $\bar{\Lambda}$-modules. But then the same is true for $\mathcal{G}_1(\Gamma)$. Note though that $\bar{\Gamma}$ is not uniserial in general.

2. The previous argument shows that for a block $B$ with cyclic defect, which is stably equivalent to its Brauer correspondent $b$, whose graph is a star, the category $\mathcal{G}_1(B)$ consists solely of projective $\bar{B}$-modules. In fact, $B$ is stably isomorphic to $b$.

3. I do not know whether the condition that $\bar{\Lambda}$ is uniserial is really necessary.

**Proof.** Since the graph $G$ of $\Gamma$ is a tree, it is derived equivalent [KöZi; 98], [Ri; 89, I], [Ri; 89, II] to an order $\Lambda$ whose graph is a star. Hence, $\Gamma$ is also stably equivalent of Morita type to $\Gamma$. And by assumption, $\bar{\Lambda}$ is uniserial. So it suffices to prove the result for $\Lambda$.

Let $U$ be an indecomposable $\bar{\Lambda}$-module, which as $\Lambda$-module has homological dimension 1. Since $\bar{\Lambda}$ is uniserial, $U$ has a projective resolution of the form:

$$0 \longrightarrow Q \longrightarrow P \longrightarrow U \longrightarrow 0$$

with $P$ indecomposable projective and $Q$ - consequently - indecomposable projective. We have $\pi \cdot P \subset Q \subset P$ and so $Q = \pi \cdot P$, since $\bar{\Lambda}$ is uniserial. But then $U$ is $\bar{\Lambda}$-projective.      **q. e. d.**

**Claim 5.8.** *In the projective resolution $\mathcal{P}_\Gamma$ every indecomposable projective occurs with multiplicity exactly 2, as is the case for $\mathcal{P}_\Lambda$. We have exact sequences*

$$0 \longrightarrow K_{i,j} \longrightarrow P_i \longrightarrow M_{i,j} \longrightarrow 0, \, 1 \leq j \leq 2$$

*such that $K \cdot M_{i,1} \simeq K \cdot K_{i,2}$ and $K \cdot M_{i,2} \simeq K \cdot K_{i,1}$.*

**Proof.** We know from the above that all projective modules in the resolution $\mathcal{P}_\Gamma$ are indecomposable. The claim will be proved if we can show that no projective can occur with multiplicity more than two.

We first note that from the structure of $\Gamma$ it follows that for all syzygies $S_i$ we have
(10)      $\underline{Hom_\Lambda(S_i, S_j) = 0}$ if and only if $i \neq j$.

The same must then hold for the syzygies $\{T_j\}$ of $\mathcal{P}_\Gamma$.

Assume that one indecomposable $\Gamma$-lattice $Q_0$ occurs with multiplicity at least three, i.e., we have three exact sequences:

$$0 \longrightarrow K_i \longrightarrow P_0 \longrightarrow M_i \longrightarrow 0, \, 1 \leq i \leq 3.$$

We shall show, that there do exist two indices $i \neq j$ such that there either is a non-zero morphism between $M_i$ and $M_j$ or a morphism between $K_i$ and $K_j$. In fact, assume that $Hom_\Gamma(M_i, M_j) = 0$ for $i \neq j$. Let $M_i$ have $n_i$ rational components and let $P_0$ have $n_0$ rational components. Then $K_i$ has $n - n_i$ rational components. Since $Hom_\Gamma(M_i, M_j) = 0$ for $i \neq j$, we conclude $\sum_{i=0}^{3} n_i \leq n$. For the kernels we conclude

$$\sum_{i=0}^{3}(n - n_i) = 3 \cdot n - \sum_{i=0}^{3} n_i \geq 2 \cdot n.$$

Consequently there must exist a pair $i \neq j$ with $Hom_\Gamma(K_i, K_j) \neq 0$.

Passing to the dual situation if necessary, we may assume that $Hom_\Gamma(M_1, M_2) \neq 0$, which implies, however, because of Equation 10, that $\underline{Hom_\Gamma(M_1, M_2)} = 0$.

If necessary, we interchange the rôle of $M_1$ and $M_2$ to arrange that we have the following situation:

1. There exists $0 \neq \phi \in Hom_\Gamma(M_1, M_2)$,

2. a $\Gamma$-lattice $M_0$ and two epimorphisms $\phi_i : M_i \longrightarrow M_0$

3. and a factorization: $\phi_1 = \phi \circ \phi_2$.

Since $\phi = 0$ in $\underline{Hom_\Gamma(M_1, M_2)}$ the epimorphism $\phi$ factorizes via a projective module. Since the projective cover of $M_0$ is also $P_0$ the epimorphism $M_1 \xrightarrow{\phi} M_0$ must factorize *via* $P_0$, a contradiction. This argument also proves the statement about the kernels and cokernels of the exact sequences involving $P_i$.        **q. e. d.**

**Lemma 5.9.**  *The order $\Gamma$ is a Green order.*

**Proof.**  Let us recapitulate what we know:

1) $\Lambda$ and $\Gamma$ have the same center;

2) they have the same number of rational components;

3) they have the same number of indecomposable projectives;

4) in the projective resolution the projectives are indecomposable and every indecomposable projective $P$ occurs exactly twice; and

5) it occurs in two projective resolutions:

$$0 \longrightarrow V_1 \longrightarrow P \longrightarrow V_2 \longrightarrow 0$$

and

$$0 \longrightarrow W_1 \longrightarrow P \longrightarrow W_2 \longrightarrow 0$$

with $K \otimes_R W_1 \simeq K \otimes_R V_2$ and $K \otimes_R W_2 \simeq K \otimes_R V_1$.

6) The above sequences give rise to the following commutative diagram with exact rows and columns

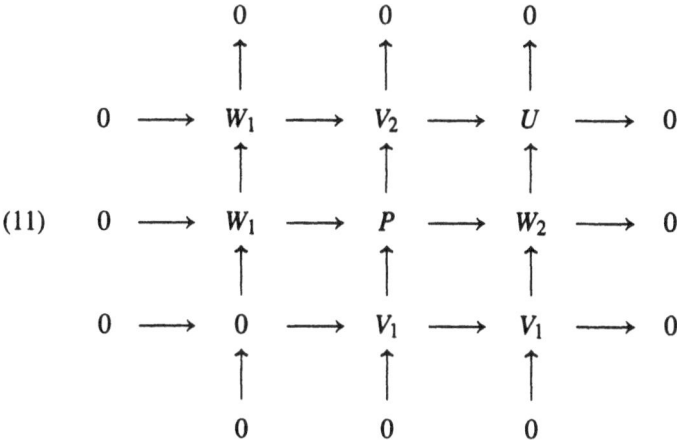

(11)

7) Moreover, if $\phi : P \longrightarrow U$ is the induced map, then $Ker(\phi) = V_1 \oplus W_2 = rad(P) \subset P$; in particular, $U$ is local as the epimorphic image of an indecomposable projective module.     **q. e. d.**

**Claim 5.10.** *Assume that one of the $\Gamma$-lattices, say $V_2$, is irreducible - note that this holds if $\Lambda$ is a block with normal cyclic defect group. Then the module $U$ from above has composition factors all isomorphic to $P/rad(P)$.*

**Proof.** Assume there is a composition factor isomorphic to $Q/rad(Q)$ for an indecomposable projective lattice $Q \not\simeq P$. Then we have a homomorphism $Q \xrightarrow{\bar{\psi}} U$, which can be lifted to a homomorphism $\psi_1 : Q \longrightarrow V_2$, which has non-zero image in $U$. Since $U$ is irreducible, the map $\psi$ factorizes *via* $\psi_2 : X \longrightarrow V_2$, where $X$ is one of the syzygies onto which $Q$ maps. This means that $\psi$ factorizes via $P$; but then $Im(\psi) \subset W_1$; a contradiction, since the map $X \longrightarrow U$ is not zero. **q. e. d.**

**Remark 5.11.** This statement also should hold if the assumption that one of the $\Gamma$-lattices $V_2$ or $W_2$ is irreducible is dropped.

**Definition 5.12.** Let $\{\Omega_i\}_{1 \leq i \leq 2 \cdot n}$ be the syzygies in a projective resolution. We then put

$$T := End_\Gamma(\oplus_1^{2 \cdot n} \Omega_i).$$

**Claim 5.13.** *The order $T$ is a triangular order which decomposes onto n inde-composable triangular orders $T_j : 1 \leq j \leq n$.*

**Proof.** $T$ is an order, and $\Omega_i$ is a projective $T$-module. We identify $\Omega_i$ with $T \cdot e_i$ for a primitive idempotent $e_i$ of $T$. From the diagram 11 we obtain the exact sequence ($T \cdot e_i = W_1$, $K_i = V_1$ and $U_i = U$):

$$0 \longrightarrow K_i \xrightarrow{\ \alpha_i\ } T \cdot e_i \longrightarrow U_i \longrightarrow 0.$$

We may assume w.l.o.g. that $\alpha_i$ is a physical inclusion and hence $J := \oplus_1^{2 \cdot n} K_i$ is a left ideal in $T$.

Though up to now there is no reason why $J$ should be a two-sided ideal, we define $J^2$ to be the image of $J$ under the isomorphism $T \simeq J$. Similarly, $J^i$ is defined for every natural number $i$.

If one follows through the proof of Theorem 1.5 in [Ro; 92], one sees the $T$ is a triangular order with ideal $J$.                                              **q. e. d.**

**Definition 5.14.** We now define THE GRAPH $G$ OF $\Gamma$ as follows: The vertices $V := \{v_i\}_{1 \leq i \leq m}$ correspond to the indecomposable orders $\{T_i\}_{1 \leq i \leq m}$. We draw an edge $v_i \xrightarrow{\ P\ } v_j$, provided there exists an indecomposable projective $\Gamma$-lattice $P$, which is defined via the pull-back

$$
\begin{array}{ccc}
P & \longrightarrow & V \\
\downarrow & & \downarrow \\
W & \longrightarrow & U
\end{array}
,
$$

such that $V$ is a projective $T_i$-lattice and $W$ is a projective $T_j$-lattice.

For further applications we fix some NOTATION

$$
(12) \qquad T_i = \begin{pmatrix}
\Omega_i & \Omega_i & \Omega_i & \cdots & \Omega_i & \Omega_i \\
\omega_i & \Omega_i & \Omega_i & \cdots & \Omega_i & \Omega_i \\
\omega_i & \omega_i & \Omega_i & \cdots & \Omega_i & \Omega_i \\
\cdots & \cdots & \cdots & \ddots & \cdots & \cdots \\
\omega_i & \omega_i & \omega_i & \cdots & \Omega_i & \Omega_i \\
\omega_i & \omega_i & \omega_i & \cdots & \omega_i & \Omega_i
\end{pmatrix}_{n_i}
,
$$

where $\omega_i = \alpha_i \cdot \Omega_i = \Omega_i \cdot \alpha_i$ is a two-sided principal ideal of the local $R$-order

$\Omega_i$. We define $M_{i,j}$ for $1 \leq j \leq n_i$ to be the $j$-th column of $T_i$

(13)
$$M_{i,j} := \begin{pmatrix} \Omega_i \\ \cdots \\ \Omega_i & j-th \text{ row} \\ \omega_1 \\ \omega_i \\ \cdots \\ \omega_i \\ \omega_i \end{pmatrix}_{n_i}$$

We put

$$S_{i,j} := M_{i,j}/rad_{T_i}(M_{i,j}) \text{ and } U_{i,j} := M_{i,j}/M_{i,j-1} \text{ where } M_{i,0} = \alpha_i \cdot M_{i,n_i}.$$

Then all composition factors of $U_{i,j}$ are the same, namely $S_{i,j}$.

With this setup we can describe the indecomposable projective $\Gamma$-lattices more explicitly. An indecomposable projective $\Gamma$-lattice $P$ is described via the commutative diagram with exact rows and columns - we use the above convention for $M_{i,0}$:

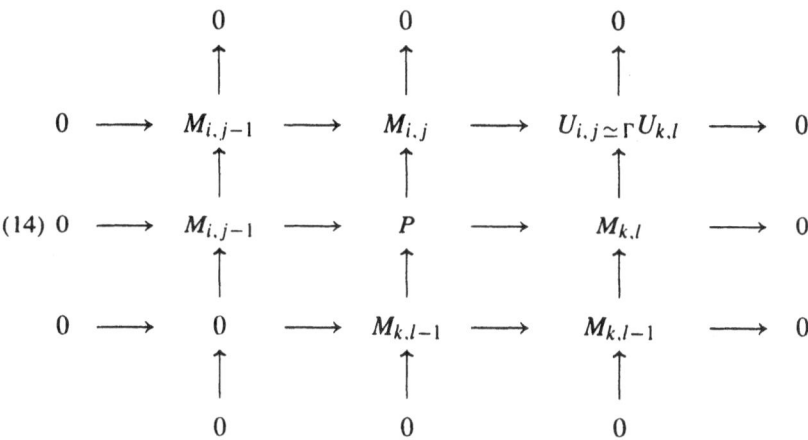

We note that either $i \neq k$ or if $i = k$, then $j \neq l$. Hence as $\mathcal{T}$-modules, $U_{i,j}$ and $U_{k,l}$ are not isomorphic, however, they are isomorphic as $\Gamma$-modules.

It remains to show that the graph $G$ of $\Gamma$ is a tree. Recall that $\Gamma$ is derived equivalent to $\Lambda$, which corresponds to a star, coming from a block. Hence it is symmetric by a result of Zimmermann [Zi; 98]. From Claim 5.6 we know that in this case $\mathbb{G}(\mathcal{G}_1(\Gamma))$ consists solely of projective $\bar{\Gamma}$-lattices.

We now assume that $G$ is not a tree, and we shall show that $\mathbb{G}(\mathcal{G}_1(\Gamma))$ does not consist solely of projective $\bar{\Gamma}$-lattices.

We note that each $T_i$ allows an outer automorphism $\rho_i$ which cyclicly permutes the indecomposable $T_i$-lattices $M_{i,j}$ and extends to an automorphism of $\Gamma$. We shall use this often to simplify the notation.

**Case 1:** $G$ contains a loop; i.e., there exists a projective indecomposable $\Gamma$-lattice $P$ which "lives" in one $T_i$; i.e., with the above notation and using $\rho_i$ we may assume that $P$ corresponds to $\{M_{1,1}, M_{1,j}1 \neq j\}$. We shall next construct a projective $\Gamma$-lattice $Q \simeq P$ with $\alpha_i \cdot P \subset Q \subset P$, and these inclusions are proper.

$$
\begin{array}{ccc}
M_{1,1} & \xrightarrow{\ P\ } & M_{1,j} \\
\cup & & \cup \\
M_{1,2} & & M_{1,j+1} \\
\cup & & \cup \\
\vdots & & \vdots \\
\cup & & \cup \\
M_{1,j} & \xrightarrow{\ Q\ } & M_{1,1} \\
\cup & & \cup \\
\vdots & & \vdots \\
\cup & & \cup \\
M_{1,n_1} & & M_{1,j-1} \\
\cup & & \cup \\
\alpha_1 \cdot M_{1,1} & \xrightarrow{\ \alpha_1 \cdot P\ } & \alpha_i \cdot M_{1,j} \\
\vdots & & \vdots
\end{array}
$$

This picture needs a word of explanation: A horizontal line between $M_{i,j}$ and $M_{k,l}$ with $X$ on top and the filtration written vertically means that $X$ is the pull-back

$$
\begin{array}{ccc}
X & \longrightarrow & M_{i,j} \\
\downarrow & & \downarrow \\
M_{k,l} & \longrightarrow & U_{i,j} \simeq_\Gamma U_{k,l}
\end{array}
,
$$

So the picture represents $P$ – omit the lines $\xrightarrow{\ Q\ }$ and $\xrightarrow{\ \alpha_i \cdot P\ }$. It also represents $Q$ – start at $M_{1,j} \xrightarrow{\ Q\ } M_{1,1}$ and omit $\xrightarrow{\ \alpha_1 \cdot P\ }$. Finally, it represents $\alpha_1 \cdot P$ – start at $\xrightarrow{\ \alpha_1 \cdot P\ }$. The picture then shows that we have proper inclusions – $\alpha_i \cdot P \subset Q \subset P$.

**Case 2:** $G$ contains a proper cycle (i.e., not a loop) consisting of the indecomposable projective modules $P_\nu : 1 \leq \nu \leq s$. Let $P := \oplus_{\nu=1}^{s} P_\nu$ and put $\Pi := \times_{\nu=1}^{s} \alpha_\nu$.

Then there is a proper inclusion: $\Pi \cdot P \subset Q \subset P$ with $Q \simeq P$, as is indicated in the following picture, which is to be interpreted as the above picture.

$$
\begin{array}{ccccccc}
M_{s,1} \xrightarrow{\ P_s\ } & & M_{1,1} \xrightarrow{\qquad P_1 \qquad} & & & & M_{2,j} \\
& {\searrow}^{P_s} & & & & & \\
\cup & M_{s,k} & \cup \xrightarrow{\ P_s\ } M_{1,i} & M_{2,1} \xrightarrow{\ P_2\ } & \cup \xrightarrow{\ P_2\ } & & \\
M_{s,2} & \cup & M_{1,2} \ \cup & \cup & M_{2,j+1} & & \\
\cup & M_{s,k+1} & \cup & M_{1,i} & M_{2,2} & \cup & \\
\vdots & \cup & \vdots & \cup & \cup & \vdots & \\
\cup & \vdots & \cup & \vdots & \vdots & \cup & \\
M_{s,j} & \cup & M_{1,i} & \cup & \cup & M_{2,l} & \\
& {}^{Q_s}\!\!\diagup & & & & & \\
\cup & M_{s,1} & \cup & M_{1,1} \xrightarrow{\ Q_1\ } M_{2,j} & \cup & & \\
\vdots & \cup & \vdots & \cup & \cup & \vdots & \\
\cup & \vdots & \cup & \vdots & \vdots & \cup & \\
M_{s,n_s} & \cup & M_{1,n_1} & \cup & \cup & M_{2,n_j-1} & \\
\cup & M_{s,k-1} & \cup & M_{1,i-1} & M_{2,n_2} & \cup & \\
\alpha_s \cdot M_{s,1} \xrightarrow{\ \alpha \cdot P_s\ } & \cup & \alpha_1 \cdot M_{1,1} \xrightarrow{\ \alpha \cdot P_1\ } & \cup \xrightarrow{\ \alpha \cdot P_1\ } & \cup \xrightarrow{\ \alpha \cdot P_1\ } \alpha_2 \cdot M_{2,j} & & \\
& {\searrow}^{\alpha \cdot P_s} & & & & & \\
\vdots & \alpha_s \cdot M_{s,k} & \vdots \xrightarrow{\ \alpha \cdot P_s\ } \alpha_1 \cdot M_{1,i} & \alpha_2 \cdot M_{2,1} \xrightarrow{\ \alpha \cdot P_2\ } & \vdots \xrightarrow{\ \pi \cdot P_2\ } & &
\end{array}
$$

Thus we have shown, that $G$ must be a tree.

# References

[CR; 82, 86] Curtis, C. and Reiner, I., *Methods of Representation theory*, Vol. I, II, John Wiley, New York 1982, 1986.

[Gr; 72] Green, J.A., Walking around the Brauer tree, *J. Austral. Math. Soc.* **17**, 197–213, 1972.

[Ka; 98] Kauer, M., *Derivierte Äquivalenzen von Graphordnungen und Graphalgebren, Ph.D. Thesis*, Stuttgart, 1998.

[KöZi; 98] König, S. and Zimmermann, A., Derived Equivalences for Group Rings, to appear as Springer Lecture Notes in Math, 1998.

[Li; 96] Linckelmann, M., Stable equivalences of Moritas type for selfinjective algebras and p-groups, *Math. Z.* **223**, 87–100, 1996.

[Re; 75] Reiner, I., *Maximal orders*, Academic Press, New York, 1975.

[Ri; 89, I] Rickard, J., Derived categories and stable equivalence. *J. pure appl. Alg.* **61**, 303–317, 1989.

[Ri; 89, II]   Rickard, J., Morita theory for derived categories, *J. London Math. Soc.* **39**, 436–456, 1989.

[Ro; 92]      Roggenkamp, K.W., Blocks of cyclic defect and Green orders, *Comm. Alg.* **20**(6), 1715–1739, 1992.

[We; 98]      Weber, H., *The global structure of the group rings of the affine groups*, Manuscript Stuttgart, 1998.

[Zi; 98]      Zimmermann, A. Tilted symmetric orders are symmetric orders, MS Amiens, 1998.

Mathematisches Institut B, Universität Stuttgart, Pfaffenwaldring 57, D-70550 Stuttgart, Germany.
E-mail: Roggenkamp@mathematik.uni-stuttgart.de

# Symmetric Elements and Identities in Group Algebras*

*Sudarshan K. Sehgal*

## 1. Introduction

Let $FG$ be the group ring of a group $G$ over a field $F$ of characteristic $p \geq 0$. Let $*$ be the natural involution, $\gamma = \sum \gamma(g)g \rightarrow \gamma^* = \sum \gamma(g)g^{-1}$. Let us denote by

$$(KG)^+ = \{\gamma \in FG : \gamma^* = \gamma\} \text{ and } (KG)^- = \{\gamma \in FG : \gamma^* = -\gamma\},$$

the sets of symmetric and skew symmetric elements respectively. We investigate whether certain identities on these and similar subsets control identities on the whole ring.

## 2. Lie Identities

Write $[x, y]$ for the Lie product $xy - yx$. We say that $FG$ is Lie nilpotent if $[FG, FG, \ldots, FG] = 0$ for some $n$. Suppose for a moment, that $G$ is finite and
$\underbrace{\qquad\qquad\qquad}_{n}$
$p > 0$. Let $FG$ be Lie nilpotent. Then for $x, y \in G$, $[x, \underbrace{y, \ldots, y}_{p^k}] = 0$ for a fixed

$k$ so that $p^k > n$. It follows that $[x, y^{p^k}] = 0$. Therefore, $y^{p^k}$ is central in $G$ for all $y \in G$. Thus by Schur's Theorem [Sehgal, 1978, p. 39], the commutator group $G'$ is a $p$-group. Moreover, since $G$/centre is a $p$-group, $G$ is nilpotent and $G'$ is a $p$-group. We have proved:-

    (1) *If $G$ is finite and $FG$ is Lie nilpotent with $F$ of characteristic $p > 0$, then $G$ is nilpotent and $G'$ is a $p$-group.*

*Conversely,*

    (2) *Suppose $F$ has characteristic $p > 0$. If $G$ is a finite nilpotent group with $G'$ a $p$-group, then $FG$ is Lie nilpotent.*

*Work supported by NSERC grant A-5300.

**Proof.** We use induction on the order of $G$. Pick a central element $z$ of order $p$ in $G$. Let $\bar{G} = G/\langle z \rangle$. By induction $[F\bar{G}, \underbrace{F\bar{G}, \ldots, F\bar{G}}_{n}] = 0$. This implies that

$[FG, \underbrace{FG, \ldots, FG}_{n}]$ is contained in $\Delta(G, \langle z \rangle)$, the kernel of the natural projection

$FG \to F\bar{G}$. We conclude that $[FG, FG, \ldots, FG] \subseteq (1-z)FG$. Thus

$$[FG, \underbrace{FG, \ldots, FG}_{2n}] \subseteq (1-z)^2 FG$$

and consequently

$$[FG, \underbrace{FG, \ldots, FG}_{pn}] \subseteq (1-z)^p FG = 0. \qquad \square$$

In general, to classify Lie nilpotent group algebras $FG$, we have to use **PI**-theory. Recall that we say a ring $R$ satisfies a polynomial identity if there exists a nonzero polynomial $f(z_1, \ldots, z_n)$ in noncommuting variables such that $f(\alpha_1, \ldots, \alpha_n) = 0$ for all $\alpha_i \in R$. We say that $R$ is a polynomial identity ring ($R$ is **PI**) or $R \in$ **PI**. Thus Lie nilpotent rings $KG$ are **PI** rings satisfying a special identity. To state the main theorem for **PI** group rings we need one more definition.

**Definition.** We say that a group $G$ is $p$-abelian if $G'$ is a finite $p$-group. We say that $G$ is 0-abelian if $G$ is abelian.

**Theorem 1. (Passman and Isaacs-Passman [Passman, 1977])**
*Suppose char $F = p \geq 0$. Then $FG \in$ PI if and only if $G$ has a $p$-abelian subgroup of finite index.*

By applying this theorem we obtain

**Theorem 2. (Passi-Passman-Sehgal, 1973)** *$FG$ is Lie nilpotent if and only if $G$ is nilpotent and $p$-abelian where $p$ is the characteristic of $F$.*

Let us write $\delta^{[1]}(FG) = [FG, FG]$, $\delta^{[i+1]}(FG) = [\delta^{[i]}(FG), \delta^{[i]}(FG)]$. We say that $FG$ is Lie solvable if $\delta^{[n]}(FG) = 0$ for some $n$.

**Theorem 3. (Passi-Passman-Sehgal, 1973)** *Necessary and sufficient conditions for Lie solvability of $FG$, char $F = p \geq 0$ are*

(i) *$G$ is $p$-abelian when $p \neq 2$ and*

(ii) *$G$ has a 2-abelian subgroup of index at most 2 when $p = 2$.*

Continuing in the same vein as above we say that $FG$ is Lie $n$-Engel if we have $[x, \underbrace{y, \ldots, y}_{n}] = 0$ for all $x, y \in FG$. A classification of Lie $n$-Engel group rings is given by

**Theorem 4. (Sehgal, 1978, p. 155)** *Let char* $F = p \geq 0$. *Then necessary and sufficient conditions for* $FG$ *to be Lie n-Engel are*

(i) $G$ *is nilpotent and contains a normal p-abelian subgroup* $A$ *with* $G/A$ *a finite p-group if* $p > 0$ *and*

(ii) $G$ *is abelian if* $p = 0$.

Let us now turn to symmetric (and skew symmetric) elements. We say that $(FG)^+$ (resp. $(FG)^-$) is Lie nilpotent if we have $[x_1, \ldots, x_n] = 0$ for all $x_i \in (FG)^+$ (resp. $(FG)^-$). We similarly define when $(FG)^+$ is Lie $n$-Engel etc. In this regard we have the following result.

**Theorem 5. (Giambruno-Sehgal, 1978)** *Suppose char* $F \neq 2$ *and that G has no 2-elements. Then*

$$(FG)^+ or (FG)^- \; Lie \; nilpotent \implies FG \; Lie \; nilpotent.$$

A related result is

**Theorem 6. (Giambruno-Sehgal, 1989)** *Let A be an additive subgroup of* $FG$. *Suppose that* $[A, FG, FG, \ldots, FG] = 0$. *Then* $[A, FG] FG$ *is an (associative) nilpotent ideal.*

By taking $A = FG$ in the above we obtain:

$$FG \; Lie \; nilpotent \implies [FG, FG]FG = \Delta(G, G') \; \text{nilpotent}$$
$$\implies G' \; \text{is a finite } p\text{-group.}$$

Similar to Theorem 5, we have

**Theorem 7. (G. Lee, 1998)** *If* $(FG)^+$ *or* $(FG)^-$ *is Lie n-Engel for some n, G has no 2-elements and char* $F \neq 2$, *then* $FG$ *is Lie m-Engel for some m.*

There is a general result due to Zalesskii-Smirnov about certain rings with involution. Again, denote by $R^-$ the subset of skew symmetric elements.

**Theorem 8. (Zalesskii-Smirnov, 1981)** *Suppose that* $R = \langle R^-, 1 \rangle$ *and that char* $R \neq 2$. *Then*
$$R^- \; Lie \; nilpotent \implies R \; Lie \; nilpotent.$$

**Remark.** In general $\langle (FG)^-, 1 \rangle \neq FG$. For example, let $G = \langle a, b : b^2 = 1, a^b = a^{-1} \rangle$ be the infinite dihedral group. Then $(FG)^- = \{\sum c_i(a^i - a^{-i}) | c_i \in F\} \subset F\langle a \rangle$. Thus $\langle (FG)^-, 1 \rangle \subseteq F\langle a \rangle \neq FG$. However, from the identity

$$\cdot \, 2g^2 = 2 + (g^2 - g^{-2}) + (g - g^{-1})^2$$

it follows that $g^2 \in \langle (FG)^-, 1 \rangle$ for $g \in G$. Thus if $G$ is a finite group of odd order and char $F \neq 2$ then $\langle (FG)^-, 1 \rangle = FG$.

In connection with Theorem 5, possibilities of 2-elements and/or $p = 2$ remain to be discussed. Let us see what may happen if we allow $G$ to have 2-elements. Let

$$G = K_8 = \langle a, b : a^4 = 1 = b^4, a^b = a^{-1}, ba = abz \rangle$$

be the quaternion group of order 8. Then

$$(FG)^+ = F(a + a^{-1}) \oplus F(b + b^{-1}) \oplus F(ab + (ab)^{-1}) \oplus F \oplus Fz$$

is clearly commutative. Moreover, for $p \neq 2$, $G'$ is not a $p$-group. Thus $FG$ is not Lie nilpotent if char $F \neq 2$.

We have seen that $(FK_8)^+$ is Lie nilpotent. Further, if $E$ is an elementary abelian 2-group and $G = K_8 \times E$, then the calculation above gives us that $(FG)^+$ is Lie nilpotent. Moreover, if $G = K_8 \times E \times P$ where $P$ is a finite $p$-group and char $F = p$, the above remarks combined with the argument in (2) of the beginning of this section give that $(FG)^+$ is Lie nilpotent. In fact this is all that can happen as seen in the following result of G. Lee.

**Theorem 9. (G. Lee, 1998)** *Suppose that $K_8 \not\subseteq G$ and char $F = p > 2$. Then $(FG)^+$ is Lie nilpotent $\iff FG$ is Lie nilpotent.*

**Corollary 10.** *Suppose that char $F = 0$ and $K_8 \not\subseteq G$. Then $(FG)^+$ is Lie nilpotent if and only is $G$ is abelian.*

**Proof.** $(FG)^+$ Lie nilpotent $\implies (\mathbb{Z}G)^+$ Lie nilpotent $\implies ((\mathbb{Z}/p\mathbb{Z})G)^+$ Lie nilpotent $\implies G'$ is a finite $p$-group. Since $p$ is arbitrary it follows that $G' = 1$ as required. $\qquad\qquad\qquad\qquad\qquad\qquad\qquad\qquad\qquad\qquad\qquad\qquad\qquad\quad$ □

**Theorem 11. (G. Lee, 1998)** *Suppose that $K_8 \subseteq G$ and char $F = p > 2$. Then $(FG)^+$ is Lie nilpotent if and only if $G = K_8 \times E \times P$ where $E^2 = 1$ and $|P| = p^m$.*

**Corollary 12.** *Suppose that char $F = 0$ and $K_8 \subseteq G$. Then $(FG)^+$ is Lie nilpotent if and only if $G = K_8 \times E$ where $E^2 = 1$.*

Recall that it is proved in Theorem 5 that in the absence of 2-elements and provided $p \neq 2$, $(FG)^-$ is Lie nilpotent if and only if $FG$ is Lie nilpotent. This is no more true if we allow 2-elements as seen by the following example. Let $G = D_8 = \langle a, b : a^4 = 1 = b^2, a^b = a^{-1} \rangle$ be the dihedral group of order 8. Then $(FG)^- \subseteq F\langle a \rangle$ is commutative. However, $FG$ is not Lie nilpotent if $p \neq 2$. Currently, Greg Lee is investigating this problem and the possibility of allowing 2-elements in the $n$-Engel property classification.

## 3. Group Identities

Let us turn to related questions in $U = U(FG)$, the group of units of $FG$. Throughout this section, $G$ is a torsion group and $F$ is of characteristic $p \geq 0$. We begin with

**Definition.** We say that $U$ satisfies a group identity if there is a non-trivial word $w(x_1, \ldots, x_m)$ in the free group generated by $x_1, \ldots, x_m$ such that $w(u_1, \ldots, u_m) = 1$ for all $u_i \in U$. In this case, we write $U \in \mathbf{GI}$. We list below special cases of $w$.

**Examples**

1)  $w = (x_1, x_2) = x_1^{-1} x_2^{-1} x_1 x_2$: commutative,

2)  $w = (x_1^m, x_2)$: torsion over the centre and

3)  $w = (x_1, \ldots, x_\ell)$: nilpotent.

In order to connect the multiplicative structure to the additive·structure, Brian Hartley made the

**Conjecture.** Let $G$ be a torsion group. Let $F$ be an infinite field. Then

$$U \in \mathbf{GI} \Longrightarrow FG \in \mathbf{PI}.$$

The first results on this conjecture were obtained by Warhurst (1981) who studied some special cases. Also, P. Menal (1981) suggested a possible solution for some $p$-groups. Goncalves and Mandel (1991) classified group algebras $FG$ of torsion groups over infinite fields whose group of units satisfies a semigroup identity ($w_1 = w_2$ where $w_1$ and $w_2$ are distinct semigroup words) proving this way Hartley's conjecture for semigroup identities. Dokuchaev and Goncalves (1997) dealt with this question for integral group rings. Giambruno-Jespers-Valenti (1994) settled in the positive the conjecture when $G$ has no $p$-elements. By using the construction suggested by Menal the authors proved.

**Theorem 12. (Giambruno-Sehgal-Valenti, 1997)** *If $G$ is a torsion group and $F$ is infinite, then $U \in \mathbf{GI} \Longrightarrow FG \in \mathbf{PI}$.*

The referee remarked that the conjecture is too weak. Sure enough, we have the

**Theorem 13. (Passman, 1997)** *Let $G$ be torsion and let $F$ be infinite. Then*

*1)  if char $F = 0$, $U(FG)$ satisfies a group identity if and only if $G$ is abelian,*

*2)  if char $F = p > 0$, $U \in \mathbf{GI} \Longleftrightarrow FG \in \mathbf{PI}$ and $G'$ is of bounded exponent $p^m$.*

It turns out that Hartley's conjecture is also true for finite fields (see Liu, *Preprint*). Moreover, the last theorem can be extended as well.

**Theorem 14. (Liu-Passman, *Preprint*)** *Let $F$ be a field of characteristic $p > 0$ and $G$ a torsion group. If $G'$ is a $p$-group, then the characterization of the last theorem holds. If $G'$ is not a $p$-group, then $U \in \mathbf{GI} \Longleftrightarrow FG \in \mathbf{PI}$, $G$ has bounded period and $F$ is finite.*

We can ask if the identities at the unit level are also controlled by symmetric elements. We use the notation:

$$U^+ = \{u \in U(FG) : u^* = u\}.$$

We say $U^+$ satisfies a group identity, $U^+ \in GI$, if there is a nontrivial word $w(x_1, \ldots, x_m)$ such that $w(u_1, \ldots, u_m) = 1$ for all $u_i \in U^+$. We should keep in mind the example, of the last section, of the quaternion group where elements of $U^+(FK_8)$ commute. We have an analogue of Hartley's Conjecture:

**Theorem 15. (Giambruno-Sehgal-Valenti, 1997)** *Let $F$ be an infinite field of characteristic $\neq 2$ and let $G$ be a torsion group. Then*

$$U^+ \in GI \Longrightarrow FG \in PI.$$

In fact, it is possible to say more.

**Theorem 16. (Giambruno-Sehgal-Valenti, 1997)** *Let $F$ be an infinite field and $G$ a torsion group. If char $F = 0$, $U^+ \in GI$ if and only if $G$ is abelian or a Hamiltonian 2-group. If char $F = p > 2$, then $U^+ \in GI$ if and only if $FG \in PI$ and either $K_8 \nsubseteq G$ and $G'$ is of bounded exponent $p^k$ or $K_8 \subseteq G$ and*

1) *the p-elements of $G$ form a subgroup $P$ and $G/P$ is a Hamiltonian 2-group; and*

2) *$G$ is of bounded exponent $4p^s$.*

## References

Dokuchaev, M.A. and Goncalves, J.Z. Semigroup identities on units of integral group rings, *Glasgow Math. J.* **39**, 1–6, 1997.

Giambruno, A. Jespers, E. and Valenti, A. Group identities on units of rings, *Arch. Math.* **63**, 291–296, 1994.

Giambruno, A. and Sehgal, S.K. A Lie property in group rings, *Proc. Amer. Math. Soc.* **105**, 287–292, 1989.

Giambruno, A. and Sehgal, S.K. Lie nilpotence of group rings, *Comm. Alg.* **21**, 4253–4261, 1993.

Giambruno, A. Sehgal, S.K. and Valenti, A. Group algebras whose units satisfy a group identity, *Proc. Amer. Math. Soc.* **125**, 629–634, 1997.

Giambruno, A. Sehgal, S.K. and Valenti, A. Symmetric units and group identities, (*Preprint*).

Goncalves, J.Z. and Mandel, A. Semigroup identities on units of group algebras, *Arch. Math.* **57**, 539–545, 1991.

Liu Chia-Hsin, Group algebras with units satisfying a group identity, (*Preprint*).

Liu Chia-Hsin and Passman, D.S. Group algebras with units satisfying a group identity II, (*Preprint*).

Menal, P. *Priv. Lett.* to B. Hartley, April 6, 1981.

Passi, I.B.S. Passman, D.S. and Sehgal, S.K. Lie solvable group rings, *Canad. J. Math.* **25**, 748–752, 1973.

Passman, D.S. *The Algebraic Structure of Group Rings*, John Wiley, New York, 1977.

Passman, D.S. Group Algebras whose units satisfy a group identity II, *Proc. Amer. Math. Soc.* **125**, 657–662, 1997.

Sehgal, S.K. *Topics in Group Rings*, Marcel Dekker, New York, 1978.

Warhurst, D.S. *Topics in group rings, Thesis* Manchester, 1981.

Zalesskii, A.E. and Smirnov, M.B. Lie algebra associated with linear group, *Comm. Alg.* **9**, 2075–2100, 1981.

Department of Mathematical Sciences, University of Alberta Edmonton, Alberta, Canada T6G 2G1.
E-mail: s.sehgal@ualberta.ca

# Serial Modules and Rings

*Surjeet Singh*

## Introduction

The fundamental theorem of abelian groups states that any finitely generated abelian group is a direct sum of cyclic groups. This theorem plays a fundamental role in the structure theory of abelian groups. It has fascinated many algebraists to look at this theorem from different points of view:-

(*a*) To find suitable generalizations of this theorem for modules over certain classes of rings, e.g. Dedekind domain, hereditary noetherian prime rings, valuation rings etc.

(*b*) To create suitable versions of this theorem in some module categories and use such versions to develop the structure theory of such module categories. For example, torsion abelian group-like modules, modules with finitely generated submodules direct sums of multiplication modules.

(*c*) To find those rings for which certain versions of the fundamental theorem of abelian groups hold. For example rings over which all finitely generated modules are direct sums of cyclic modules.

(*d*) To examine the structure of certain classes of abelian groups and to try to find modules with similar structures.

(*e*) To find roles of answers to some of above types of questions in the general theory of modules.

There are a host of other questions raised that have been inspired by the fundamental theorem of abelian groups. Some of the questions raised above are surveyed here in some details.

## §1. Serial Rings

Throughout all rings are with identity and all modules are unital right modules, unless otherwise stated. Let $R$ be a ring. $J(R)$ denotes the Jacobson radical of $R$.

For any $n \geq 1$, $R^{n \times n}$ denotes the ring of all $n \times n$ matrices over $R$. This ring is known as the **full matrix ring** over $R$.

**Definition 1.1.** Let $M$ be any $R$-module. $M$ is said to be **uniserial** if for any two submodules $A$ and $B$ of $M$, either $A \subseteq B$ or $B \subseteq A$. A module that is a direct sum of uniserial modules is called a serial module.

**Example 1.** For any prime number $p$, any cyclic $p$-group or the quasi-cyclic $p$-group $C(p^\infty)$ is a uniserial $Z$-module, $Z$, the ring of integers. Any finitely generated torsion abelian group, being a direct sum of cyclic $p$-groups, where $p$ ranges over some finite set of primes, is a serial $Z$-module.

**Definition 1.2.** A non zero module $M$ is called a **uniform** module if intersection of any two non zero submodules of $M$ is non zero.

Any non zero uniserial module is uniform. However, $Z$ as a $Z$-module is uniform but not uniserial.

**Definition 1.3.** A ring $R$ is called a right (left) serial ring if $R_R(_R R)$ is a serial module. A ring that is both sided serial is called a serial ring (Faith, 1976).

Commutative indecomposable serial rings are known as **valuation** rings. For the structure theory of valuation ring, one may consult Krull (1968), Larsen and McCarthy (1971) or Nagata (1975). Every local, principal ideal domain, other than a field, is a valuation ring, called a **discrete valuation ring of rank one**. Some other serial rings are:-

(1) Any full matrix ring over a valuation ring.

(2) Let $R$ be a discrete valuation ring of rank one. Consider the full matrix ring $R^{n \times n}$. The subring $S$ of $R^{n \times n}$ consisting of those matrices $[a_{ij}]$ for which $a_{ij} \in J(R)$ for $i > j$.

(3) For any division ring $D$, the subring of $D^{n \times n}$, consisting of all upper triangular matrices.

(4) Let $\sigma$ be an automorphism of a field $F$, and

$$R = \left\{ \sum_0^\infty X^i a_i : a_i \in F \right\}$$

be the set of all right power series over $F$. In $R$ define addition as usual, but define multiplication by using distributivity and the law:

$$aX = Xa^\sigma, a \in F.$$

Then $R$ is a serial ring. We write $R = F[[X, \sigma]]$. Some of the one sided serial rings are:

(5) For any field $F$, the matrix subring

$$\begin{bmatrix} F & 0 & 0 \\ 0 & F & F \\ 0 & 0 & F \end{bmatrix}$$

of $F^{3 \times 3}$ is right serial, but not left serial.

(6) Let $\sigma$ be an endomorphism of a field $F$ such that $0 \neq \sigma(F) \neq F$. Then analogous to (4), $R = F[[X, \sigma]]$ is right serial but not left serial.

There are many constructions of valuation rings. For general valuation rings, see Nagata (1975) and for finite, local, principal ideal rings, see McDonald (1974). Notable among the later are, so called Galois rings. These are very close to Galois Fields, in construction as well as theory (see Krull, 1968 and McDonald, 1974).

## §2. Decomposition of Modules

Let $G$ be an abelian $p$-group, where $p$ is some prime number. Then $G$ is said to be **bounded** if $p^n G = 0$ for some $n \geq 1$. Any bounded abelian $p$-group is a direct sum of cyclic $p$-groups [Fuchs (1970)]; so it is a serial $Z$-module.

**Question.** *When is every module over a ring $R$ serial?*

It was shown by Köthe (1935) that every module over a both sided artinian and both sided principal ideal ring is serial. Nakayama (1941) gave a complete answer for the above question. He proved the following:

**Theorem 2.1.** *For a ring $R$ the following are equivalent:*

*(a) $R$ is an artinian serial ring;*

*(b) Every right $R$ module is serial;*

*(c) Every left $R$ module is serial.*

**Definition 2.2.** A module $M_R$ is said to be finitely presented if there exists an exact sequence

$$0 \to Q \to P \to M \to 0$$

with $P$ a finitely generated projective module and $Q$ a finitely generated module.

Any finitely presented module is finitely generated. However a finitely generated module need not be finitely presented. If $R$ is right noetherian, then any finitely generated $R$-module is finitely presented. A module $M$ is called a **local** module in case it has a unique maximal submodule $N$ and every proper submodule is contained in $N$. Analogous to the above definition we can define **locally presented** module and a **cyclically presented** module. A locally presented module is finitely presented. Recall that a ring $R$ is a local ring if $R/J(R)$ is a division ring; i.e., $R_R$ is a local module. Every valuation ring is a local ring. Kaplansky (1949) proved that every finitely presented module over a valuation ring is a direct sum of locally presented (hence cyclically presented) modules. One likes to know other local rings, if any, with this property. The following Theorem by Warfield (1970) is worth noticing.

**Theorem 2.3.** *Let $R$ be a commutative local ring which is not a valuation ring. Then for any $n > 0$ there exists a finitely presented indecomposable module $M$ which cannot be generated by fewer than $n$ elements.*

Thus if a commutative local ring $R$ has to have every finitely presented module a direct sum of cyclic modules, $R$ must be a valuation ring. The study of artinian serial rings was initially done by Nakayama (1940, 1941). He called them generalized uniserial rings. Kupisch (1959) proved that given an indecomposable generalized uniserial ring $R$, and any representative set $e_1 R, e_2 R, \ldots, e_n R$ of non-isomorphic indecomposable summands of $R_R$, we can index this set in such a way that for $J = J(R)$

(a) $e_i J/e_i J^2 \cong e_{i+1} R/e_{i+1} J$ and $d(e_i R) \leq d(e_{i+1} R) + 1$ for $1 \leq i \leq n - 1$; and

(b) either $e_n R$ is simple or $e_n J/e_n J^2 \cong e_1 R/e_1 J$ and $d(e_n R) \leq d(e_1 R) + 1$.

Here $d(M)$ denotes the composition length of a module $M$ admitting a composition series. This result plays a central role in the structure theory of generalized uniserial rings. The detailed structure theory of these rings was essentially developed by Murase (1963*a&b* and 1964).

The study of non-artinian, non-commutative serial rings was started by Warfield (1975). A ring $R$ is called a semi-perfect ring, if $R/J(R)$ is semisimple artinian ring and for any idempotent $\bar{g} = g + J(R)$ in $R/J(R)$, there exists an idempotent $e$ in $R$, such that $\bar{g} = \bar{e}$. For any indecomposable idempotent $e$ in a semiperfect ring $R$, $eR$ is a local module. Warfield (1975) proved the following:

**Theorem 2.4.** *The following properties of a semiperfect ring $R$ are equivalent:*

(i) *$R$ is serial;*

(ii) *Every finitely presented left $R$-module is a direct sum of locally presented modules;*

(iii) *Every finitely presented right $R$-module is a direct sum of locally presented modules.*

In the same paper Warfield also showed that any both sided noetherian serial ring is a direct sum of prime rings and an artinian ring. However a serial ring, which is one sided noetherian, may be indecomposable, but neither artinian nor prime.

**Example 2.** Let $Q$ be the field of rational numbers, $p$ be a prime number and

$$Z_p = \{a/b : a, b \in Z, p \nmid b\}.$$

Then the matrix ring

$$R = \begin{bmatrix} Z_p & Q \\ 0 & Q \end{bmatrix}$$

is serial, right noetherian and indecomposable but $R$ is neither artinian nor prime (see Singh, 1984).

Warfield gave a structure theorem of a noetherian, prime serial ring. This result along with the known structure of artinian serial, describes to very great extent all both sided noetherian serial rings. One sided noetherian serial rings have been studied by Singh (1984). But still not much is known about the other type of serial rings. Serial rings with Krull dimension one have been studied by Upham (1987).

It follows from Theorem 2.4 that any indecomposable finitely presented module over any serial ring is local. But this assertion is false for one sided serial rings.

**Example 3.** Let $F$ be any field admitting an endomorphism $\sigma$ such that $1 < [F : \sigma(F)] < \infty$. Consider

$$R = \{a + Xb : a, b \in F\}$$

in which $(a + Xb)(c + Xd) = ac + X(a^\sigma d + bc)$. Then $R$ is right serial local but $R$ is not left serial. This ring is both sided artinian. The injective hull $E = E(R_R)$ is finitely presented and uniform. $E/Soc(E)$ is a direct sum of $n$ simple modules where $n = [F : \sigma(F)]$. So $E$ is not a local module.

Infact not much is known about finitely generated modules over both sided artinian, but one sided serial rings. One can easily construct finitely generated indecomposable non-uniform modules over certain artinian right serial rings. One such construction is in Singh (1997).

## §3. Serial Modules

Any uniform torsion $Z$-module is either isomorphic to a cyclic $p$-group or to quasi-cyclic $p$-groups, for some prime number $p$; so it is uniserial. It follows from Theorem 2.4 that any uniform, finitely presented module over a serial ring is uniserial. However, a finitely generated uniform module over a serial ring need not be uniserial.

**Example 4.** Let $R = Z_{(p)} \times C(p^\infty)$, where $p$ is a prime number. $C(p^\infty)$ is a $Z_{(p)}$-module. $R$ is a valuation ring under component-wise addition, and multiplication given by

$$(x, a)(y, b) = (xy, bx + ay).$$

Consider the injective hull $E = E(R_R)$. Then $E_R$ is uniform, but not uniserial. The reason being the following. Let $\hat{Z}_{(p)}$ be the $p$-adic completion of $Z_{(p)}$. Then $Z_{(p)}$ is isomorphic to the ring of endomorphism of $C(p^\infty)$. (See Matlis [16].) So $C(p^\infty)$ is a $\hat{Z}_{(p)}$-module. Analogous to $R$, we get a ring

$$S = \hat{Z}_{(p)} \times C(p^\infty)$$

containing $R$. Then $R$ is an essential submodule of $S_R$, so $S$ embeds in $E$. There is a lattice isomorphism between $R$-submodules of $S/R$ and $Z_{(p)}$-submodules of $\hat{Z}_{(p)}/Z_{(p)}$. But $\hat{Z}_{(p)}/Z_{(p)}$ contains infinite direct sums of $Z_{(p)}$-submodules. So $S_R$ is not uniserial and hence $E$ is not uniserial. We can find $u, v \in E$ such that $uR \nsubseteq vR$ and $vR \nsubseteq uR$. Then $M = uR + vR$ is finitely generated, uniform but not uniserial.

Observe that in the above example $E = E(R/J(R))$. Serial ring over which uniform modules are uniserial are not known. But for valuation rings an answer is known (see Theorem 3.2).

**Definition 3.1.** A valuation ring $R$ is called a **maximal valuation** ring, if given any family of ideals $(I_\alpha)_A$ of $R$, and congruene equations

$$x \equiv a_\alpha (mod \, I_\alpha), a_\alpha \in R$$

then this system has a solution for $x \in R$ whenever every finite subsystem has a solution for $x \in R$. A valuation ring $R$ is called an **almost maximal valuation** ring if $R/I$ is maximal for every nonzero ideal $I$.

For Example $Z_{(p)}$ is almost maximal, but it is not maximal. $\hat{Z}_{(p)}$ is a maximal valuation ring. Gill (1971) gives the following:

**Theorem 3.2.** *For any commutative local ring $R$, the following are equivalent:*

*(i) $R$ is an almost maximal valuation ring;*

*(ii) the injective hull $E(R/J(R))$ is uniserial;*

*(iii) any finitely generated $R$-module is a direct sum of cyclic modules.*

Thus in the above theorem condition $(iii)$ shows that fundamental theorem of abelian groups generalizes to modules over almost valuation rings. A classification of all commutative rings over which every finitely generated $R$-module is a direct sum of cyclic modules have been established. The details of this development can be found in Brandal (1979). But for non-commutative non-noetherian rings nothing much is known.

Let $N \subset' M$ denote that $N$ is an essential submodule of $M$.

**Definition 3.3.** For any module $M_R$, $Z(M) = \{x \in M : xA = 0$ for some $A \subset' R_R\}$ is called the singular submodule of $M$. $Z(R_R)$ is an ideal of $R$, called (right) singular ideal of $R$. A module $M$ with $Z(M) = 0$ is called a non singular module.

Given a uniform module over a serial ring, can we isolate some uniserial submodules? Again not much is known. But some information is available. We start with the following due to Upham (1987).

**Theorem 3.4.** *Any non singular uniform module over a serial ring is uniserial.*

Upham (under the name Wright) continued the study of above question, for serial rings with Krull dimensions (Wright, 1989 & 1990). All the results proved by her do not depend on Krull dimension but on a concept close to non-singularity. This was seen by Mueller and Singh (1991). First of all they proved the following.

**Theorem 3.5.** *For an ideal I of a serial ring R, the following are equivalent:*

(i) *For every indecomposable idempotent $e \in R$, either $eI = 0$ or $eR/eI$ is a non-singular $R/I$-module;*

(ii) *For every indecomposable idempotent $e \in R$, and every $x \in eR \backslash eI$, $xI = eI$;*

(iii) *For every uniserial module $M_R$, and $x \in M \backslash MI$, $xI = MI$.*

Call an ideal $I$ of a serial ring $R$, a non-singular ideal, if $R/I$ is a non-singular ring. Any non-singular ideal $I$ satisfies the equivalent conditions of the above theorem. In view of this fact we define the following.

**Definition 3.6.** An ideal $I$ of a serial ring $R$ is called (right) almost non-singular, if it satisfies the three equivalent conditions of Theorem 3.5.

**Definition 3.7.** For any ideal $I$ of a ring $R$, a module $M_R$ is called an $I$-non-singular module if $ann_M(I)$ is a non-singular $R/I$-module.

Mueller and Singh (1991) have proved the following:

**Theorem 3.8.** *Let $R$ be a serial ring and $I$ an almost non-singular ideal of $R$. Let $M$ be a uniform $R$-module, such that it is $I$-non-singular. Then the annihilator $K$ of $\cap_1^\infty I^n$ in $M$ is a uniserial module.*

For example, if $R$ is a valuation ring, then $J(R)$ is almost non-singular. If $M = E(R/J(R))$, then $M$ is $J$-non-singular for $J = J(R)$. So $K = ann_M(\cap_1^\infty J^n)$ is a uniserial submodule of $M$. In general consider any serial ring $R$, for any ordinal $\alpha$ define $J^\alpha$ as follows. $J^0 = R$. If $\alpha = \beta + 1$, $J^\alpha = J^\beta.J$. If $\alpha$ is a limit ordinal $J^\alpha = \cap_{\beta<\alpha} J^\beta$. One can prove that every $J^\alpha$ is an almost non-singular ideal. So the above theorem can be applied to $J^\alpha$-non-singular uniform $R$-modules. Such applications give various results for uniserial submodules proved by Wright (1989).

## §4. Serial Homomorphic Images

Let $D$ be a Dedekind domain. Then for any non zero ideal $A$ of $D$, $D/A$ is an artinian PIR, so $D/A$ is artinian, commutative serial ring. The converse of this result also holds. Possibly there are many other rings that can be described in terms of serial rings. We outline some of such results.

A ring $R$ is called left (right) hereditary ring of all of its left (right) ideals are projective. A prime ring which is both sided noetherian and hereditary is called an hereditary noetherian prime ring (in short (hnp)-ring). A Dedekind domain is an (hnp)-ring. Eisenbud and Robson (1970) proved the following:

**Theorem 4.1.** *For any non zero ideal $A$ of an (hnp)-ring $R$, $R/A$ is an artinian serial ring.*

Given a simple ring $R$, hypothesis, that for any non zero ideal $A$ of $R$, $R/A$ is an artinian serial ring, vacuously holds. But a simple ring need not be an (hnp)-ring. So unlike for commutative integral domains, converse of the above result does not hold. We put some extra conditions on $R$ to get a converse of the result in Theorem 4.1.

**Definition 4.2.** A ring $R$ is said to be right bounded if for any $A \subset' R_R$ there exist a non zero ideal $B$ of $R$ contained in $A$.

Any commutative ring is left and right bounded. Singh (1975) proved the following:

**Theorem 4.3.** *Let $R$ be a prime, right bounded, right Goldie ring, such that for each non zero ideal $A$ of $R$, $R/A$ is an artinian serial ring. Then $R$ is right hereditary.*

Singh (1984) further proved the following two results:

**Theorem 4.4.** *Let $R$ be a right noetherian semi-prime ring all of whose proper homomorphic images are serial rings. Then either $R$ is a serial noetherian ring or a prime ring.*

**Theorem 4.5.** *Let $R$ be a prime, right bounded, right noetherian ring, such that for each ideal $A \neq 0$, $R/A$ is serial. Then for each ideal $A \neq 0$, $R/A$ is artinian.*

Next step will be to consider rings with Krull dimensions, with proper homomorphic images serial rings.

## §5. Some Other Aspects

Consider the following two conditions on a module $M_R$:
   (I) Any finitely generated submodule of any homomorphic image of $M_R$ is a direct sum of finite·length uniserial modules.

(II) Given any two finite length uniserial submodules $U$ and $V$ of a homomorphic image $M/K$ of $M$, and a submodule $W$ of $U$, then any homomorphism $\sigma : W \to V$ can be extended to a homomorphism $v : U \to V$, provided $d(U/W) \le d(V/\sigma(W))$.

Any torsion abelian group as a Z-module satisfies (I) and (II). More generally given any bounded (hnp)-ring $R$, any torsion $R$ module satisfies these conditions. These conditions were introduced by Singh (1976). They arose out of the study of torsion modules over bounded (hnp)-rings. In a series of papers, it has been seen that the structure theory of modules satisfying (I) and (II) is very similar to that of torsion abelian groups. Because of this similarity, they are named torsion abelian groups-like modules (in short TAG modules) in Benabdallah and Singh (1983). The concepts of basic subgroups, Ulm-Kaplansky sequences, neat and pure subgroups, have been extended to those for (TAG)-modules (see Mehran & Singh, 1985; Singh, 1979; and Singh & Al-Zaid, 1996). Most of the basic results of torsion abelian groups known for these concepts have been extended to (TAG)-modules. In 1987 Singh showed that (II) can be partially deduced from (I). A module satisfying (I) is called QTAG-module. Structure theory of torsion abelian groups is highly developed (see Fuchs, 1970 & 1973). This theory can act as a motivating force for the development of theory of TAG-modules or of QTAG-modules. One more crucial step to be taken is to find a set of axioms extending those for TAG-modules, which can satisfactorily cover non-torsion abelian groups.

**Definition 5.1.** A module $M$ over a commutative ring $R$ is called a multiplication module if for any submodule $N$ of $M$, $N = MA$ for some ideal $A$ of $R$.

Any cyclic module, any ideal of a Dedekind domain, and any ideal of a von-Neumann regular commutative ring, are some examples of multiplication modules. For detailed information on multiplication modules one may consult Barnard (1981), El-Bast and Smith (1988) or Larsen and McCarthy (1971). The fundamental theorem of abelian groups can be reworded to say that any finitely generated Z-module is a direct sum of multiplication modules. Consider a module $M_R$ ($R$ commutative) satisfying:

(Mu): Any finitely generated submodule of $M_R$ is a direct sum of multiplication modules.

One can hope to develop the structure theory of a module satisfying (Mu) on the lines of abelian groups. As a first step, Singh (*Preprint*) has recently studied a commutative ring $R$ such that any ideal of any homomorphic image $R/A$ is a finite direct sum of multiplication modules. Such rings are called WPIR. He has proved the following:

**Theorem 5.2.** *Let $R$ be a commutative noetherian ring. Then $R$ is WPIR if and only if $R = R_1 \oplus R_2 \oplus \ldots \oplus R_n$, where each $R_i$ is an indecomposable ring of one of the following types:-*

(a)  $R_i$  has a semi-simple ideal  $B_i$ , with  $R_i/B_i$  a PIR.

(b)  $R_i$  has unique minimal prime ideal  $P_i$ . This  $P_i$  is semi-simple and  $R/P_i$  is a Dedekind domain.

# References

Barnard, A., Multiplication modules, J. Alg., **71**, 174–178, 1981.

Benabdallah, K. and Singh, S., On torsion abelian groups-like modules, *Proc. Conf. Abelian Groups, Hawaii, LNM*, Springer Verlag, **1006**, 639–653, 1983.

Benabdallah, K., Bouanane, A. and Singh, S., On sums of uniserial modules, *Rockey Mount. J. Math.*, **20**, 15–29, 1990.

Brandal, W., Commutative rings whose finitely generated modules decompose, *LNM*, **723**, Springer Verlag, 1979.

Eisenbud, D. and Robson, J.C., Hereditary noetherian prime rings. *J. Alg.* **16**, 86–104, 1970.

El-Bast, Z.A. and Smith, P.F., Multiplication modules, *Comm. Alg.*, **16**, 755–779, 1988.

Faith, C. *Algebra, II, Ring Theory*, Grundlehren der mathematischen Wissenchaften **191**, Springer Verlag, 1976.

Fuchs, L. Infinite Abelian Groups, Vol I, *Pure appl. Math.*, **No. 36**, Academic Press, 1970.

Fuchs, L. Infinite Abelian Groups, Vol II, *Pure appl. Math.*, **No. 36-II**, Academic Press, 1973.

Gill, D.T., Almost maximal valuation rings, *J. London Math. Soc.*, **4**, 140–146, 1971.

Kaplansky, I., Elemantary divisors and modules, *Trans. Amer. Math. Soc.*, **66**, 464–491, 1949.

Köthe, G., Verallgemeinerte abelsche gruppen mit hyperkomplexem operatoren ring, *Math. Z.*, **39**, 31–44, 1935.

Kupisch, H., Beitrage zur theorie nitchalbein facher ringe mit minimal bedingung, *J. Reine Angew. Math.*, **201**, 100–112, 1959.

Krull, W., *Idealtheorie*, Ergebnisse der Mathematik und ihrer Grenzgebiete, **46**, Springer Verlag, 1968.

Larsen, M.D. and McCarthy, P.J., *Multiplicative Theory of Ideals*, Academic Press, 1971.

Matlis, E., Injective modules over noetherian rings, *Pacific. J. Math.*, **8**, 511–528, 1958.

McDonald, B.R., Finite Rings with Identity, *Pure appl. Math.*, **28**, Marcel Dekker, 1974.

Muller, B.J. and Singh, S., Uniform modules over serial rings, *J. Alg.*, **144**, 94–108, 1991.

Mehran, H.A. and Singh, S., Ulm-Kaplansky invariants for TAG modules, *Comm. Alg.*, **13**, 355–373, 1985.

Mehran, H.A. and Singh, S., On  $\sigma$ - pure submodules of QTAG-modules, *Arch. Math.*, **46**, 501–510, 1986.

Murase, I., On the structure of generalized uniserial rings I, *Sci. Pap. Coll. Gen. Edu. Univ. Tokyo*, **13**, 1–22, 1963a.

Murase, I., On the structure of generalized uniserial rings II, *ibid*, **13**, 131–158, 1963b.

Murase, I., On the structure of generalized uniserial rings III, *ibid*, **14**, 12–25, 1964.

Nagata, M., *Local Rings*, Robert E. Kreiger Publishing Company, 1975.

Nakayama, T., Note on uniserial and generalized uniserial rings, *Proc. Imp. Acad., Tokyo*, **16**, 285–289, 1940.

Nakayama, T., On Frobeniusean algebra I, *Ann. Math.*, **40**, 611–133, 1939.

Nakayama, T., On Frobeniusean algebra II, *ibid.*, **42**, 1–21, 1941.

Nakayama, T., On Frobeniusean algebra III, *Japan J. Math.*, **18**, 49–65, 1942.

Singh, S., Modules over hereditary noetherian prime rings, *Can. J. Math.*, **27**, 867–883, 1975.

Singh, S., Some decomposition theorems in abelian groups and their generalizations, *Proc. Ohio Univ. Conf.* (Ed. S. K. Jain) LNP, Marcel Dekker **25**, 183–189, 1976.

Singh, S., Some decomposition theorems in abelian groups and their generalizations II. *Osaka J. Math.*, **16**, 45–55, 1979.

Singh, S., Serial right noetherian rings, *Can. J. Math.*, **36**, 22–37, 1984.

Singh, S., Abelian groups-like modules, *Acta Mathematica Hungarica*, **50**, 85–95, 1987.

Singh, S., On generalized uniserial rings, *Chinese J. Math.*, **17**, 117–137, 1989.

Singh, S., Indecomposable modules over artinain right serial rings, *Ad. Ring Theory*, (Eds. S.K. Jain and S. Tariq Rizvi) Birkhauser, 295–304, 1997.

Singh, S., Direct sums of multiplication modules (*Preprint*).

Singh, S. and Al-Zaid, H., On h-pure and pure submodules of a QTAG-module, *Kuwait J. Sci. Eng.*, **23**, 7–11, 1996.

Singh, S. and Ansari, W.A., On Ulm's Theorem, *Comm. Alg.*, **10**, 2031–2042, 1982.

Upham, M.H., Serial rings with right Krull dimension one, *J. Alg.*, **109**, 319–333, 1987.

Warfield, R.B., Decomposability of finitely presented modules, *Proc. Amer. Math. Soc.*, **25**, 167–172, 1970.

Warfield, R.B., Serial rings and finitely presented modules. *J. Alg.*, **37**, 187–222, 1975.

Wright, M.H., Certain modules over serial rings are uniserial, *Comm. Alg.*, **17**, 441–469, 1989.

Wright, M.H., Uniform modules over serial rings with Krull dimensions, *Comm. Alg.*, **18**, 2541–2557, 1990.

Department of Mathematics, King Saud University, P.O. Box 2455, Riyadh 11451, Kingdom of Saudi Arabia

# On Subgroups Determined by Ideals of an Integral Group Ring

*L. R. Vermani*

## 1. Introduction

Let $G$ be a group, $ZG$ the integral group ring of $G$ and $I(G)$ its augmentation ideal. Recall that $I(G)$ is the kernel of the ring homomorphism $\in : ZG \to Z$ given by $\in (\sum n_i g_i) = \sum n_i$, $n_i \in Z$, $g_i \in G$, and it is generated as a free $Z$-module by the elements $g - 1$, $g \in G$, $g \neq e$. For $n \geq 1$, let $I^n(G)$ denote the $n$th associative power of $I(G)$. For an ideal $J$ of $ZG$, let $G \cap (1 + J) = \{x \in G/x - 1 \in J\}$. Observe that for $x, y \in G \cap (1 + J)$, $z \in G$,

$$xy - 1 = x(y - 1) + x - 1 \in J$$

$$x^{-1} - 1 = -x^{-1}(x - 1) \in J,$$

and

$$z^{-1}xz - 1 = z^{-1}(x - 1)z \in J,$$

which imply that $G \cap (1 + J)$ is a normal subgroup of $G$. This subgroup is called the subgroup of $G$ determined by the ideal $J$ of $ZG$. When $J = I^n(G)$, $n \geq 1$, the subgroup $G \cap (1 + I^n(G)) = D_n(G)$ is called the $n$th integral dimension subgroup of $G$ and has been well studied during the last forty years—but we donot discuss this problem here (*cf*. Gupta, 1987; Gupta *et al.*, 1984; Gupta and Kuzmin-*unpublished*; Passi *et al.*, 1968, 1974, 1979, 1987, 1983; and Sandling 1972a & b; the list of references for dimension subgroups is by no means exhaustive).

Let $F$ be a free group and $R$ be a normal subgroup of $F$. For $n \geq 1$, the subgroup $F(n, R) = F \cap (1 + I^n(F)I(R))$ of $F$ is called the $n$th Fox subgroup of $F$ relative to $R$. The Fox subgroups of $F$ relative to a normal subgroup $R$ have been completely identified by I.A. Yunus (1984). Refer N. Gupta (1987) for details regarding the Fox subgroup problem.

Lately there has been interest in the identification of the subgroup $G \cap (1 + I^n(G)I(H))$ of an arbitrary group $G$ when $H$ is a normal subgroup of $G$. A related problem is the study of the relative dimension subgroups

$$D_n(G, H) = G \cap (1 + I^n(G) + I(G)I(H)) \text{ for } n \geq 1,$$

where $H$ is a normal subgroup of $G$. Both these problems are very closely related to the study of quotient groups $Q_n(G) = I^n(G)/I^{n+1}(G), n \geq 1$, and $I^n(G)I(H)/I^{n+1}(G)I(H), n \geq 1$, when $H$ is a normal subgroup of $G$.

## 2. Group $G \cap (1 + I^n(G) + I(G)I(H))$.

Let $G$ be a group and $H$ be a subgroup of $G$. Let $S$ be a left transversal of $H$ in $G$ with $e \in S$. Then every element of $G$ can be uniquely written as $sh, s \in S, h \in H$. Let $\theta : G \to H$ be the map defined by $\theta(sh) = h$. Extending $\theta$ by linearity to $ZG$ we get a $Z$-linear map $\theta : ZG \to ZH$. An application of the homomorphism $\theta$ gives the following simple but extremely useful observation of R. Sandling.

**Proposition 2.1. (Sandling, 1972b)** *If $J$ is a left ideal of $ZH$ contained in $I(H)$, then*

$$G \cap (1 + ZGJ) = H \cap (1 + J).$$

*The same argument, in fact, also gives*

**Proposition 2.2. (Ram Karan and Vermani, 1986)** *If $J$ is a left ideal of $ZH$ contained in $I(H)$, then*

$$G \cap (1 + I(G)J) = H \cap (1 + I(H)J).$$

**Corollary 2.3. (12)** $G \cap (1 + I(G)I^n(H)) = D_{n+1}(H)$ *for all* $n \geq 1$.

**Lemma 2.4.** *Let $G = HK$, where $H, K$ are subgroups of a group $G$ with $H$ normal in $G$. Then, for all $n \geq 2$,*

$$I^n(G)I(H) \subseteq I^{n+1}(H) + I(K)I(H) + \sum_{i=1}^{n-1} I([H, iK])I^{n-i}(H),$$

*where for $i \geq 1$, $[H, iK]$ is the commutator subgroup $[H, \underbrace{K, ..., K}_{i \text{ trems}}]$ of $G$.*

**Proposition 2.5.** *Let $G = HK$, where $H, K$ are subgroups of $G$ with $H$ normal in $G$. Then, for all $n \geq 2$,*

$$G \cap (1 + I^n(G)I(H)) \subseteq H \cap (1 + I^{n+1}(H) + I(H \cap K)I(K)$$
$$+ \textstyle\sum_{i=1}^{n-1} I([H, iK])I^{n-i}(H)).$$

*In particular, if $H \cap K \subseteq [H, K]$, then*

$$G \cap (1 + I^2(G)I(H)) = H \cap (1 + I^3(H) + I([H, K])I(H)).$$

Thus the study of the subgroups $G \cap (1 + I^n(G)I(H))$ is very much related to and influenced by the study of the relative dimension subgroups $G \cap (1 + I^n(G) + I(G)I(H))$ of $G$, $H$ a normal subgroup of $G$.

For any subgroup $H$ of $G$,

2.6. $\qquad\qquad \gamma_2(H)\gamma_3(G) \subseteq G \cap (1 + I^3(G) + I(G)I(H)),$

where, for any group $M$ and integer $n \geq 1$, $\gamma_n(M)$ denotes the $n$th term in the lower central series of $M$.

Equality holds in (2.6) if

(a) $H$ is normal and $G/H$ is cyclic (I.B.S. Passi, 1974; Passi, in fact, proves this result with 3 replaced by any $n \geq 3$ on both sides of this relation);

(b) $[H, G] \subseteq \gamma_3(G)$ (Passi and Sharma, 1974; also cf. Sandling, 1972a);

(c) $G = HK$, where $H$, $K$ are subgroups of $G$ with $H$ normal in $G$ and $H \cap K$ a central subgroup of $G$ (Ram Karan and L.R. Vermani, 1988);

(d) $HG'/G'$ is divisible (we write $G'$ for the derived group $\gamma_2(G)$ of $G$);

(e) One of the subgroups $G/H'\gamma_3(G)$, $[G, H]\gamma_3(G)/H'\gamma_3(G)$, $G/HG'$ is torsion-free.

(One part of (e) is a result of Ram Karan and Vermani, 1988, while (d) and the other two parts of (e) follow from a result of M. Hartl (Theorem 2.8) to follow).

*2.7. Equality Does Not Hold in (2.6) Always:* For example M. Hartl (*Preprint*): Let $p$ be a prime and $r > 2$. The group

$$G = \langle x, y | x^{p^r} = y^{p^r} = [x, y, x] = [x, y, y] = 1 \rangle$$

is nilpotent of class 2. For some $s$, $0 < s < r - 1$, consider the subgroup

$$H = \langle x^{p^s}, y^{p^{r-1}}, [x, y] \rangle$$

of $G$. Since $[x, y]$ is a central element of $G$ contained in $H$, modulo $I^3(G) + I(G)I(H)$,

$$
\begin{aligned}
[x, y]^{p^{r-1}} - 1 &\equiv p^{r-1}([x, y] - 1) \\
&\equiv p^{r-1}((x - 1)(y - 1) - (y - 1)(x - 1)) \\
&\equiv (x - 1)(y^{p^{r-1}} - 1) - p^{r-s-1}(y - 1)(x^{p^s} - 1) \\
&\equiv 0.
\end{aligned}
$$

Therefore, $[x, y]^{p^{r-1}} \in G \cap (1 + I^3(G) + I(G)I(H))$ but $[x, y]^{p^{r-1}}$ has order $p$ modulo $H'\gamma_3(G)$.

The structure of the subgroup $G \cap (1 + I^3(G) + I(G)I(H))$ for any normal subgroup $H$ of $G$ has been determined by Hartl using homological methods.

**Theorem 2.8. (Hartl, *Preprint*)** *Let G be a group and H be a normal subgroup of G. Then*

$$G \cap (1 + I^3(G) + I(G)I(H)) = \gamma_3(G)V,$$

*where* $V = \langle [x^m, y] \mid x^m, y^m \in HG' \text{ for some } m \geq 1, x, y \in G \rangle.$

An independent proof of this result, using free group ring techniques, has very recently been obtained by K. Tahara, Vermani and A. Razdan (1998).
    Passi (1974) proved.

**Theorem 2.9.** *If G is the semidirect product H|K of a normal subgroup H by a subgroup K, then*

$$G \cap (1 + I^n(G) + I(G)I(H)) = H'D_n(K)[H, (n-1)G].$$

**Theorem 2.10.** *If H is a central subgroup of G, then*

$$G \cap (1 + I^4(G) + I(G)I(H)) = \gamma_4(G),$$

*provided G is a finite p-group, p an odd prime.*

Ram Karan and Vermani are able to say a little more regarding Theorem 2.10.

**Theorem 2.11. (Ram Karan and Vermani, 1988)** *Let G be a finite p-group of odd order with normal subgroups H and K such that G = HK and H ∩ K is contained in the centre of G. Then*

$$G \cap (1 + I^4(G) + I(G)I(H)) = H'\gamma_4(G).$$

Passi (1974) proved that
2.12. If $H$ is normal subgroup of a group $G$ with $G/H$ cyclic, then

$$G \cap (1 + I^n(G) + I(G)I(H)) = H'\gamma_n(G).$$

Let $G$ be a group with a normal subgroup $H$ such that $G/H$ is free Abelian. Let $X = \{x_\delta \mid \delta \in \Delta\}$ be a set of elements of $G$ such that $\{x_\delta H \mid \delta \in \Delta\}$ is a basis of $G/H$. We may suppose that the index set $\Delta$ is well ordered. We can then choose a set $S$ of representatives of $H$ in $G$ consisting of elements of the form

$$x_{\delta_1}^{t_1} \ldots x_{\delta_n}^{t_n}, t_i \in Z, n \geq 1, \delta_1 < \delta_2 < \ldots < \delta_n.$$

Let $S_n$ be the $Z$-submodule of $I(G)$ generated by elements of the form

$$(x_{\delta_1}^{\epsilon_1} - 1) \ldots (x_{\delta_n}^{\epsilon_n} - 1), \epsilon_i = 1 \text{ or } -1, \text{ for every } i \text{ and } \delta_1 \leq \delta_2 \leq \ldots \leq \delta_n.$$

For $m \geq 2$, let $S^{(m)} = \sum_{n \geq m} S_n.$ Then

**Theorem 2.13. (Vermani et al., 1993)** *For* $n \geq 2$,

$$I^n(G) = I^{n-1}(G)I(H) + \sum_{i=2}^{n-1} I^{n-i}(G)I(\gamma_i(G)) + I(\gamma_n(G)) \oplus S^{(n)}.$$

*Using the above decomposition of* $I^n(G)$ *Vermani, Razdan and Ram Karan prove*

**Theorem 2.14.** *Let G be a group with a normal subgroup H such that G/H is free Abelian. Then*

$$G \cap (1 + I^n(G) + I(G)I(H)) = H'\gamma_n(G) \text{ for all } n \geq 1.$$

*The decomposition of* $I^n(G)$ *as in Theorem 2.13 has been extensively used by Vermani and Razdan to obtain subgroups determined by certain ideals which we come to in the next section.*

## 3. Some Intersection Theorems

For studying the subgroups $G \cap (1+I^n(G)+I(G)I(H))$ and $G \cap (1+I^n(G)I(H))$ when $H$ is a normal subgroup of a group $G$, we are, in general, led to consider intersections of certain ideals of $ZG$ contained in $I(G)$. A few such intersections have been obtained by Ram Karan, Vermani and Razdan. For example, Ram Karan and Vermani (1989, 1990) prove that

3.1. (*i*) if $G$ is a group with subgroups $H, K$ such that $HK$ is a subgroup of $G$, then
$$ZH \cap I(K)I(H) = I(H \cap K)I(H);$$

(*ii*) if $H$ is a normal subgroup and $K$ a subgroup of $G$ with $G = HK$, $H \cap K \subseteq [H, K]$, then
$$I^2(H) \cap I^2(G)I(H) = I^3(H) + I([H, K])I(H);$$

(*iii*) if $G$ is Abelian and $H$ is a subgroup of $G$, then
$$ZH \cap I^2(G) = I^2(H);$$

(*iv*) if $H$ is a central subgroup of $G$ with $G/H$ Abelian, then
$$I(H) \cap (I(H)I(G) + I^3(G)) = I^2(H);$$

(*v*) (a) $I(G') \cap I^3(G) = I^2(G') + I(\gamma_3(G));$
(b) $ZGI(G') \cap I^3(G) = I(G)I(G') + I(\gamma_3(G)).$

Let $p$ be a given prime and let $c$, $n$ be natural numbers. B. Hartley (1982) proves:

**Theorem 3.2.** *There exists a function $\phi\,(a)$ depending on $p, c$ and $n$ such that if $G = HK$ is a nilpotent group of class atmost $c$, where $H$ is a normal subgroup of $G$ and $K$ has finite exponent dividing $p^n$, then*

$$I^{\phi(a)}(G) \cap ZH \subseteq I^a(H) \text{ for all } a \geq 1.$$

When $H$ is a normal subgroup of $G$ with $G/H$ torsion-free, Hartley has another intersection theorem.

**Theorem 3.3. (Hartley, 1982)** *If $G$ is a nilpotent group of class $n$, $H$ is a normal subgroup of $G$ such that $G/H$ is torsion-free and $\phi(b, n) = (b-1)\, n\, (n+1)+1$, then*

$$I^{\phi(b,n)}(G) \cap ZH \subseteq I^b(H) \text{ for all } n \geq 1.$$

This result of Hartley is useful in the study of intersections of all integral powers of the augmentation ideal and has also been used by Passi and Vermani (1983) and by Passi, Sucheta and Tahara (1987) for studying the subgroup $P_n H^2(G, T)$ of the Schur multiplicator $H^2(G, T)$ of $G$.

Using the decomposition of $I^n(G)$ as in Theorem 2.13, Vermani and Razdan prove

**Theorem 3.4. (Vermani and Razdan, 1995)** *Let $G$ be a group and $H$ be a normal subgroup of $G$ such that $G/H$ is free Abelian. Then, for $n \geq 2$,*

$$I^n(G) \cap ZGI(H) = I^{n-1}(G)I(H) + \sum_{i=2}^{n-1} I^{n-i}(G)I(\gamma_i(G)) + I(H \cap \gamma_n(G)).$$

For free groups we have

**Theorem 3.5. (Vermani and Razdan, 1995)** *If $F$ is a free group with normal subgroups $R$ and $S$ and $A = R \cap S$, then for all $n \geq 1$,*

$$I^n(R)I(S)I(F) \cap I^n(R)I(A) = I^n(R)I(A)I(R) + I^n(R)I(R \cap S').$$

*In Particular, for all $n \geq 1$,*

$$I^n(R)I^2(F) \cap I^{n+1}(R) = I^{n+2}(R) + I^n(R)I(R \cap F').$$

**Proposition 3.6. (Vermani and Razdan, 1995)** *For all $n \geq 1$,*

$$I(R)I^n(F) \cap I^{n+2}(F) = I(R)I^{n+1}(F) + I(R \cap F')I^n(F).$$

**Theorem 3.7. (Vermani and Razdan, 1995)** *If the normal subgroup $R$ of a free group $F$ is such that $F \cap (1 + I^k(F) + I(F)I(R)) = R'\gamma_k(F)$ for all $k \geq 1$, then for all $m \geq n + 3$ and $n \geq 1$,*

$$I(R)I^n(F) \cap I^m(F) = \sum_{i=1}^{m-n} I(R \cap \gamma_i(F))I^{m-i}(F).$$

**Corollary 3.8. (Vermani and Razdan, 1995)** *For $n \geq 1, m \geq n + c + 1$ and $c \geq 2$,*

$$I(\gamma_c(F))I^n(F) \cap I^m(F) = \sum_{i=c}^{m-n} I(\gamma_i(F))I^{m-i}(F).$$

## 4. The subgroup $G \cap (1 + I^2(G)I(H))$

Khambadkone (1985) gives the following

**Theorem 4.1.** *If $G$ is the semi direct product $H|K$ of a finitely generated nilpotent normal subgroup $H$ of class atmost 2 by a subgroup $K$, and if $\bar{x}_1, \ldots, \bar{x}_m$ is a basis of $H/H'$ adapted to $[H, K]$ i.e., $[H, K]$ is generated modulo $H'$ by $x_1^{e_1}, \ldots, x_m^{e_m}$ where $e_i$ are non-negative integers with $e_1 \mid e_2 \mid \ldots \mid e_m$, then*

$$G \cap (1 + I^2(G)I(H)) = \langle [x_i, x_j^{e_j}] \mid i < j \rangle.$$

Ram Karan and Vermani (1989) prove

**Proposition 4.2.** *Let $G = HK$ where $H, K$ are both normal subgroups of $G$ with $H \cap K$ a central subgroup of $H$. Then*

$$G \cap (1 + I^2(G)I(H)) = \gamma_3(H).$$

The argument leading to the above result also shows that if the subgroup $H$ is in addition a finite $p$-group, $p$ an odd prime, then

(4.3) $$G \cap (1 + I^3(G)I(H)) = \gamma_4(H).$$

Vermani, Razdan and Ram Karan (1993) prove

**Proposition 4.4.** *Let $G$ be a group and $H, K$ be subgroups of $G$ with $H$ normal in $G$, $G = HK$, $H \cap K \subseteq [H, K]$ and $H/[H, K]$ free Abelian. Then*

$$G \cap (1 + I^2(G)I(H)) = \gamma_2(H \cap G')\gamma_3(H).$$

With the result of Hartl (Theorem 2.8) now being available, we can improve upon Propositions 4.2 and 4.4 as follows:

**Proposition 4.5.** *Let $G = HK$, where $H, K$ are subgroups of $G$ with $H$ normal in $G$ and $H \cap K \subseteq [H, K]$. Then*

$$G \cap (1 + I^2(G)I(H)) = \gamma_3(H)V,$$

*where $V = \langle [x^m, y] \mid x^m, y^m \in [H, K] \text{ for some } m \geq 1, x, y \in H \rangle$.*

A variation of this result has recently been obtained by Tahara, Vermani and Razdan:

**Theorem 4.6. (Tahara et al., 1998)** *Let $G = HK$ where $H, K$ are subgroups of $G$ such that $H$ is normal in $G$ and $H \cap K \subseteq G'$. Then*

$$G \cap (1 + I^2(G)I(H)) = \gamma_3(H)V,$$

*where $V = \langle [x^m, y] \mid x^m, y^m \in G' \text{ for some } m \geq 1, x, y \in H \rangle$.*

**Proof.**

$$I^2(G) \quad = \quad I^3(H) + I([H, K])I(H) + I(K)I^2(H) + I^2(K)I(H)$$
$$\subseteq \quad I^3(H) + I([H, K])I(H) + I(K)I(H).$$

Therefore,

$$G \cap (1 + I^2(G)I(H)) \quad = \quad H \cap (1 + I^2(G)I(H))$$
$$\subseteq \quad H \cap (1 + I^3(H) + I([H, K])I(H) + I(K)I(H))$$
$$\subseteq \quad H \cap (1 + I^3(H) + I(H \cap G')I(H)).$$

For the last step we need to use the right $ZH$-module homomorphism: $ZG \to ZH$ which is the extension of the map $sh \to h, s \in S, h \in H$, where $S$ is a set of coset representatives of $H \cap K$ in $K$ and, hence, also of $H$ in $G$. The reverse inclusion being trivially true, we have

$$G \cap (1 + I^2(G)I(H)) = H \cap (1 + I^3(H) + I(H \cap G')I(H)).$$

The result then follows from Theorem 2.8.

Observe that the hypothesis of the theorem is satisfied if the exact sequence

$$1 \to HG'/G' \to G/G' \to G/HG' \to 1$$

splits. This is true, for example, in the following cases:

(a) if $G/H$ is free Abelian;

(b) if $H$ is a divisible subgroup of $G$; and

(c) if $H$ splits over $G'$.

For a finitely generated group $G$ with normal subgroup $H$, the subgroup $G \cap (1 + I^2(G)I(H))$ has been identified by Curzio and Gupta (1995). Hartl has recently considered a unification of the third relative dimension subgroup and the second Fox subgroup for arbitrary groups. He proves

**Theorem 4.7. (Hartl, *Preprint*)** *Let $G$ be a group and $H$, $K$ be subgroups of $G$. Then*

$$G \cap (1 + I(K)I(H) + I^2(G)I(H))$$
$$= \langle \pi[h, k]^{a_{hk}} | a_{hk} \in Z \text{ and for all } k \in H \text{ there exists } d_k \geq 0$$
$$\text{such that } k^{d_k} \in H' \text{and } \underset{h \in H}{\pi} h^{a_{hk} - a_{kh}} \in KG'G^{d_k} \rangle.$$

Observe that if $H = G$, the above result gives the third relative dimension subgroup of $G$ relative to $K$ while if $K = \{1\}$, it gives the second Fox subgroup of $G$ relative to the subgroup $H$.

## 5. Subgroups determined by Some Other Ideals

Bergman and Dicks prove

**Theorem 5.1. (Bergman and Dicks, 1975)** *If $H$ is a normal subgroup of $G$ and $K$ is any subgroup of $G$, then*

$$G \cap (1 + ZGI(H)ZGI(K) = \gamma_2(H \cap K).$$

When $H$ is not necessarily normal, we have

**Theorem 5.2. (Ram Karan and Vermani, 1988)** *If $H, K$ are subgroups of a group $G$, then*

(i) $G \cap (1 + ZGI(H)ZGI(K)) = \gamma_2(H^G \cap K)$; and

(ii) $G \cap (1 + ZGI(H)I(K)) = \gamma_2(H^K \cap K)$;

*where $H^G$ denotes the normal closure of $H$ in $G$ and $H^K$ is the subgroup of $G$ generated by all elements $k^{-1}hk, k \in K, h \in H$.*

**Proposition 5.3. (Vermani, 1990)** *Let $G$ be a group and $H, K$ be normal subgroups of $G$ with $G = HK$. Then*

$$G \cap (1 + I(K)I^n(H) + I(H)I(K)) = [K, nH]\gamma_2(H \cap K)$$

*for all $n \geq 1$.*

Taking $K = G$ in this result gives the following unpublished result of Passi.

**Corollary 5.4.** *If $H$ is a normal subgroup of $G$, then*

$$G \cap (1 + I(H)I^n(G) + I(G)I(H)) = [H, nG]H'$$

*for all $n \geq 1$.*

If $F$ is a free group and $R, S$ are normal subgroups of $F$, then C.K. Gupta (1978, 1983) (also cf. Gupta, 1987) proved

5.5. (i) $F \cap (1 + I(R)I(F)I(S)) = \gamma_2(R' \cap S)\gamma_2(R \cap S')[R' \cap S', R \cap S]$;

(ii) $F \cap (1 + I^2(R)I^2(F)) = \gamma_3(R \cap F')\gamma_4(R)$;

(iii) $F \cap (1 + I(R)I^2(F)I(R)) = [\gamma_2(R \cap F'), R]\gamma_4(R)$;

while Stohr (Gupta, 1987) proved

(iv) $F \cap (1 + I(F)I^n(R)I(F)) = \sqrt{[\gamma_{n+1}(R), F]}$ for all $n \geq 1$,

where for a subgroup $H$ of a group $G$, $\sqrt{H} = \{x \in G | x^m \in H \text{ for some } m \geq 1\}$. Vermani, Razdan and Ram Karan prove

**Proposition 5.6. (Vermani *et al.*, 1993)** *If H is a normal subgroup of a group G, then*

(a) $G \cap (1 + I^2(G)I(H) + I(G)I(H)I(G) + I(H)I^2(G)) = [H, G, G]$, *if* $H' \leq [H, G, G]$; *and*

(b) $G \cap (1 + I(G)I^2(H) + I(H)I(G)I(H) + I^2(H)I(G)) = [G, H, H]$.

For free groups we have

**Theorem 5.7. (Vermani and Razdan, 1996)** *Let F be a free group and R be a normal subgroup of F. Then for any $n \geq 1$,*

$$F \cap \left(1 + \sum_{i+j=n} I^i(R)I(F)I^j(R)\right) = [F, nR].$$

*Using the decomposition of $I^n(G)$ as in Theorem 2.13, Vermani and Razdan prove*

**Theorem 5.8. (Vermani and Razdan, 1996)** *Let G be a group with a normal subgroup H such that $G/H$ is free Abelian. Then*

(a) $G \cap (1 + I^n(G) + I^2(G)I(H) + I(G)I(H)I(G) + I(H)I^2(G)) = [H, G, G] \gamma_n(G)$;

(b) $G \cap (1 + I^n(G) + I^2(G)I(H) + I(G)I(H)I(G)) = [G, H, H]\gamma_n(G)V$;

(c) $G \cap (1 + I^n(G) + I^2(G)I(H)) = \gamma_n(G)\gamma_3(H)V$;

   *where $V = \langle [x^m, y] \mid x^m, y^m \in G'$ for some $m \geq 1, x, y \in H \rangle$; and*

(d) $G \cap (1 + I^n(G) + I(G)I^2(H) + I(H)I(G)I(H) + I^2(H)I(G)) = [G, H, H] \gamma_n(G)$;

   *provided $H/G'$ is also assumed free Abelian when $n = 4$.*

Let $H$ be a normal subgroup of a group $G$. For $i \geq 1$, define

$$K_i = \begin{cases} \gamma_{(i+1)/2}(G) & \text{if } i \text{ is odd,} \\ \gamma_{i/2}(H)\gamma_{(i+2)/2}(G) & \text{if } i \text{ is even, and} \end{cases}$$

$W_i = \langle [x^m, y] \mid x^m, y^m \in K_{i+1}$ for some $m \geq 1, x, y \in K_i \rangle$.
We then have the following

**Theorem 5.9. (Vermani and Razdan, 1996)** *If the quotient groups $K_i/K_{i+1}$ are free Abelian for $1 \leq i \leq n - 1$, then*

$$G \cap (I + I^{n+2}(G) + I^n(G)I(H)) = \gamma_{n+1}(H)\gamma_{n+2}(G)W_n$$

*for $n = 3, 4$ and $5$.*

We have now been able to prove the above result for $n = 6$ as well. Since the computations involved are very tedius and the original paper does not contain a proof for the case $n = 5$, we include here partial details of the proof for the case $n = 6$.

**Theorem 5.10.** *If the quotient groups $K_i/K_{i+1}$ are free Abelian for $1 \leq i \leq 5$, then*

$$G \cap (1 + I^8(G) + I^6(G)I(H)) = \gamma_7(H)\gamma_8(G)W_6.$$

**Proof. (outline)** Let $G \cap (1 + I^8(G) + I^6(G)I(H)) = X$ (say).

Let $x, y \in \gamma_3(H)\gamma_4(G)$ such that $x^m, y^m \in \gamma_4(G)$ for some $m \geq 1$. Then, modulo $I^3(\gamma_3(H)\gamma_4(G))$, we have

$$
\begin{aligned}
[x^m, y] - 1 &\equiv (x^m - 1)(y - 1) - (y - 1)(x^m - 1) \\
&\equiv (x^m - 1)(y - 1) - (y^m - 1)(x - 1)
\end{aligned}
$$

and, as $I(\gamma_4(G))I(\gamma_3(H)\gamma_4(G)) \subseteq I^8(G) + I^6(G)I(H)$, it follows that $W_6 \subseteq X$. That $\gamma_7(H)\gamma_8(G) \subseteq X$ is clear and so $\gamma_7(H)\gamma_8(G)W_6 \subseteq X$.

We now proceed to prove the reverse inclusion.

Let $\tau^1 : ZG \to ZH$ be the linear extension of the map from $G$ to $H$ defined by mapping $g = sh$ to the element $h$, $h \in H$, $s \in S(= $ a left transversal of $H$ in $G)$. Then $\tau^1$ is a homomorphism of right $ZH$-modules. We may write $\tau^r$ if a right transversal of $H$ in $G$ is chosen (so that $\tau^r$ is a left $ZH$-homomorphism). By our choice of left transversal of $H$ in $G$ made just before Theorem 2.13 and with $S^{(n)}$ as defined there we find that $\tau^1 \mid S^{(n)} = 0$. Similarly, $\tau^r \mid S^{(n)} = 0$ in the other case. Observe that $\tau^1$ (or $\tau^r$) is the identity map on $I(H)$.

Using Theorem 2.13 a number of times, we find that

$$I^8(G) + I^6(G)I(H)$$
$$= I(G)I^6(H) + I(G')I^5(H) + I(G)I(G')I^4(H) + I(\gamma_3(G))I^4(H)$$
$$\quad + I^2(G')I^3(H) + I(G)I(\gamma_3(G))I^3(H) + I(\gamma_4(G))I^3(H)$$
$$\quad + I(G)I^2(H)I(G')I^2(H) + I(G')I(H)I(G')I^2(H) + I(G)I^2(G')I^2(H)$$
$$\quad + I(\gamma_3(G))I(G')I^2(H) + I(G')I(\gamma_3(G))I^2(H) + I(G)I(\gamma_4(G))I^2(H)$$
$$\quad + I(\gamma_5(G))I^2(H) + I(G)I^3(H)I(G')I(H) + I(G')I^2(H)I(G')I(H)$$
$$\quad + I(G)I(G')I(H)I(G')I(H) + I(\gamma_3(G))I(H)I(G')I(H)$$
$$\quad + I(G)I(H)I^2(G')I(H) + I^3(G')I(H) + I(G)I(\gamma_3(G))I(G')I(H)$$
$$\quad + I(\gamma_4(G))I(G')I(H) + I(G)I^2(H)I(\gamma_3(G))I(H)$$
$$\quad + I(G')I(H)I(\gamma_3(G))I(H) + I^2(\gamma_3(G))I(H)$$
$$\quad + I(G')I(\gamma_4(G))I(H) + I(G)I(\gamma_5(G))I(H) + I(\gamma_6(G))I(H)$$

$$+I(G)I^4(H)I(\gamma_3(G)) + I(G')I^3(H)I(\gamma_3(G))$$

$$+I(G)I(G')I^2(H)I(\gamma_3(G)) + I(\gamma_3(G))I^2(H)I(\gamma_3(G))$$

$$+I(G)I(H)I(G')I(H)I(\gamma_3(G)) + I^2(G'))I(H)I(\gamma_3(G))$$

$$+I(G)I(\gamma_3(G))I(H)I(\gamma_3(G)) + I(G)I^2(H)I(G')I(\gamma_3(G))$$

$$+I(G')I(H)I(G')I(\gamma_3(G)) + I(G)I^2(G')I(\gamma_3(G))$$

$$+I(\gamma_3(G))I(G')I(\gamma_3(G)) + I(G)I(H)I^2(\gamma_3(G)) + I(G')I^2(\gamma_3(G))$$

$$+I(G)I(\gamma_4(G))I(\gamma_3(G)) + I(\gamma_5(G))I(\gamma_3(G))$$

$$+I(G)I(H)I(G'))I(\gamma_4(G)) + I^2(G'))I(\gamma_4(G)) + I(G)I^3(H)I(\gamma_4(G))$$

$$+I(G')I^2(H)I(\gamma_4(G)) + I(\gamma_3(G))I(H)I(\gamma_4(G)) + I(G)I(\gamma_7(G))$$

$$+I(\gamma_8(G)) + I(\gamma_4(G))I(H)I(\gamma_3(G)) + \sum_{i=2}^{8} S^{(i)} A_i,$$

where $A_i$ are certain ideals contained in $I(H)$.

Applying the right $ZH$-homomorphism $\tau^1 : ZG \to ZH$ which vanishes on $S^{(m)}$, we get (after certain simplifications)

$$\begin{aligned}
X = \ & G' \cap (1 + I^7(H) + I(G')I^5(H) + I(\gamma_3(G))I^4(H) + I^2(G')I^3(H) \\
& + I(\gamma_4(G))I^3(H) + I(\gamma_3(G))I(G')I^2(H) + I(\gamma_5(G))I^2(H) + I^3(G')I(H) \\
& + I^2(\gamma_3(G))I(H) + I(G')I(\gamma_4(G))I(H) + I(\gamma_6(G))I(H) \\
& + I(\gamma_5(G))I(\gamma_3(G)) + I^2(\gamma_4(G)) + I(\gamma_8(G)).
\end{aligned}$$

Since $H/G'$ is free Abelian, decomposing powers of $I(H)$ and applying the left $ZG'$-homomorphism $\tau^r : ZH \to ZG'$, we get

$$\begin{aligned}
X = \ & G' \cap (1 + I(\gamma_7(H)) + I(G')I(\gamma_5(H)) + I(\gamma_3(G))I(\gamma_4(H)) \\
& + I^2(G')I(\gamma_3(H)) + I(\gamma_4(G))I(\gamma_3(H)) + I(\gamma_3(G))I(G'))I(H') \\
& + I(\gamma_5(G))I(H') + I^4(G') + I^2(\gamma_3(G))I(G') \\
& + I(G'))I(\gamma_4(G))I(G') + I(\gamma_6(G))I(G') \\
& + I(\gamma_5(G))I(\gamma_3(G)) + I^2(\gamma_4(G)) + I(\gamma_8(G))).
\end{aligned}$$

The group $G'/H'\gamma_3(G)$ being free Abelian, we can decompose powers of $I(G')$ using Theorem 2.13 and applying the right $Z(H'\gamma_3(G))$-homomorphism $\tau^1 : ZG' \to Z(H'\gamma_3(G))$ we get

$$\begin{aligned}
X = \ & H'\gamma_3(G) \cap (1 + I(\gamma_7(H)) + I(H'\gamma_3(G))I(\gamma_5(H)) \\
& + I(\gamma_3(G))I(\gamma_4(G)) + I^2(H'\gamma_3(G))I(\gamma_3(H))
\end{aligned}$$

$$+I(\gamma_4(G))I(\gamma_3(H)) + I(H'\gamma_3(G))I(\gamma_3(G))I(H')$$
$$+I(\gamma_5(G))I(H') + I^4(H'\gamma_3(G)) + I(H'\gamma_3(G))I^2(\gamma_3(G))$$
$$+I^2(H'\gamma_3(G))I(\gamma_4(G)) + I(H'\gamma_3(G))I(\gamma_6(G))$$
$$+I(\gamma_5(G))I(\gamma_3(G)) + I^2(\gamma_4(G)) + I(\gamma_8(G))).$$

Next, we expand powers of $I(H'\gamma_3(G))$ using Theorem 2.13 the group $H'\gamma_3(G)/\gamma_3(G)$ being free Abelian and apply the right $Z(\gamma_3(G))$-homomorphism $\tau^1$ : $Z(H'\gamma_3(G)) \to Z(\gamma_3(G))$. We get

$$X = H'\gamma_3(G) \cap (1 + \gamma_7(H) + I(\gamma_3(G))I(\gamma_4(H))$$
$$+I(\gamma_4(G))I(\gamma_3(H)) + I^3(\gamma_3(G)) + I(\gamma_5(G))I(\gamma_3(G))$$
$$+I^2(\gamma_4(G)) + I(\gamma_8(G))).$$

We next expand $I^3(\gamma_3(G))$ the group $\gamma_3(G)/\gamma_3(H)\gamma_4(G)$ being free Abelian and apply the right $Z(\gamma_3(H)\gamma_4(G))$-homomorphism $\tau^1$ : $Z(\gamma_3(G)) \to Z(\gamma_3(H)\gamma_4(G))$. This leads us to

$$\begin{aligned}
X &= \gamma_3(H)\gamma_4(G) \cap (1 + I(\gamma_7(H)) + I(\gamma_3(H)\gamma_4(G))I(\gamma_4(H)) \\
&\quad + I(\gamma_4(G))I(\gamma_3(H)) + I^3(\gamma_3(H)\gamma_4(G)) + I(\gamma_3(H)\gamma_4(G))I(\gamma_5(G)) \\
&\quad + I^2(\gamma_4(G)) + I(\gamma_8(G)) \\
&= \gamma_3(H)\gamma_4(G) \cap (1 + I(\gamma_7(H)) + I(\gamma_3(H)\gamma_4(G))I(\gamma_4(H)) \\
&\quad + I(\gamma_4(G))I(\gamma_3(H)) + I^2(\gamma_4(G)) + I^2(\gamma_4(G))I(\gamma_3(H)) \\
&\quad + I^3(\gamma_3(H)\gamma_4(G)) + I(\gamma_3(H)\gamma_4(G))I(\gamma_5(G)) + I(\gamma_8(G))). \\
&= \gamma_3(H)\gamma_4(G) \cap (1 + I(\gamma_7(H)) + I(\gamma_3(H)\gamma_4(G))I(\gamma_4(H)) \\
&\quad + I(\gamma_4(G))I(\gamma_3(H)\gamma_4(G)) + I^3(\gamma_3(H)\gamma_4(G)) \\
&\quad + I(\gamma_3(H)\gamma_4(G))I(\gamma_5(G)) + I(\gamma_8(G))). \\
&= \gamma_3(H)\gamma_4(G) \cap (1 + I(\gamma_7(H)) + I(\gamma_4(G))I(\gamma_3(H)\gamma_4(G)) \\
&\quad + I^3(\gamma_3(H)\gamma_4(G)) + I(\gamma_8(G))). \\
&= \gamma_7(H)\gamma_8(G)\{\gamma_3(H)\gamma_4(G) \cap (1 + I(\gamma_4(G))I(\gamma_3(H)\gamma_4(G)) \\
&\quad + I^3(\gamma_3(H)\gamma_4(G))\} \\
&= \gamma_7(H)\gamma_8(G)W_6,
\end{aligned}$$

by Theorem 2.8. This completes the proof.

Using the results of Theorems 5.9 and 5.10, we are in a position to improve the result 5.5($ii$) of Gupta as follows.

**Theorem 5.11.** *Let $F$ be a free group and $R$ be a normal subgroup of $F$ Then*

(a) *(Vermani and Razdan, 1996)* $F \cap (1 + I^n(R)I^2(F)) = \gamma_{n+1}(R \cap F')\gamma_{n+2}(R)$, *for n = 3 and 4;*

(b) *(Vermani and Razdan, 1996)* $F \cap (1 + I^5(R)I^2(F)) = \gamma_6(R \cap F')\gamma_7(R)V$, *where* $V = \langle [x^m, y] \mid x^m, y^m \in \gamma_3(R \cap F')\gamma_4(R)$ *for some* $m \geq 1, x, y \in \gamma_3(R) \rangle$; *and*

(c) *if the quotient group* $\gamma_3(R)/\gamma_3(R \cap F')\gamma_4(R)$ *is free Abelian,* $F \cap (1 + I^6(R)I^2(F)) = \gamma_7(R \cap F')\gamma_8(R)W$, *where* $W = \langle [x^m, y] \mid x^m, y^m \in \gamma_4(R \cap F')$ *for some* $m \geq 1, x, y \in \gamma_3(R \cap F')\gamma_4(R) \rangle$.

Identification of the subgroup $G \cap (1 + I(H)I(G)I(H))$ of a group $G$ where $H$ is any normal subgroup of $G$ corresponding to the result 5.5 $(i)$ of Gupta for free groups (the case $S = R$) is still evading a solution although we have the following variant of it.

**Proposition 5.12. (Tahara *et al.*, 1998)** *If $H$ is a normal subgroup of a group $G$, then* $G \cap (1 + I(H)I(G)I(H) + I([H, G])I(H)) = \gamma_3(H)V$, *where* $V = \langle [x^m, y] \mid x^m, y^m \in [H, G]$ *for some* $m \geq 1, x, y \in H \rangle$.

Subgroups determined by some other ideals of $ZF$, $F$ a free group, have been obtained by Kuzmin and Stöhr but our main interest being in the subgroups determined by ideals in $ZG$, $G$ an arbitrary group, the same have not been touched upon here. The interested reader may refer to Gupta (1987) for details.

The subgroup $G \cap (1 + I(H)I(G)I(H))$ for any subgroup $H$ of a group $G$ has since been identified by Vermani (unpublished).

## Acknowledgement

The work is partly supported by the National Board for Higher Mathematics.

## References

Bergman, G.M. and Dicks, W. On universal derivations, *J. Alg.*, **36**, 193–211, 1975.

Curzio, M. and Gupta, C.K. Second Fox subgroup of arbitrary groups, *Canad. Math. Bull.*, **38**, 177–181, 1995.

Gupta, C.K. Subgroups of free groups induced by certain products, *Comm. Alg.*, **6**, 1231–1238, 1978.

Gupta, C.K. Subgroups induced by certain ideals of free group rings, *Comm. Alg.*, **11**(22), 2519–2525, 1983.

Gupta, N. Free Group Rings, *Contemporary Math.* **66**, *Amer. Math. Soc.*, 1987.

Gupta, N. Hales, A.W. and Passi, I.B.S. Dimension subgroups of matabelian groups, *J. reine. angew. Math.*, **346**, 194–198, 1984.

Gupta, N. and Kuzmin, Y. On varietal quotients defined by ideals generated by Fox derivatives, *J. pure appl. Alg.*, *(to appear)*.

Hartl, M. Some successive quotients of group ring filtrations induced by N-series, *Comm. Alg.*, **23**(10), 3831–3853, 1995.

Hartl, M. Polynomiality properties of group extensions with torsion-free abelian kernel, *J. Alg.*, **179**, 380–415, 1996.

Hartl, M. The third relative Fox dimension subgroup (*Preprint*).

Hartley, B. Powers of the augmentation ideal in group rings of infinite nilpotent groups, *J. London Math. Soc.*, **25**(2), 43–61, 1982.

Hartley, B. An intersection theorem for powers of the augmentation ideal in group rings of certain nilpotent $p$-groups, *J. London Math. Soc.*, **25**(2), 425–434, 1982.

Ram Karan and Vermani, L.R. A note on augmentation quotients, *Bull. London Math. Soc.*, **18**, 5–6, 1986.

Ram Karan and Vermani, L.R. A note on polynomial maps, *J. pure appl. Alg.*, **51**, 169–173, 1988.

Ram Karan and Vermani, L.R. Augmentation quotients of integral group rings, *J. Indian Math. Soc.*, **54**, 107–120, 1989.

Ram Karan and Vermani, L.R. Augmentation quotients of integral group rings II, *J. pure appl. Alg.*, **65**, 253–262, 1990; A corrigendum: *J.pure appl. Alg.*, **77**, 229–230, 1992.

Ram Karan and Vermani, L.R. Augmentation quotients of integral group rings III, *J. Indian Math Soc.*, **58**, 19–32, 1992; A corrigendum: *J. Indian Math. Soc.*, **59**, 261–262, 1993.

Khambadkone, M. Subgroup ideals in group rings I, *J. pure appl. Alg.*, **30**, 261–276, 1983.

Khambadkone, M. On the structure of augmentation ideals in group rings, *J. pure appl. Alg.*, **35**, 35–45, 1985.

Magnus, W. On a theorem of Marshal Hall, *Ann. Math.*, **40**, 764–776, 1939.

Passi, I.B.S. Dimension subgroups, *J. Alg.*, **9**, 152–182, 1968.

Passi, I.B.S. Polynomial maps on groups II, *Math. Z.*, **135**, 137–141, 1974.

Passi, I.B.S. Group rings and their Augmentation Ideals, *LNM.*, **715**, Springer-Verlag, Berlin, 1979.

Passi, I.B.S. and Sharma, S. The third dimension subgroup mod $n$, *J. London Math. Soc.*, **9**(2), 176–182, 1974.

Passi, I.B.S. and Sucheta, Dimension subgroups and Schur multiplicator-II, *Topol. Appl.*, **25**, 121–124, 1987.

Passi, I.B.S. Sucheta and Tahara, K. Dimension subgroups and Schur multiplicator-III, *Japan J. Math. (N.S.)* **13**(2), 371–379, 1987.

Passi, I.B.S. and Vermani, L.R. Dimension subgroups and Schur multiplicator, *J. pure appl. Alg.*, **30**, 61–67, 1983.

Sandling, R. The dimension subgroup problem, *J. Alg.*, **21**, 216–231, 1972*a*.

Sandling, R. Note on integral group ring problem, *Math. Z.*, **124**, 255–258, 1972*b*.

Schumann, H.-G. Über modulen und Gruppenbilder, *Math. Ann.*, **114**, 385–413, 1937.

Tahara, K. On the structure of $Q_3(G)$ and the fourth dimension subgroup, *Japan J. Math. (N.S)* **3**, 381–396, 1977.

Tahara, K. Vermani, L.R. and Razdan, A. On generalized third dimension subgroups, *Canad. Math. Bull.*, **41**(1), 109–117, 1998.

Vermani, L. R. Dimension subgroups, In: *Second Biennial Conference, Allahabad Math. Soc.*, a9–a19, 1990.

Vermani, L.R. and Razdan, A. Some intersection theorems and subgroups determined by certain ideals in group rings, *Alg. Colloq.*, **2** : 1, 23–32, 1995.

Vermani, L.R. and Razdan, A. Some remarks on augmentation ideals in group rings, In: *Proc. 4th Ramanujan Symp. Alg. Appl.*, Ramanujan Institute for Advanced Study in Mathematics, **5**, 58–66, 1996.

Vermani, L.R. and Razdan, A. On subgroups determined by certain ideals in group rings, *Alg. Colloq.*, **3**: 4, 301–306, 1996.

Vermani, L.R. and Razdan, A. and Ram Karan, Some remarks on subgroups determined by certain ideals in integral group rings, *Proc. Indian Acad. Sci. (Math. Sci.)* **103**, 249–256, 1993.

Yunus, I.A. On a problem of Fox, *Sov. Math. Dokl.*, **30**, 346–350, 1984.

Department of Mathematics, Kurukshetra University, Kurukshetra - 136 119, India

# A Complex Irreducible Representation of the Quaternion Group and a Non-free Projective Module over the Polynomial Ring in Two Variables over the Real Quaternions

*R. Sridharan*\*†

## Introduction

It is a classical result of Frobenius and Schur ([F], p. 20–22 or [Se], p. 121–124) that any finite dimensional complex irreducible representation of a finite group, whose character is real, either descends to a real representation or can be extended to a representation of the group over the real quaternion algebra. The simplest example where the latter phenomenon holds is the standard 2-dimensional complex representation of the group of integral quaternions. An application of some general results of Barth and Hulek shows that this representation leads to a canonical rank 2 (stable) vector bundle over the complex projective plane. It can be shown that the restriction of this bundle to the affine plane gives rise to a non-free projective module of $H[X, Y]$, isomorphic to the one constructed in [OS] in another context in a different manner. (The existence of this projective module led, incidentally, to the construction in ([P1]) of non diagonalisable, (in fact indecomposable), non singular symmetric $4 \times 4$ matrices of determinant one over the polynomial ring in two variables over the field of real numbers, producing remarkable counter examples to the so called quadratic analogue of Serre's conjecture and opening up a new and fruitful area of research (cf. [P3]). On the other hand, it was shown in ([KPS]) that any non-free projective module over $D[X, Y]$, where $D$ is a finite dimensional division algebra over a field, extends (and essentially uniquely) to a vector bundle over the projective plane over this field, with a $D$-structure.

The aim of this article is to explain in a general setting the rather non obvious connection between the existence of the non extended irreducible complex

---

\*To Parimala, who wove golden garments with the gossamer strands of my thoughts, for the jubilee year

†I would like to thank R. Preeti and V. Suresh for their invaluable help in the preparation of this article

representation and the existence of a non-free projective module over the polynomial ring in two variables over the real quaternion algebra.

## 1. The Canonical Representation of the Quaternion Group and the Pauli Matrices

Let $G$ be the group $\{\pm 1, \pm i, \pm j, \pm k\}$ of integral quaternions. Then, there is a representation of $G$ on $\mathbb{C}^2$ given by $i \mapsto \alpha_0 = \begin{pmatrix} i & 0 \\ 0 & -i \end{pmatrix}, j \mapsto \alpha_1 = \begin{pmatrix} 0 & 1 \\ -1 & 0 \end{pmatrix}, k \mapsto \alpha_2 = \begin{pmatrix} 0 & i \\ i & 0 \end{pmatrix}$. This leads to an $\mathbb{R}$-algebra injection $\mathbb{H} \hookrightarrow M_2(\mathbb{C})$, $\mathbb{H}$ denoting the algebra of real quaternions, inducing an isomorphism $\mathbb{C} \otimes_{\mathbb{R}} \mathbb{H} \xrightarrow{\sim} M_2(\mathbb{C})$. (The above $2 \times 2$ matrices are just the complex quaternions.) Obviously, this representation of $G$ on $\mathbb{C}^2$ is irreducible. Since $\mathbb{H}$ is not isomorphic to $M_2(\mathbb{R})$, it follows that this representation cannot descend to a real representation of $G$. It is a classical result of Frobenius and Schur that an irreducible finite dimensional complex representation of a finite group $G$ with real character $\chi$ descends to a real representation if $\frac{1}{|G|} \sum_{x \in G} \chi(x^2) = 1$ and it extends to a representation of $G$ over the real quaternions if $\frac{1}{|G|} \sum_{x \in G} \chi(x^2) = -1$. It is clear that for the representation above, the character is real and the latter condition holds. We note that the triple of matrices $(i\alpha_0, i\alpha_1, i\alpha_2)$ were considered by W. Pauli in ([Pa]).

## 2. A Barth-Hulek Construction of Vector Bundles Over the Projective Plane, Associated to Certain Triples of Matrices

In this section, we mention a result (which can be proved on similar lines as some results of Barth and Hulek ([B], [BH], [H])), which establishes a bijection between isomorphism classes of certain vector bundles over the projective plane over any field, and equivalence classes of triples $n \times n$ matrices with entries in a finite dimensional central division algebra over the field, satisfying a non degeneracy condition called *pre-stability*.

Let $k$ be an algebraically closed field. A triple $\alpha = (\alpha_0, \alpha_1, \alpha_2), \alpha_i \in M_n(k)$, is called *pre-stable*, if for any non zero $v \in k^n$, the $k$-subspaces spanned by $\{\alpha_0(v), \alpha_1(v), \alpha_2(v)\}$ and $\{\alpha_0^t(v), \alpha_1^t(v), \alpha_2^t(v)\}$ are both at least 2 dimensional. If $k$ is an arbitrary field, and $D$ a finite dimensional central division algebra over $k$ (of degree $r$), a triple $\alpha = (\alpha_0, \alpha_1, \alpha_2), \alpha_i \in M_n(D)$ is called *pre-stable*, if the triple $1 \otimes \alpha = (1 \otimes \alpha_0, 1 \otimes \alpha_1, 1 \otimes \alpha_2), 1 \otimes \alpha_i \in \bar{k} \otimes M_n(D) = M_n(\bar{k} \otimes D) = M_{rn}(\bar{k})$ is pre-stable, $\bar{k}$ denoting the algebraic closure of $k$. For example, it can be seen that, if $\alpha_0$, and $\alpha_2 \alpha_0^{-1} \alpha_1 - \alpha_1 \alpha_0^{-1} \alpha_2$ are both invertible, then the triple $\alpha$ is pre-stable.

**Example 2.0.** For $k = \mathbb{C}$, the triple $\alpha = (\alpha_0, \alpha_1, \alpha_2)$ of matrices $\alpha_0 = \begin{pmatrix} i & 0 \\ 0 & -i \end{pmatrix}, \alpha_1 = \begin{pmatrix} 0 & 1 \\ -1 & 0 \end{pmatrix}, \alpha_2 = \begin{pmatrix} 0 & i \\ i & 0 \end{pmatrix}$ in $M_2(\mathbb{C})$, considered in the previous section is obviously pre-stable; infact $\alpha_2 \alpha_0^{-1} \alpha_1 - \alpha_1 \alpha_0^{-1} \alpha_2$ is two times the identity matrix! For $k = \mathbb{R}$ and $D = \mathbb{H}$, the triple $(i, j, k)$ of matrices in $M_1(\mathbb{H}) = \mathbb{H}$ is pre-stable by definition, since the triple $(1 \otimes i, 1 \otimes j, 1 \otimes k)$ under the isomorphism $\mathbb{C} \otimes \mathbb{H} \simeq M_2(\mathbb{C})$ maps to $(\alpha_0, \alpha_1, \alpha_2)$ considered above.

Let $\mathcal{O}$ denote the structure sheaf of $\mathbb{P}_k^2$. Let $\mathcal{E}$ be a vector bundle over $\mathbb{P}_k^2$ and let $\Gamma(\mathcal{E})$ denote the space of global sections. Following Hulek ([H]), we say that $\mathcal{E}$ is *s-stable* if $\Gamma(\mathcal{E})$ and $\Gamma(\mathcal{E}^*)$ are zero, where $\mathcal{E}^* = \mathcal{H}om(\mathcal{E}, \mathcal{O})$. Let $D$ be a finite-dimensional central division algebra over $k$. We say that a vector bundle over $\mathbb{P}_k^2$ has a *D-structure*, if it is a module over the constant sheaf associated to $D$.

For any vector bundle $\mathcal{F}$ over $\mathbb{P}_k^2$, let $H^m(\mathcal{F})$ denote the cohomology group $H^m(\mathbb{P}_k^2, \mathcal{F})$ and $\mathcal{F}(n) = \mathcal{F} \otimes \mathcal{O}(n)$. For a vector bundle $\mathcal{F}$ over $\mathbb{P}_k^2$ with a *D*-structure, the rank of $\mathcal{F}$ is divisible by the dimension of $D$ over $k$, and we write $\mathrm{rk}_D \mathcal{F} = \mathrm{rk}(\mathcal{F})/\dim_k D$. Note that if $\mathcal{F}$ has a *D*-structure, then there is an induced *D*-structure on $\mathcal{F}(n)$. Further, all the cohomology groups $H^m(\mathcal{F})$ are *D*-vector spaces. Let $\mathcal{E}$ be an *s*-stable vector bundle over $\mathbb{P}_k^2$ with a *D*-structure. We then have $\dim_D H^1(\mathcal{E}(-2)) = \dim_D H^1(\mathcal{E}(-1)) = n$ and the second Chern class $c_2(\mathcal{E})$ of $\mathcal{E}$ is $n \dim_k D$ ([H], 1.4.2). Let

$$\alpha_i(\mathcal{E}) : H^1(\mathcal{E}(-2)) \rightarrow H^1(\mathcal{E}(-1)) \qquad (*)$$

be the *D*-linear maps induced by the multiplications by $Z_i$, for $i = 0, 1, 2$, $Z_i$ denoting the homogeneous coordinates of $\mathbb{P}_k^2$. By fixing *D*-bases for $H^1(\mathcal{E}(-1))$ and $H^1(\mathcal{E}(-2))$, $\alpha_i(\mathcal{E})$ can be identified with elements of $M_n(D)$. It can be proved as in ([H], 1.5.2) that $\alpha(\mathcal{E}) = (\alpha_0(\mathcal{E}), \alpha_1(\mathcal{E}), \alpha_2(\mathcal{E}))$ is pre-stable.

Following the classification of *s*-stable vector bundles over $\mathbb{P}_k^2$ given in [H], we now indicate a classification of *s*-stable vector bundles over $\mathbb{P}_k^2$ with a *D*-structure, in terms of equivalence classes of pre-stable triples of matrices over $D$ ([PSiSSu], cf. [OPS1] and [OPS2]).

We begin by associating to any pre-stable triple $\alpha = (\alpha_0, \alpha_1, \alpha_2)$ an *s*-stable vector bundle $\mathcal{E}(\alpha)$.

The set $\{Z_0, Z_1, Z_2\}$ is a *k*-basis for $\Gamma(\mathcal{O}(1))$, where $Z_0, Z_1, Z_2$ denotes, as before, the homogeneous coordinates of $\mathbb{P}_k^2$. Let $V = \Gamma(\mathcal{O}(1))^* = \mathrm{Hom}_k(\Gamma(\mathcal{O}(1)), k))$ and $\{v_0, v_1, v_2\}$ be the basis of $V$ dual to $\{Z_0, Z_1, Z_2\}$. Let $s : \mathcal{O}(-1) \rightarrow V \otimes \mathcal{O}$ be the map induced by the multiplication $\Gamma(\mathcal{O}(1)) \otimes \mathcal{O}(-1) \rightarrow \mathcal{O}$ and $s^t : V^* \otimes \mathcal{O} \rightarrow \mathcal{O}(1)$ be the transpose of $s$. Let $\alpha = (\alpha_0, \alpha_1, \alpha_2)$ be a triple, with $\alpha_i \in M_n(D)$. Let $A(\alpha) : D^n \otimes V \rightarrow D^n \otimes V^*$ be the *D*-linear map defined for $\phi \in D^n$ by $A(\alpha)(\phi \otimes v_i) = \alpha_{i+1}(\phi) \otimes Z_{i-1} - \alpha_{i-1}(\phi) \otimes Z_{i+1}$, where $i = 0, 1, 2$ mod 3. The map $A(\alpha)$ can be represented by the matrix

$$\begin{pmatrix} 0 & \alpha_2 & -\alpha_1 \\ -\alpha_2 & 0 & \alpha_0 \\ \alpha_1 & -\alpha_0 & 0 \end{pmatrix}$$

with respect to the canonical basis of $D^n$ and the bases for $V$, $V^*$ defined as above. Let $U$ be the image of $A(\alpha)$ and $a : D^n \otimes \mathcal{O}(-1) \rightarrow U \otimes \mathcal{O}$ be the composite map

$D^n \otimes \mathcal{O}(-1) \xrightarrow{1 \otimes s} D^n \otimes V \otimes \mathcal{O} \xrightarrow{A(\alpha) \otimes 1} U \otimes \mathcal{O}$. Let $c : U \otimes \mathcal{O} \to D^n \otimes \mathcal{O}(1)$ be the restriction of the map $1 \otimes s^t : D^n \otimes V^* \otimes \mathcal{O} \to D^n \otimes \mathcal{O}(1)$ to $U \otimes \mathcal{O}$. We then have a complex

$$M(\alpha) : D^n \otimes \mathcal{O}(-1) \xrightarrow{a} U \otimes \mathcal{O} \xrightarrow{c} D^n \otimes \mathcal{O}(1)$$

of vector bundles. We have the following proposition, for whose proof we refer to ([PSiSSu]).

**Proposition 2.1.** *Let $\alpha$ be a pre-stable triple. Then the homology $\mathcal{E}(\alpha)$ of the complex $M(\alpha)$ is a vector bundle over $\mathbb{P}_k^2$ with a D-structure.*

Let $\alpha = (\alpha_0, \alpha_1, \alpha_2)$ be a pre-stable triple in $M_n(D)$. Let $\mathcal{E}(\alpha)$ be the homology sheaf of the complex $M(\alpha)$. Then, by (2.1), $\mathcal{E} = \mathcal{E}(\alpha)$ is a vector bundle over $\mathbb{P}_k^2$ with a $D$-structure. We have $\mathrm{rk}_D \mathcal{E} = \mathrm{rk}_D A(\alpha) - 2n$, $\Gamma(\mathcal{E}) = \Gamma(\mathcal{E}^*) = 0$, i.e., $\mathcal{E}$ is $s$-stable in the sense of Hulek. Further, the first Chern class $c_1(\mathcal{E}) = 0$ and the second Chern class $c_2(\mathcal{E}) = n \dim_k D$ ([H], 1.4.2).

We note that if $k = \mathbb{C}$ and $\alpha$ is the triple of $2 \times 2$ matrices in $M_2(\mathbb{C})$ considered in (2.0), then $\mathcal{E}(\alpha)$ is a rank 2, $s$-stable bundle over $P_{\mathbb{C}}^2$ with $c_1(\mathcal{E}(\alpha)) = 0$ and $c_2(\mathcal{E}(\alpha)) = 2$ ([KPS]). If $k = \mathbb{R}$ and $D = \mathbb{H}$, then the triple $(i, j, k)$ of (2.0) gives rise to a rank 4, $s$-stable bundle $\mathcal{E}$ over $P_{\mathbb{R}}^2$ with $c_1(\mathcal{E}) = 0$ and $c_2(\mathcal{E}) = 4$. In fact $\mathcal{E}$ is the bundle $\mathcal{E}(\alpha)$ over $P_{\mathbb{C}}^2$ considered as a bundle over $P_{\mathbb{R}}^2$. ([OPS1], [OPS2]).

Let $\alpha = (\alpha_0, \alpha_1, \alpha_2)$ and $\beta = (\beta_0, \beta_1, \beta_2)$ be two pre-stable triples. We say that $\alpha$ and $\beta$ are *equivalent* if there exist $u_1, u_2 \in GL_n(D)$ such that $\beta_i = u_1 \alpha_i u_2$ for $i = 0, 1, 2$. Let $T(n)$ denote the set of equivalence classes of pre-stable triples $\alpha = (\alpha_0, \alpha_1, \alpha_2)$, $\alpha_i \in M_n(D)$, and $V(n)$ the set of isomorphism classes of $s$-stable vector bundles $\mathcal{E}$ over $\mathbb{P}_k^2$ with $D$-structures for which $c_1(\mathcal{E}) = 0$ and $c_2(\mathcal{E}) = n \dim_k D$. We then have the following

**Theorem 2.2. ([PSiSSu])** *The assignment $\alpha \mapsto \mathcal{E}(\alpha)$ induces a bijection $T(n) \to V(n)$.*

We note that the inverse of the above correspondence is obtained by associating to the isomorphism class of any $s$-stable vector bundle $\mathcal{E}$, the equivalence class of the triple $\alpha(\mathcal{E}) = (\alpha_0(\mathcal{E}), \alpha_1(\mathcal{E}), \alpha_2(\mathcal{E}))$ defined by $(*)$.

Let $\mathcal{E}$ be a vector bundle over $\mathbb{P}_k^2$ with a $D$-structure. Since all its cohomology groups are $D$-vector spaces, and $c_2(\mathcal{E})$ is divisible by the dimension of $D$ over $k$, the theorem above gives a classification of *all* $s$-stable vector bundles over $\mathbb{P}_k^2$ with a $D$-structure.

## 3. Non-Free Projective Modules Over $D[X, Y]$

Let $D$ be a non-commutative division ring. We begin by recalling a construction given in ([OS]) of a non-free projective $D[X, Y]$-module of rank 1. Let $\mu, \nu \in D$ with $\mu \nu - \nu \mu \neq 0$. Let $P_{\mu, \nu}$ be the kernel of the $D[X, Y]$-linear map $D[X, Y]^2 \to$

$D[X, Y]$ defined by $(1,0) \mapsto X - \mu$ and $(0, 1) \mapsto Y - \nu$. In [OS], it was shown that $P_{\mu,\nu}$ is a non-free projective $D[X, Y]$-module of rank 1.

Let now $D$ be a finite-dimensional central division algebra over $k$. Suppose $D$ is non-commutative. Let $\alpha = (\alpha_0, \alpha_1, \alpha_2)$ be a triple with $\alpha_0, \alpha_1, \alpha_2 \in D$. It is easy to check that $\alpha$ is pre-stable if and only if $\alpha_i \in D^*, 0 \le i \le 2$ and $\alpha_0 \alpha_2^{-1} \alpha_1 - \alpha_1 \alpha_2^{-1} \alpha_0 \in D^*$.

Let $\alpha = (\alpha_0, \alpha_1, \alpha_2)$ be a pre-stable triple with $\alpha_i \in D$. Let $\mathcal{E} = \mathcal{E}(\alpha)$ be the associated $s$-stable vector bundle over $\mathbb{P}_k^2$. The restriction of $\mathcal{E}$ to an affine plane in $\mathbb{P}_k^2$ gives a projective $D[X, Y]$-module $P$ of rank 1.

**Lemma 3.0** *The $D[X, Y]$-module $P$ is not free*

**Proof.** Suppose, if possible $P \simeq D[X, Y]$. Since $D$ is a finite dimensional central division algebra, the projective module $P$ has an extension to $\mathbb{P}_k^2$ as a vector bundle with a $D$-structure, which is *unique* up to a line bundle ([cf. [KPS]]). Since $P \simeq D[X, Y]$, we have $\mathcal{E} \simeq D \otimes \mathcal{O}(m)$ for some $m$. Thus either $\Gamma(\mathcal{E})$ or $\Gamma(\mathcal{E}^*)$ is not zero (since $\Gamma(\mathcal{O}(r)) \neq 0$ for $r \ge 0$). However, since $\mathcal{E}$ is $s$-stable, $\Gamma(\mathcal{E})$ and $\Gamma(\mathcal{E}^*)$ are both zero, leading to a contradiction. Hence, $P$ is not free. $\square$

The following proposition compares the above two constructions of non-free projective $D[X, Y]$-modules of rank 1.

**Proposition 3.1.** *Let $\alpha = (\alpha_0, \alpha_1, \alpha_2)$ be a pre-stable triple with $\alpha_0, \alpha_1, \alpha_2 \in D^*$ and $\mathcal{E}(\alpha)$ the associated $s$-stable vector bundle over $\mathbb{P}_k^2$. Let $P$ be the projective $D[X, Y]$-module given by the restriction of $\mathcal{E}(\alpha)$ to the affine plane defined by $Z_2 \neq 0$ in $\mathbb{P}_k^2$, $Z_i, 0 \le i \le 2$, denoting the homogeneous coordinates of $\mathbb{P}_k^2$. Let $\mu = \alpha_2^{-1} \alpha_0$ and $\nu = \alpha_2^{-1} \alpha_1$. Then $P \simeq P_{\mu,\nu}$.*

To prove the proposition, we first interpret the complex $M(\alpha)$ of §2 as a complex of graded $D[Z_0, Z_1, Z_2]$-modules. We note that given a quasi-coherent sheaf on $\mathbb{P}_k^2$, there is a graded $k[Z_0, Z_1, Z_2]$-module, canonically associated to it. For any graded $k[Z_0, Z_1, Z_2]$-module $N$ and for any integer $m$, let $N(m)$ denote the graded $k[Z_0, Z_1, Z_2]$-module defined by $N(m)_d = N_{d+m}$. Let $A(\alpha)$ and $U$ be as in §2. We consider the following complex of graded modules

$$\Lambda(\alpha) : D^n \otimes k[Z_0, Z_1, Z_2](-1) \xrightarrow{a} U \otimes k[Z_0, Z_1, Z_2]$$
$$\xrightarrow{c} D^n \otimes k[Z_0, Z_1, Z_2](1),$$

where $a$ and $c$ are the graded $D \otimes k[Z_0, Z_1, Z_2]$-linear maps defined for any $v \in D^n$ by

$$a(v) = (A(\alpha) \otimes 1) \begin{pmatrix} v \otimes Z_0 \\ v \otimes Z_1 \\ v \otimes Z_2 \end{pmatrix} = \begin{pmatrix} \alpha_2(v) \otimes Z_1 - \alpha_1(v) \otimes Z_2 \\ -\alpha_2(v) \otimes Z_0 + \alpha_0(v) \otimes Z_2 \\ \alpha_1(v) \otimes Z_0 - \alpha_0(v) \otimes Z_1 \end{pmatrix}$$

and for $w = (w_0, w_1, w_2) \in U \subset (D^n) \otimes V^* = (D^n)^3$, by

$$c(w) = w_0 \otimes Z_0 + w_1 \otimes Z_1 + w_2 \otimes Z_2.$$

The complex $\Lambda(\alpha)$ is the complex associated to $M(\alpha)$ for the choice of bases $V$ and $V^*$ mentioned in §2.

Let $a_1 : D^n \otimes k[X, Y] \to D^{3n} \otimes k[X, Y]$ and $c_1 : D^{3n} \otimes k[X, Y] \to D^n \otimes k[X, Y]$ be the $D[X, Y]$-linear maps defined respectively by

$$a_1(v) = A(\alpha) \begin{pmatrix} X_v \\ Y_v \\ v \end{pmatrix}$$

and $c_1(w_0, w_1, w_2) = Xw_0 + Yw_1 + w_2$ for all $w_0, w_1, w_2 \in D^n$. Since $\mathcal{E}(\alpha)$ is the homology of the complex $M(\alpha)$, its restriction to $D(Z_2)$ can be identified with the homology $P$ of the complex of $D[X, Y]$-modules

$$\Lambda_1(\alpha) : D^n \otimes k[X, Y] \overset{a_1}{\to} U \otimes k[X, Y] \overset{c_1}{\to} D^n \otimes k[X, Y].$$

**Proof of 3.1.** Since $\alpha$ is a pre-stable triple with $\alpha_0, \alpha_1, \alpha_2 \in D^*$, it follows that $A(\alpha)$ is non-singular, so that $U = A(\alpha)(D^3) = D^3$. Let $a_1 : D[X, Y] \to D[X, Y]^3$ and $c_1 : D[X, Y]^3 \to D[X, Y]$ be the $D[X, Y]$-linear maps defined as above. Since $P$ is the restriction of $\mathcal{E}(\alpha)$ to the affine plane given by $Z_2 \neq 0$, it follows that $P$ is isomorphic to the $D[X, Y]$-module

$$\frac{\ker(c_1)}{\text{image}(a_1)} = \frac{\{(f, g, h) \in D[X, Y]^3 \mid Xf + Yg + h = 0\}}{\{(Y\alpha_2 f - \alpha_1 f, -X\alpha_2 f + \alpha_0 f, X\alpha_1 f - Y\alpha_0 f) \mid f \in D[X, Y]\}},$$

and the $D[X, Y]$-module on the right is clearly isomorphic to

$$\frac{\{(f, g) \mid f, g \in D[X, Y]\}}{\{(Y\alpha_2 f - \alpha_1 f, -X\alpha_2 f + \alpha_0 f) \mid f \in D[X, Y]\}}.$$

Therefore, we have the exact sequence

$$0 \to D[X, Y] \overset{\theta}{\to} D[X, Y]^2 \to P \to 0,$$

where $\theta$ is defined by $f \mapsto (Y\alpha_2 f - \alpha_1 f, -X\alpha_2 f + \alpha_0 f)$. Let $\psi : D[X, Y]^2 \to D[X, Y]$ be the $D[X, Y]$-linear map given by $(f, g) \mapsto f(X - \mu) + g(Y - v)$. Then it is easy to see that $\psi\theta$ is the identity map on $D[X, Y]$, i.e., $\psi$ is a section to $\theta$. Therefore, $P$ is isomorphic to the kernel of $\psi$ which is $P_{\mu,v}$. $\qquad\square$

With the notation as above, for $k = \mathbb{R}$ and $D = \mathbb{H}$, $\mu = -j$ and $v = i$, we obtain a non-free projective module $P = P_{\mu v}$ over $\mathbb{H}[X, Y]$, which, in view of the above proposition, is isomorphic to the restriction of the vector bundle $\mathcal{E}$ over $\mathbb{P}^2_{\mathbb{R}}$ corresponding to the triple $(i, j, k)$. Since $(i, j, k)$, treated as a triple of elements of $M_4(\mathbb{R})$ under the regular representation of $\mathbb{H}$ is a triple of skew symmetric matrices, it can be shown ([OPS2]) that $\mathcal{E}$ carries a non-singular quadratic form. Thus $P$ considered as an $\mathbb{R}[X, Y]$-module carries a non-singular quadratic form. Since $P$ is a free $\mathbb{R}[X, Y]$-module of rank 4, the quadratic form on $P$ gives rise to a non-singular symmetric $4 \times 4$ matrix and this is indeed the example of Parimala ([P1]) of a non-singular, non-diagonalisable $4 \times 4$ symmetric matrix over $\mathbb{R}\,[X, Y]$.

We end this article by remarking that the non-free projective modules constructed above led eventually to a study of non-trivial principal bundles over the plane for groups of classical types as well as for the exceptional groups of type $G_2$ and $F_4$. (cf. [P2], [P3] for further references, [PSuT], [PST] and [PSiSSu] for some recent work).

## References

[B]       Barth W.: Moduli of vector bundles on the projective plane, *Invent Math.* **42**, 63–91, 1977.

[BH]      Barth W. and Hulek K.: Monads and moduli of vector bundles, Manuscripta Math. **25**, 323–347, 1978.

[F]       Feit W.: *Characters of Finite Groups*, Benjamin, New York, 1967.

[H]       Hulek K.: On the classification of stable rank-r vector bundles over the projective plane, In: *"Vector Bundles and Differential Equations"*, Proc. Nice, 1979, Prog. Math. **7**, 113–144, 1980.

[KPS]     Knus M.A., Parimala R. and Sridharan R.: Non-free projective modules over $H[X, Y]$ and stable bundles over $\mathbb{P}^2_{\mathbb{C}}$, *Invent. Math.* **65**, 13–27, 1981.

[OPS1]    Ojanguren M., Parimala R. and Sridharan R.: Indecomposable Quadratic Bundles of rank $4n$ over the Real Affine Plane, *Invent, Math.* **71**, 643–653, 1983.

[OPS2]    Ojanguren M., Parimala R. and Sridharan R.: Anisotropic Quadratic spaces over the affine plane, In: *"Vector Bundles on Algebraic Varieties"*, Bombay 1984, Oxford University Press, 465–489, 1987.

[OS]      Ojanguren M. and Sridharan R.: Cancellation of Azumaya algebras, *J. Alg.* **18**, 501–505, 1971.

[P1]      Parimala R.: Failure of quadratic analogue of Serre's conjecture, *Amer. J. Math.* **100**, 913–924, 1978.

[P2]      Parimala R.: Indecomposable quadratic spaces over the affine plane, *Adv. Math.* **62**, 1–6, 1986.

[P3]      Parimala R.: Study of quadratic forms — some connections with geometry, in *Proc. ICM 94*, Birkhauser Verlag, Basel, Switzerland, 1995.

[PSiSSU]  Parimala, R., Sinclair, P., Sridharan R. and Suresh V.: Anisotropic hermitian spaces over the plane, to appear in Math. Zeitschreft.

[PST]     Parimala R., Sridharan R. and Thakur M.L.: Tits' constructions of Jordan algebras and $F_4$ bundles on the plane, to appear in Compositio Mathematics.

[PSuT]    Parimala R., Suresh V. and Thakur M.L.: Jordan Algebras and $F_4$ Bundles over the Affine Plane, *J. Alg.* **198**, 582–607, 1997.

[Pa]      Pauli W.: Zur Quantenmechanik des magnetischen Elektrons, *Z.f. Phys.* **23**, 601–623, 1927.

[Se]      Serre J.P.: *Representations Lineaire des Groupes Finis*, Herman, Paris, 1967.

School of Mathematics, Tata Institute of Fundamental Research, Homi Bhabha Road, Bombay 400 005, India